Aircraft Propulsion

Pitman Aeronautical Engineering Series

The books in this series are intended primarily as textbooks for students at universities and technical colleges who are studying aeronautical engineering in one or more of its various branches with the aim of obtaining degrees or diplomas in aeronautical engineering or allied subjects. Their contents, however, are not confined to the syllabus of any particular university or college, and they may profitably be read by anyone engaged in aircraft engineering, or interested in aeronautical problems, who wishes to further his acquaintance with the subjects covered by the series.

A knowledge of mathematics and physics to approximately the A-level standard of the General Certificate of Education is assumed, but an intelligent reader who has not reached that standard should have no difficulty in following the main parts of the text. The majority of books in the series contain worked examples, and other questions for the student to work out for himself, and also references for further study—thus the books should be as helpful to those who wish to teach themselves as to those engaged on recognized courses.

It is unfortunate that at the present time, when co-operative efforts are being made between the aircraft industries of the United States, Great Britain and the Continent of Europe, there still exist differences in the systems of units and symbols and even in the aeronautical technical terms, used in these countries; but, while the units and terms usually accepted in aeronautical circles in Great Britain have so far been used in this series, the problem has been borne in mind with a view to making later editions of the books more international in character.

It is intended that the series will eventually cover all subjects which can be classified under the general heading of Aeronautical Engineering, e.g. aerodynamics, stability and control, aircraft propulsion, aircraft materials and structural design, performance, rockets and space flight, and even hovercraft.

Air Vice-Marshal A. C. KERMODE, *R.A.F.* (*Ret'd.*)
Advisor Editory

Aircraft Propulsion

P J McMAHON
B.Sc., A.F.R.Ae.S.
School of Engineering
Bath University of Technology

with a Foreword by
S G Hooker

PITMAN PUBLISHING

First published 1971

SIR ISAAC PITMAN AND SONS LTD
Pitman House, Parker Street, Kingsway, London, WC2B 5PB
P.O. Box 6038, Portal Street, Nairobi, Kenya

SIR ISAAC PITMAN (AUST.) PTY. LTD
Pitman House, Bouverie Street, Carlton, Victoria 3053, Australia

PITMAN PUBLISHING COMPANY S.A. LTD
P.O. Box 9898, Johannesburg, S. Africa

PITMAN PUBLISHING CORPORATION
6 East 43rd Street, New York, N.Y. 10017, U.S.A.

SIR ISAAC PITMAN (CANADA) LTD
Pitman House, 381-383 Church Street, Toronto, 3, Canada

THE COPP CLARK PUBLISHING COMPANY
517 Wellington Street, Toronto, 2B, Canada

ISBN: 0 273 42324 X

Made in Great Britain at the Pitman Press, Bath
G1—(A.198)

Foreword

By Dr. S. G. HOOKER, formerly Technical Director, Rolls-Royce Ltd. (Bristol Engines Division)

In the beginning, manned flight was only possible when the Wright Brothers produced an aero engine developing about 15 hp for a weight of about 180 lb. With this engine, on 3 December 1903, Orville Wright staggered into the air, and a new epoch of human communications was borne.

Since that day, dramatic advances have been made, and aero engines have always been the pacemakers in aircraft performance. Today the Concorde flies with jet engines producing, under cruising conditions, nearly 150000 hp, and giving 6 hp for every pound of of engine weight.

No one can foresee the end of the possible improvements to aero engines. More power, lower fuel consumption, less noise, more reliability, are the continuous objectives of the aero engineer, and the rate of progress in achieving these aims is greater today than ever before.

Because of the international nature of the aircraft business, home-produced engines and aircraft give tremendous benefits to the national economy. In the defence of this country they are a paramount weapon, and we need continuously to remind ourselves of the supreme importance which the victory of the Battle of Britain had on the survival of the civilization of the world as we know it today.

The life of an aero engineer is interesting and arduous. It has its triumphs and failures, and its disappointments when economical and political decisions override the application of new technical advances.

Nonetheless, I believe that there is no branch of engineering which calls for a wider and more diverse application of the latest results of applied science. The wide scope of Mr. McMahon's excellent book bears witness to this.

I commend it to those ambitious young engineers who wish to use their brains in these enthralling developments, and are not afraid to compete with men of considerable ability, whom they will find in plenty in the Aircraft Industry.

6 *August* 1970

Preface

When first starting to write a textbook, an author must try to envisage the type of person who, he hopes, will read his work and who in so doing will gain knowledge and some enjoyment. In this particular instance an attempt has been made to produce a work which will be suitable as an undergraduate text for engineering students who are particularly interested in the aero-engine field. But in addition, the large and growing body of engineers already working with engines for aircraft have been kept in mind. For these men, with their specialized interests in such lines as design, testing or manufacture, the main need is to look again at the fundamentals and to see how their knowledge fits into the overall picture. The operators of aircraft too, come into this class.

With these thoughts in mind I have tried first to give a physical picture of the more important aspects of engineering science as applied to propulsion, without reproducing too many of the proofs which are given in standard textbooks. This emphasis on the physical picture has been carried over into the subsequent chapters. In these, the main classes of engine and their components have been considered in greater detail. My other main aim has been to emphasize the design aspect of the work. By this I mean that in designing real machinery, compromises must always be made between conflicting requirements. Although achieving good performance is important, other factors such as reliability and economy are of almost equal importance.

There is nothing static about the aircraft propulsion field and things that are at the limits of knowledge and ability at one time are commonplace in a few short years. For conventional aero-engines, the performance has improved steadily over the years at the price of increased complexity of design but without sacrificing reliability or life. Two examples can illustrate this rise. Cycle pressure ratios have increased from 4 or 5 in the early days to 25 or 30 in modern advanced engines; turbine entry temperatures have risen from 1 050 K to 1 400 K and more. For rocket engines the advances have been even more outstanding. Beneath all these changes though, the fundamental principles do not alter—it will be as important to understand the principles in twenty years time as it is now.

A word should be said on the decision to use SI units in this book. This has been done because the advantages of SI are such that it is bound to be the system of the future. Those who have grown up on SI will never know what they have missed in trying to decide whether to multiply or divide by 32·2! The change in units is bound to be more difficult for those used to the British system but, from my personal experience, it is easier than at first seems likely and is well worth the effort.

A book like this, although the work of one author, in fact involves contributions from many people. I would like to acknowledge firstly my indebtedness to my friends and former colleagues at what is now the Bristol Engine Division of Rolls-Royce Ltd. Together we learnt about engines. There are too many to mention all by name, but I must single out Pierre Young who taught me so much. Then my colleagues at Bath University of Technology under Professor Black and Professor Wallace have given me much helpful advice and comment. The students too, with their questions and discussions, have contributed greatly to my understanding. I must also thank the firms who have helped by allowing me to use their information. Where necessary I have tried to show the source of such information.

On a more personal basis I must express my thanks to Miss Joyce Codrington, Mrs. B. Grigg and Mrs. P. Foley who between them typed the manuscript and to Mrs. Jean Dyer for all her help with the illustrations. My wife and children know what their help and encouragement has meant to me but my appreciation of all they have done should not go unrecorded.

Backwell 1970 P.J.McM.

Contents

1 Fundamentals of Engineering Science Applied to Propulsion

1.1 INTRODUCTION

Before we can start to study the mechanisms of aero-engines, it is important to have a sound understanding of some of the basic disciplines of Engineering Science. In this introductory chapter an attempt is made to review concisely the relevant parts of Applied Mechanics, Thermodynamics, and Mechanics of Fluids, which are needed for this understanding. It is obviously impracticable to give full proofs in a book of this kind and reference is made where necessary to standard textbooks. The real attempt made here is to give a physical understanding of the various processes involved in propulsion and to present sufficient information and design data to allow the reader to carry out representative calculations.

1.2 APPLIED MECHANICS

So much of the study of aircraft propulsion is based on Newton's Laws of Motion that no apology is needed for beginning by stating these laws and discussing their meaning.

Newton's First Law states that "Every body continues in its state of rest or of motion in a straight line unless compelled by an external force to change that state."

The Second Law states that "The rate of change of velocity of a body is proportional to the applied force and takes place in the direction of application of the force."

The Third Law states that "For every force there is an equal and opposite reaction force."

Looking at the *First Law of Motion*, we see that the main concept it introduces is that of the inertia inherent in all matter. Concurrently, there appears the idea of the momentum of a body, its tendency to keep going in a straight line at uniform speed once it is moving. From this it is clear, too, that velocity is a vector quantity and that changes in *direction* require applied forces just as much as do changes in the *magnitude* of a velocity. We will assume that the mass of a given body is constant at all times and in all places and that the variations in mass due to changes in velocity (as required by relativity) are never significant.

The importance of the *Second Law of Motion* is that it allows us to define our unit of force. Writing the Law as

$$\text{Force} \propto \text{Mass} \times \text{Acceleration}$$
$$F \propto ma$$
$$F = kma \tag{1.1}$$

the question becomes one of selecting the correct units and a suitable value of k to give a consistent equation. Throughout this book the SI system of units is used. In this system the basic unit of FORCE is the *Newton*, that force which will accelerate unit mass of 1 kilogramme at a rate of 1 metre per second per second. With this, therefore, the constant k is unity so we can write directly

$$F = ma \tag{1.2}$$

(When compared with the most commonly used British system, the advantages of SI will be apparent. In the British system the unit of force, the pound force (lbf), is that force which will accelerate unit pound mass (lb) at a rate of 32·2 ft/s^2. A constant of proportionality of 1/32·2 is therefore needed if masses are measured in pounds and accelerations in feet per second per second.)

It is often found convenient to express equation 1.1 as

$$\begin{aligned}\text{Force} &= \text{Rate of change of momentum}\\ &= \text{Mass flow rate times change in velocity}\end{aligned}$$

The validity may be checked by writing the equation as follows:

$$F = ma = m\frac{dV}{dt} = \left(\frac{m}{dt}\right) dV \tag{1.3}$$

This alternative form is particularly useful when considering jet propulsion engines.

The last part of the second law of motion as stated above—the proviso that changes in velocity take place in the direction of the applied force—emphasizes that forces are vector quantities and that their direction is important as well as their magnitude. Very often only certain components of a force will be of use or of interest. It may then be necessary to use the theorem of the parallelogram of forces in order to resolve a simple force into two components.

The total force exerted by the gases on a turbine blade, for example, can be split up into a useful component in the direction of motion of the blade and one at right angles which merely has the undesirable bad effect of applying a load to the turbine bearings.

One particular case of non-linear motion which comes up frequently in dealing with turbo-machinery is motion in a circular path. If the angular velocity is ω radians per second and the radius is r, then, as shown in any standard textbook on applied mechanics, there will be an acceleration towards the centre of curvature given by

$$\text{Centripetal acceleration} = \omega^2 r \tag{1.4}$$

Associated with this acceleration there must be a force to keep a mass moving in a curved path.

Turning now to the *Third Law of Motion*, this defines the fundamental principle underlying all means of aircraft propulsion. Whereas with transport on the surface of the earth it is always possible to make physical contact between a vehicle and a solid surface so that straightforward friction effects can be used to provide a propelling force, this is not possible for a vehicle wholly immersed in a fluid medium. Propulsion must involve some indirect effect, the most convenient being to accelerate backwards a mass of fluid and utilize the *reaction* force for forward propulsion.

The one other fundamental principle in applied mechanics which is of great interest to the propulsion engineer is that of *conservation of energy*. In dealing with gases, for example, the question arises of energy absorption by the molecules causing an increase in kinetic energy of the particles. The equation is

$$\text{Kinetic energy} = \tfrac{1}{2}\,mV^2 \tag{1.5}$$

In equation 1.5 if the mass is in kilogrammes and the velocity is in metres per second the ENERGY will be in *newton-metres*. This unit of energy is known as the *joule* (1 J = 1 N-m).

Those, then, are the main principles of applied mechanics which are of interest when considering the propulsion process. There are a few other topics, in solid mechanics for example, which arise when dealing with the design of engines—mainly with stressing and vibration aspects—but these are dealt with as they arise.

1.3 PRINCIPLES OF THERMODYNAMICS

Throughout our study of aircraft engines the principles of thermodynamics will be invoked to explain in detail the behaviour of components and of complete units. It is of the great importance to have a clear understanding of the main *ideas* behind this branch of science as the explanations it can give are so complete and satisfying. In recent years there has been a real renaissance in the teaching of thermodynamics as writers have tried to give a sound physical understanding of a wide range of different types of engines without concentrating too closely on any one type. This rigorous approach is usually associated with the name of Keenan (ref. 1) and a large number of excellent texts have followed the lead set by Keenan. As examples, refs. 2 and 3 may be cited as standard textbooks which give clear, detailed expositions of thermodynamic principles. The aim in this section is simply to state the main results which are of interest to us and to give a qualitative picture of what is going on.

As a start, the main thermodynamic terms needed in our discussion of the basic principles will be defined.

Definitions

A *system* is a region of space with boundaries which may be flexible and across which energy movements in the form of heat transfer or work transfer may take place. *A closed system* is one with boundaries impervious to the passage of matter; we will be more interested in an *open system* where matter can cross the boundaries. If there is a *steady flow* through a system, the rates of flow into and out of the system are equal and are invariant with time so that the mass of matter within the system is constant.

The matter crossing the boundaries of an open system is known as the *working fluid. Heat* energy is a form of energy which is transferred to the working fluid whenever there is a temperature difference between the fluid and the surroundings. *Work* is energy which is transferred to the system if there is a movement of the boundaries or if a shaft passes through a rigid boundary and is rotated by a couple applied to the shaft. A special case of work known as *paddle wheel work* which is a stirring of the working fluid with energy dissipation due to viscous friction. In general we will ignore the transfer of paddle wheel work unless special circumstances make it necessary to take account of it.

Keenan (ref. 1) defines a *property* as an observable quantity. For the working fluids of interest to us, we will be interested in such well-known properties as pressure, temperature and density, as well as with the rather more specialized thermodynamic ones of internal energy, enthalpy and entropy. Following the usual practice, we will work mainly in terms of *specific* properties which are properties per unit mass—in our case, per kilogramme. With a given pure substance, each specific property is a unique property of the *state* of the substance and for substances in the gaseous phase the state is defined completely be defining any two independent thermodynamic properties. This means that for gases, property charts may be prepared using any two convenient properties as the variables. All required properties can then be plotted on these graphs. (See, for example, Figures 1.1 and 1.2.)

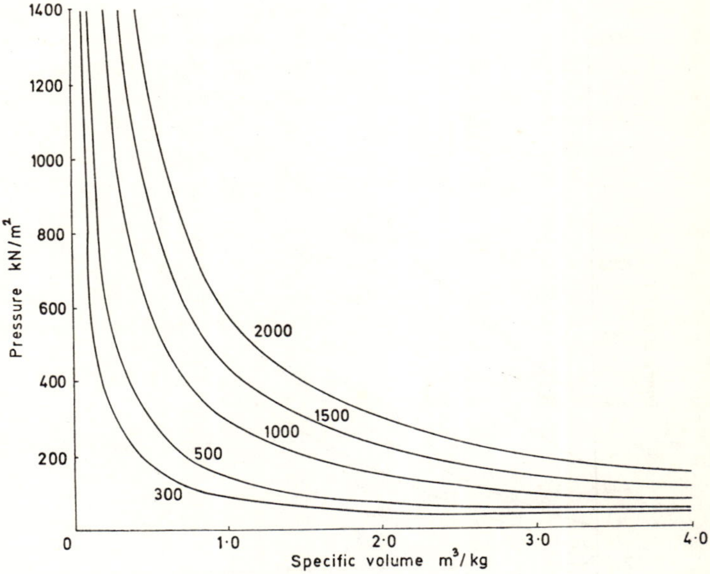

FIGURE 1.1 *Pressure–specific volume chart for 1 kg of dry air*
Temperatures marked are in kelvins

One of the simplest but most important equations of state for a perfect gas is the *characteristic equation* which is derived from a combination of Boyle's and Charles' Laws. It is

$$pv = RT \tag{1.6}$$

where p is the pressure of gas, v is the volume per unit mass, R is a characteristic constant and T is the absolute temperature. In this book the *kilonewton per square metre* is used as the unit of PRESSURE*

* An alternative pressure unit, the *bar*, is often used, where 1 bar = 100 kN/m^2. Although the bar has advantages in that standard atmospheric pressure is very nearly equal to 1 bar, it falls out of line of the preferred system of units in SI. With pressure in bars equation 1.6 must be written as $100\,pv = RT$.

and the *cubic metre per kilogramme* as the unit of SPECIFIC VOLUME. The characteristic constant must therefore have units of kilonewton-metres per kilogramme-kelvin. But kilonewton-metres are the same as kilojoules so R is quoted in kJ/kg-K. For air R has a value of 0·2871 kJ/kg-K and this value has been used in plotting the constant temperature lines on Figure 1.1.

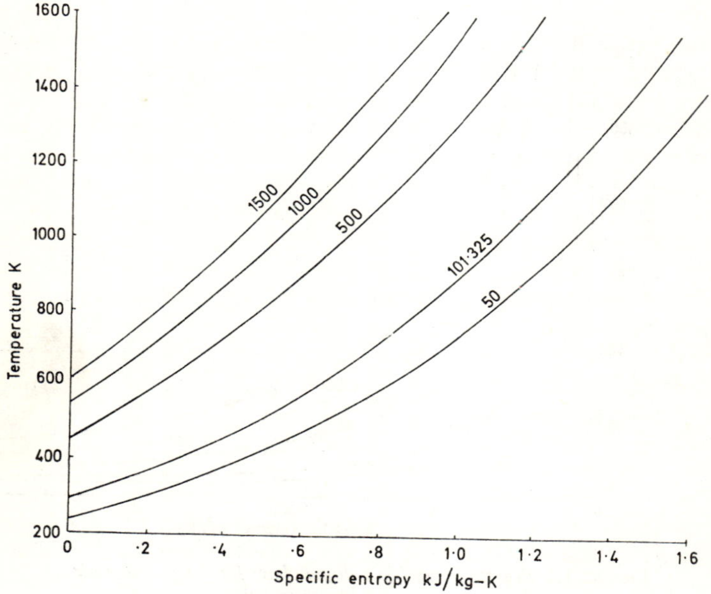

FIGURE 1.2 *Temperature–specific entropy chart for 1 kg of dry air*
Pressures marked are in kN/m² (bars × 10²)

When a working fluid changes from one state to another, it is said to undergo a *process*. It is shown in all standard thermodynamic textbooks that the change in the value of a property of a pure substance between start and finish of any process is independent of the path taken between the end states. This fact is of very great importance as it allows us to take an overall look at a component to get a measure of its performance and to neglect the details of its behaviour between the end states.

Heat and work are *not* properties, so the quantities of heat and work transferred during any change of state depend on the path taken between the original and final state points. The truth of this may best be seen by taking a purely mechanical analogy where a change of height results in a change in the property, the potential energy. In climbing a mountain, more work is done in taking a circuitous path than in climbing directly. The change in potential energy, though, is the same in each case.

Finally in these introductory remarks on thermodynamics, the *specific heats* of a substance must be defined. If the temperature of unit mass of matter rises this is evidence that some energy has been transferred to it. By the definition of heat as "energy in process of of being transferred" due to a temperature difference it is clearly wrong to talk about heat being "stored" by the substance, but this misnomer is in common use and people talk quite freely of the "heat capacity" of materials and objects. The misuse of the word "heat" occurs again in the term "specific heat", which is defined as the energy required to alter the temperature of unit mass of the substance by one degree. For the particular case of a gas, the value of the specific heat will depend on how the gas is allowed to expand as the energy is supplied. Two main specific heats are generally defined, that for constant volume heat addition c_v and that for constant pressure heat addition c_p. Each is a property of the gas which varies with temperature. The relationships between c_p and c_v are discussed in greater detail below (p. 8).

The First Law of Thermodynamics

The principle of conservation of energy is one of the keystones of engineering science and the first law of thermodynamics is basically a statement of this principle with particular reference to work and heat transfers. As stated for a closed system (Figure 1.3), the law says that if quantities of heat and work energy are transferred to a system then

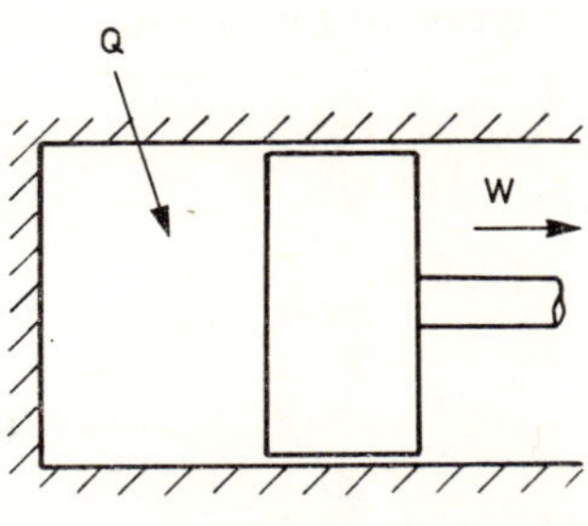

FIGURE 1.3 *Energy transfer to a closed system*

any difference between the net quantities supplied and rejected must be stored as *internal energy* in the system. This internal energy is a *property* of the system. It is possible to write the first law for a closed system undergoing a process from state 1 to state 2 in the form of the equation

$$Q = \Delta U_{1-2} + W \tag{1.7}$$

Here Q is the net heat supplied and W is the net work done by the system while ΔU_{1-2} is the internal energy change between states 1 and 2.

If the volume of a system is fixed and some heat is transferred to it then no work is done (no movement of the boundaries) and all the heat goes to increase the internal energy of the system. Taking a system composed of unit mass* of perfect gas which goes from T_1 to T_2 we can write

$$Q = c_v(T_2 - T_1) = u_2 - u_1 \tag{1.8}$$

From the differential form of this equation we can write the definition of specific heat at constant volume as

$$c_v = \left(\frac{du}{dT}\right) \tag{1.9}$$

Suppose now that the same system is taken through another process from 1 to 2 but with the pressure kept constant so that work is done due to the change in volume. Then

$$\begin{aligned} Q = c_p(T_2 - T_1) &= \Delta u + p(v_2 - v_1) \\ &= c_v(T_2 - T_1) + R(T_2 - T_1) \end{aligned}$$

Therefore

$$c_p - c_v = R \tag{1.10}$$

This equation may be rearranged in terms of the specific heat ratio $\gamma = c_p/c_v$ to give

$$c_v = \frac{R}{\gamma - 1} \tag{1.11}$$

and

$$c_p = \frac{R\gamma}{\gamma - 1} \tag{1.11a}$$

* It is customary to use capital letters to denote properties of a *system* and small letters to denote properties per unit mass, that is *specific* properties. For example an internal energy U will be in kJ while a specific internal energy u will be in kJ/kg.

Consider now the first law of thermodynamics applied to an open system (Figure 1.4) with steady flow conditions so that the mass

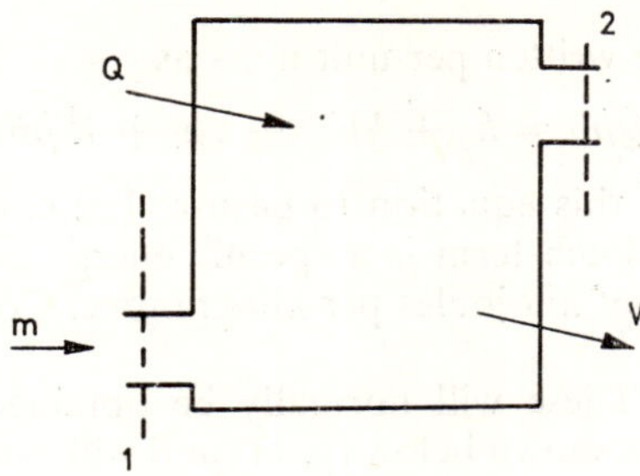

FIGURE 1.4 *Energy transfer to a steady flow open system*

flow rate into the system at station 1 is equal to the mass flow rate out of the system at station 2. By the first law the net energy flow *in* must equal the net energy flow *out* so we can write the *steady flow energy equation* (s.f.e.e.) as

Energy into system = Energy out of system
Energy into system = Internal energy at 1
+ Work of admission*
+ Kinetic energy at 1
+ Potential energy at 1
+ Heat transfer

Energy out of system = Internal energy at 2
+ Work of expulsion*
+ Kinetic energy at 2
+ Potential energy at 2
+ Work transfer

which can be written symbolically as

$$u_1 + p_1 v_1 + \tfrac{1}{2}V_1^2 + z_1 g + \dot{Q}$$
$$= u_2 + p_2 v_2 + \tfrac{1}{2}V_2^2 + z_2 g + \dot{W}$$

This basic form of the s.f.e.e. may be simplified by defining a new

* The ideas of works of admission and expulsion are sometimes a little difficult to grasp. Suppose unit mass of a gas at pressure p and with a volume v enters the system in unit time. This may be thought of as occuring behind a piston of area A which moves across the boundary of the system through a distance L. The work done by the gas on the piston will be force × distance which must equal pv.

composite property, the *enthalpy*. The specific enthalpy of a substance is defined by the equation

$$h = u + pv$$

so that the equation may be written per unit mass as

$$h_1 + \tfrac{1}{2}V_1^2 + z_1 g + \dot{Q}/\dot{m} = h_2 + \tfrac{1}{2}V_2^2 + z_2 g + \dot{W}/\dot{m} \quad (1.12)$$

A little care is needed with this equation to ensure that consistent units are used throughout. Each term is a specific energy and it is convenient to work in units of kilojoules per kilogramme. Considering each term in turn:

1. *Enthalpies h_1 and h_2*. These will normally be obtained from tables or equations of state as shown below (p. 11) and will usually be quoted directly in kJ/kg.

2. *Kinetic Energy*. The kinetic energy of 1 kg moving at 1 m/s will be 1 joule. It will therefore be necessary to divide the $\frac{1}{2}V^2$ terms by 1 000 to convert them from J/kg to kJ/kg.

3. *Potential Energy*. Similarly the potential energy of 1 kg at a height of 1 m above datum in a gravitational field of 1 m/s^2 will be 1 joule. Again the zg terms must be divided by 1 000 to convert them from J/kg to kJ/kg.

4. *Heat Transfer Rate*. This will normally be quoted in kilowatts which are equivalent to kilojoules per second. If divided by a mass flow rate $\dot{m}$ kg/s the units of $\dot{Q}/\dot{m}$ will be the desired ones of kJ/kg.

5. *Work Transfer Rate*. Again the energy transfer rate will be measured in kilowatts so dividing by the mass flow rate in kg/s will give the required specific energy term of kJ/kg.

(In passing, the advantage of SI in using the same unit of kilowatts for both heat and work transfers will be seen. In British units it is customary to use British Thermal Units or Centigrade Heat Units to measure heat quantities and foot pounds-force or horsepower seconds to measure work. Conversion factors are then needed.)

Writing in the various factors the s.f.e.e. becomes

$$h_1 + \frac{V_1^2}{2 \times 1\,000} + \frac{z_1 g}{1\,000} + \frac{\dot{Q}}{\dot{m}}$$
$$= h_2 + \frac{V_2^2}{2 \times 1\,000} + \frac{z_2 g}{1\,000} + \frac{\dot{W}}{\dot{m}} \quad (1.13)$$

If required Q/m can be written for $\dot{Q}/\dot{m}$ and W/m for $\dot{W}/\dot{m}$.

Note that the sign convention is such that heat transferred *to* the working fluid and work transferred *from* the working fluid are the positive values. It will be found that in very many cases one or more of the terms in the s.f.e.e. may be neglected. As mentioned above, the potential energy terms are rarely significant in thermodynamic work and, in addition, for many cases changes in kinetic energy are small compared with other energy transfers. Each case must be considered on its own merits, however, and it would be dangerous to neglect certain terms automatically every time.

Properties of Real Gases

As a prerequisite for using the energy equation, information must be available on the main thermodynamic properties of the working fluid. In this book the most important working fluid approximates to air and we will want to know how its properties vary with temperature and pressure. The main information needed will be on specific enthalpy and specific volume. Table 1.1 shows the variation of enthalpy of air with temperature at low pressures. For calculating volumes the characteristic equation (equation 1.6) will be used.

TABLE 1.1 *Enthalpy and Specific Heat and Specific Heat Ratio of Dry Air at Low Pressures*

Temperature K	*Specific enthalpy h* kJ/kg	*Specific heat c_p* kJ/kg-K	*Specific heat ratio γ*
200	200	1·002	1·402
400	401	1·013	1·395
600	607	1·051	1·373
800	822	1·098	1·354
1 000	1 046	1·141	1·336
1 200	1 278	1·172	1·324
1 400	1 515	1·200	1·315
1 600	1 758	1·220	1·308
1 800	2 003	1·237	1·302
2 000	2 252	1·250	1·298

Based on Keenan and Kaye, *Gas Tables*, by kind permission of the publishers, J. Wiley & Co.

Table 1.1 also shows the variation of c_p and γ with temperature. Looking at the equation for the first law for a closed system again, written in differential form, it is seen that specific enthalpy can be introduced so that for a constant pressure heat addition

$$dQ = c_p dT = d(u + pv) = dh$$

$$c_p = \frac{dh}{dT} \tag{1.14}$$

Although derived from a particular process this equation is true for *any* process because c_p, h and T are fluid properties.

Example 1.1 illustrates the application of the s.f.e.e. and shows how powerful it can be in giving an overall picture of unit performances.

Example 1.1 Air at a temperature of 15°C passes through a heat exchanger at a velocity of 30 m/s where its temperature is raised to 800°C. It then enters a turbine with the same velocity of 30 m/s and expands until the temperature falls to 650°C. On leaving the turbine the air is taken at a velocity of 60 m/s to a nozzle where it expands until the temperature has fallen to 500°C. If the air flow rate is 2 kg/s calculate.

(i) The rate of heat transfer to the air in the heat exchanger.
(ii) The power output from the turbine assuming no heat loss.
(iii) The velocity at outlet from the nozzle again assuming no heat losses during the expansion process.

Solution: In each case the steady flow energy equation is applied to find the overall effects of a process. Changes in potential energy will always be negligible.

(i) For the heat exchanger no work is done and there is no change in kinetic energy. The equation reduces therefore to

$$\dot{Q} = \dot{m}\,(h_2 - h_1)$$

From Table 1.1 the enthalpies at 15°C and 800°C are found by interpolation as 281 kJ/kg and 1 120 kJ/kg. Therefore

$$Q = 2\,(1\,120 - 281) = 1\,678 \text{ kJ/s} = 1\,678 \text{ kW}$$

(ii) Coming to the turbine, no heat is lost so $Q = 0$ but account must be taken of the kinetic energy changes:

$$h_1 + \frac{V_1^2}{2 \times 1\,000} = h_2 + \frac{V_2^2}{2 \times 1\,000} + \frac{\dot{W}}{\dot{m}}$$

Therefore

$$\dot{W} = \dot{m}\left(h_1 - h_2 + \frac{V_1^2 - V_2^2}{2 \times 1\,000}\right)$$

At 800°C, $h = 1\,120$ kJ/kg and at 650°C (again by interpolation from Table 1.1) $h = 960$ kJ/kg; also $V_1 = 30$ m/s and $V_2 = 60$ m/s. Therefore

$$\dot{W} = 2\left(1\,120 - 960 + \frac{30^2 - 60^2}{2 \times 1\,000}\right)$$

$$= 317{\cdot}3 \text{ kW}$$

Note that kinetic energy changes are very small compared with the enthalpy changes in this particular instance.

(iii) In the nozzle no external work is done and the heat loss, we are told, is negligible. There is therefore an interchange of enthalpy and kinetic energy with the energy equation reducing to

$$V_2 = \sqrt{[2 \times 1\,000\,(h_1 - h_2) + V_1^2]}$$

Interpolation of Table 1.1 shows that the enthalpy at 500°C is 791 kJ/kg, so that V_2 becomes

$$V_2 = \sqrt{[2 \times 1\,000\,(960 - 791) + 60^2]}$$
$$= 585 \text{ m/s}$$

Some Special Processes

The actual path followed by a working substance in going from one state to another may be of a complex form but it is convenient to define certain standard processes to which many real processes approximate. Many of the standard processes have names which are self-explanatory—constant pressure, isothermal, constant volume—but a few have traditional names whose meaning is less clear. For example, an *adiabatic process* is one in which no heat is transferred to the working fluid.

There is one qualifying term for processes which has a very particular meaning in thermodynamics and which deserves close consideration and that is to say that a process is *reversible.* To appreciate the meaning of reversibility, consider first the simple mechanical case of a pendulum. A pendulum will be designed with the lowest possible friction at the pivot and, if accurate results are needed, with the least possible air resistance. In the limit, one can envisage a pendulum where the friction and air resistance losses become zero and each half-swing would be the exact reversal of the previous half-swing—the motion of the ideal pendulum would be a succession of reversible processes.

This analogy can be extended to a compressor coupled directly to a turbine (Figure 1.5). Assuming steady flow through the two units,

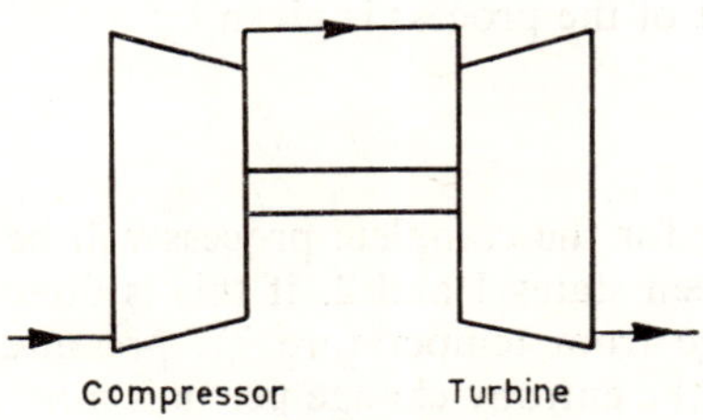

FIGURE 1.5 *The ideal compressor–turbine set*

work must be supplied to drive the compressor while work can be got out from the turbine by feeding it with the air delivered from the compressor. If heat losses are neglected, then an ideal set-up could be reached where there are no losses (due to internal friction of the air for example) in the compressor and no losses in the turbine. With these conditions, the turbine expansion would be an exact reversal of the compressor compression and the air would leave the turbine at an identical state to that at compressor inlet. In any *real* case there are bound to be friction losses so the compressor work requirement will be higher than in the ideal case and the turbine work will be lower. It will, therefore, be necessary to supply some additional energy from outside if the compressor–turbine unit is to run.

These two examples have been chosen to illustrate the point that a reversible process is the ultimate ideal which real processes might approach but which they will never reach. Reversibility gives, therefore, a standard of comparison for real processes and a target at which to aim. The work of power engineers must be devoted to reducing irreversibilities to the economic minimum and so obtaining better and more efficient machines. Unfortunately, when the conditions for reversibility are examined it is found that a reversible process is one which is negligibly removed from a state of equilibrium and so a machine which was to approach reversibility would be very slow and very large. It is for this reason that the qualification "to the economic minimum" was introduced above!

Entropy

The time has now come to introduce the thermodynamic property *entropy*. Suppose that a system undergoes a reversible process from state 1 to state 2 and that during an infinitesimal part of the process an amount of heat dQ is transferred to the system at an absolute temperature T (dQ may be positive or negative). A property, the entropy, can then be defined such that the change in entropy dS during the infinitesimal part of the process is given by

$$dS = \left(\frac{dQ}{T}\right)_{rev} \tag{1.15}$$

The total change in entropy for the complete process will be found by integrating (dQ/T) between states 1 and 2. If this is done it can be shown that for a change from temperature T_1, pressure p_1 to temperature T_2, pressure p_2 the entropy change per unit mass is

$$s_2 - s_1 = c_p \ln (T_2/T_1) - R \ln (p_2/p_1) \tag{1.16}$$

The units of SPECIFIC ENTROPY used in this book are *kilojoules per kilogramme-kelvin* (kJ/kg-K)

Any convenient arbitrary datum can be taken for the zero point of entropy. In preparing Figure 1.2, the *T–s* chart for air, the datum was taken as 0°C and a pressure of 101·3 kN/m^2 (standard atmospheric pressure).

So much for the mathematical definition of entropy. What now is the significance and physical meaning of this property?

In discussing reversibility above it appeared that the ideal process in many cases would be one where friction losses fell to zero so that the process was reversible. With real plant (such as compressors and turbines) losses must be kept to a minimum and these will include both frictional losses and any heat transfers from the working medium. A second qualification arises, therefore, for the ideal process, that it should be *adiabatic* as well as reversible. Now, by our definition of entropy in equation 1.15, if the process is adiabatic $dQ = 0$ and if it is also reversible $dS = (dQ/T)_{rev} = 0$. The ideal process with which real compressions and expansions can be compared is therefore the constant entropy or *isentropic* process. Chapter 3 will show how the "isentropic efficiency" of various components is defined and used.

The characteristics of an isentropic process can be found by writing $s_2 - s_1 = 0$ in equation 1.16 above. When this is done we find

$$c_p \ln (T_2/T_1) = R \ln (p_2/p_1)$$

Therefore

$$\frac{p_2}{p_1} = \left(\frac{T_2}{T_1}\right)^{c_p/R}$$

But from the equation 1.11(a)

$$\frac{c_p}{R} = \frac{\gamma}{\gamma - 1}$$

For an isentropic process therefore

$$\frac{p_2}{p_1} = \left(\frac{T_2}{T_1}\right)^{\gamma/\gamma - 1} \qquad (1.17)$$

This very important equation shows that for a fixed value of the ratio γ of the specific heats, (T_2/T_1) depends only on the pressure ratio of a process. Equation 1.17 can be combined with the characteristic equation $pv = RT$ to show that, for an isentropic process

$$pv^\gamma = \text{constant}$$

As the isentropic process is so useful as defining the ideal, it is convenient to use entropy as one of the ordinates in a chart of fluid properties. When dealing with air we will also be interested in the temperature so a temperature–entropy plot is particularly handy for illustrating the processes making up an engine cycle. The temperature–entropy chart drawn as Figure 1.2 covers the range of conditions of greatest interest in gas turbine work and can be used to show real engine cycles as will be described in Chapter 3. This type of chart gives an added advantage as areas represent heat transfers in much the same way as areas on a pressure–volume diagram represent work transfers.

Although the preceding discussion explains some of the *uses* of entropy it does not really give a physical meaning for this property. To explain entropy fully is far beyond the scope of this book as it involves a fairly detailed study of statistical thermodynamics, but a basic feeling for entropy can be obtained if it is equated to *randomness* or *disorder*. It is common experience that throughout the world the natural order of things is towards increasing disorder. On the domestic scale, two children let loose in a tidy room can soon produce confusion! On an engineering scale, fluid which flows smoothly into a pipe will soon be full of whirls and vortices arising from the friction of the fluid at the pipe wall. The universal rule is that all real processes undergone by an isolated system (such as fluid flowing through an insulated pipe or gas expanding in a turbine) result in an *increase* in entropy between the initial and the final states. The great thermodynamicist Clausius could say

> "The Energy of the Universe remains constant.
> The Entropy of the Universe tends to a maximum."

The thermodynamic penalties of increases in entropy can be seen by looking at the property known as the *Gibb's function*. This function, like enthalpy, is a compound property and is defined as

$$g = h - Ts \tag{1.18}$$

Physically, the Gibb's function gives a measure of the maximum amount of work which could possibly be obtained from a steady flow system in expanding down from a given initial condition to a certain *dead state* where no further change would be possible. It is clear from equation 1.18 that any increase in entropy will lead to a reduction in the possible work output.

The Second Law of Thermodynamics

Much of the discussion above on ideal processes and entropy is relevant to an understanding of the second law of thermodynamics although this might not be immediately clear from any given statement of the law—a number of possible definitions are available and are given in various textbooks. A typical statement is that known as the Kelvin–Planck statement which goes: "It is impossible to construct an engine which, operating in a cycle, will produce no other effect than the extraction of heat from a single heat reservoir and the performance of an equivalent amount of work." This statement bars the possibility of ever constructing a perpetual motion machine which could operate by picking up "free" energy from its surroundings and converting it completely into work. Of the energy taken in by a heat engine from a high temperature source, only a proportion can be converted into useful work and the remainder must be rejected to a low temperature sink.

In essence, the second law comes down to saying which processes are possible and which are impossible. As an example, it is well known that, when a vehicle is brought to rest by the brakes, its initial kinetic energy goes to heat up the braking system. This is a natural dissipation process of the type we encounter every hour of every day. Experience tells us, though, that the reverse process never occurs spontaneously—the brakes never cool down and give out energy which can be used to accelerate the car to its original condition. The first law tells us what rate of exchange holds if energy interchange occurs, but the second law says whether a given interchange *can* ever take place. In fact, a process which converts work completely into an equivalent amount of heat energy is always possible but the reverse can never take place.

These thoughts on the second law can be tied in with the work done above on entropy if heat is thought of as being a "randomized" form of energy and work as being a more "directed" form. (Always with mechanical work it is possible to discern some direction in which the work takes place.) With this in mind, the realization that the direction of all natural processes is towards increasing disorder or randomness will give the pointer to which processes are possible and which are impossible. It will always be possible to convert work into heat as this entails an increase in randomness. The opposite can never be carried out completely, although, by using a carefully designed machine, some heat energy can be converted to useful work. As was said earlier, power engineering consists in trying to keep down as far as possible the downgrading of energy into more random forms.

The Kinetic Theory of Gases and the Mol

During the nineteenth century, the kinetic theory was developed to explain a number of the observed properties of gases. Subsequent discoveries have made detailed modifications to the theory necessary but, in its simplest form, it can still give satisfying explanations in many cases. Above all, the kinetic theory is useful for giving a firm physical picture of the behaviour of gases.

Basically the theory treats a gas as a cloud of molecules which are in constant motion and which behave like completely rigid spheres so that no energy is lost in collisions. Impact of the molecules against the walls of the chamber exerts a force on the walls which we record as the gas pressure. If the volume occupied by the gas is reduced the impacts become more frequent, so the pressure goes up. The total volume of the molecules is small compared with the complete volume of the space they occupy, at least at normal pressures, so gases can be compressed to a comparatively large extent.

When energy is stored by a gas as internal energy the mechanism is that energy transfers result in changes in the sum of kinetic energy of all the molecules. Clearly, therefore, what is observed as an increase in temperature of a gas represents in fact an increase in the average speed of the molecules. If the energy supplied comes in as a work transfer, the general increase in velocity takes place in a directed fashion, whereas if the energy is transferred as heat the increase in molecular motion will be completely random. This again emphasizes the difference in nature between heat and work.

(In his book *Thermostatics and Thermodynamics* (ref. 4) Tribus draws an analogy between the supply of energy to the molecules of a gas and to the supply of energy to a number of balls on a billiard table. He likens a *work* input to a lifting of the entire table while a *heat* input might be simulated by releasing a number of fast moving balls on the table—that is, mixing in some hot gas. With this analogy in mind it is easy to see how energy put in as a work transfer may be recovered with comparative ease, while that put in as heat transfer can only be got out with considerable difficulty.)

Avogadro's Hypothesis (put forward in 1811) states that equal volumes of all gases contain the same number of molecules when the gases are at some standard conditions of temperature and pressure. This hypothesis predicts that the density of a gas when measured at standard conditions will vary directly with its molecular weight—a prediction which is closely borne out in practice. Further, if the *mol* is defined as the amount of gas whose mass in kilogrammes is numerically equal to its molecular weight, the hypothesis says that 1 mol of any gas will occupy some standard volume when measured at standard pressure and temperature. In fact, at a pressure of $101{\cdot}3\ kN/m^2$ and a temperature of $0°C$ most gases occupy a volume of $22{\cdot}4\ m^3/kg\text{-}mol$.

Because of this property of the mol, a *universal gas constant* can be derived from the characteristic equation. Writing V_{mol} as the volume of 1 mol of a gas of molecular weight M the characteristic equation becomes

$$pV_{mol} = MRT$$

$$MR = \bar{R} = \frac{pV_{mol}}{T} = \text{constant} \tag{1.19}$$

$\bar{R}$ the universal gas constant has a value of 8·314 kJ/kg-mol-K. From these values, R for any gas may be found provided the molecular weight is known.

At times it is more convenient to work with specific heats of a gas based on mols rather than on the more usual basis of unit mass. For a gas of molecular weight M the *molal specific heats* will be $K_v = Mc_v$ and $K_p = Mc_p$, so the units are kJ/mol-K. From the Kinetic Theory of Gases it can be shown that for a monatomic gas (such as helium) K_v will have a value of about 12, for a diatomic gas (such as hydrogen or oxygen) K_v will have a value of about 21, while for a polytomic gas (such as carbon dioxide) K_v will have a value of about 29. In each case of course

$$K_p - K_v = \bar{R}$$

It is in the field of combustion work that the mol proves most useful. Very often here, mixtures of gases will occur and it will be necessary to write chemical equations for reactions involving these mixtures. For a mixture of gases, the percentage composition by *volumes* will be the same as the percentage composition by *mols*; further the composition by *mols* will be the same as the percentage composition by *molecules*. As we will see in Example 1.2 of the following section, this simplifies greatly many combustion calculations.

Reacting Systems

One type of process which is of very great interest in engineering thermodynamics is that involving a reacting system wherein heat is supplied to a working fluid by means of the combustion of a fuel. The fundamental laws of thermodynamics will still apply despite any change in composition of the system during the reaction. Combustion work *can* get very complex if detailed analysis is wanted but it is sufficient for the present to look at the overall picture as shown by the conservation laws. A small amount of elementary chemistry is needed for this type of work and it can best be revised by working through a specific problem such as the following.

Example 1.2 Calculate the correct amount of air required for the complete combustion of unit mass of a fuel whose chemical composition is represented by the formula C_7H_{16}. Find also the volumetric composition of the exhaust products if this fuel is burned with 400% of the chemically correct amount of air. Assume in each case that the fuel burns to carbon dioxide and water vapour.

Solution: The first step is to write the equation for complete combustion bearing in mind that the number of *atoms* of each element must balance on each side of the equation. The volumetric analysis of air can be taken as 21% oxygen, 79% nitrogen; from the work on the mol on p. 19 we know that the volumetric analysis is the same as the molal analysis, and that the molal analysis is the same as the analysis by molecules.

Writing the equation we get

$$C_7H_{16} + 11O_2 + 11 \times \frac{79}{21} N_2 \rightarrow$$

$$7CO_2 + 8H_2O + 11 \times \frac{79}{21} N_2$$

To find the air/fuel ratio by weight we write in the atomic weights for each of the reactants, C = 12, H = 1, 0 = 16, N = 14 and get

$$C_7H_{16} + 11O_2 + 11 \times \frac{79}{21} N_2 \text{ as follows}$$

$$(7 \times 12 + 16 \times 1) + (11 \times 32) + \left(11 \times \frac{79}{21} \times 28\right)$$

$$100 + 352 + 1\,160$$

Therefore Air/fuel ratio by weight = 1 512/100 = 15·12:1

This is known as the *stoichiometric* air/fuel ratio.

For the second case, with 400% of the stoichiometric air supply the equation becomes

$$C_7H_{16} + 4 \times 11O_2 + 4 \times 11 \times \frac{79}{21} N_2 \rightarrow$$

$$7CO_2 + 8H_2O + 3 \times 11O_2 + 4 \times 11 + \frac{79}{21} N_2$$

To get the volumetric analysis of the products we work out each constituent as a percentage of the total number of *molecules*.

$$\text{Total molecules of products} = 7 + 8 + 33 + 44 \times \frac{79}{21} = 214$$

Percentage CO_2 = 3·27   Percentage O_2 = 15·41
Percentage H_2O = 3·74   Percentage N_2 = 77·58

Turning now to the first law analysis of a reacting system, consider the same case as the second part of Example 1.2, namely a steady flow system where, in unit time, one mol of C_7H_{16} enters with four times the stoichiometric quantity of air and burns at constant pressure to give the products as shown above. Looking at the reactants first, it is possible to plot H_R, the total enthalpy of all the

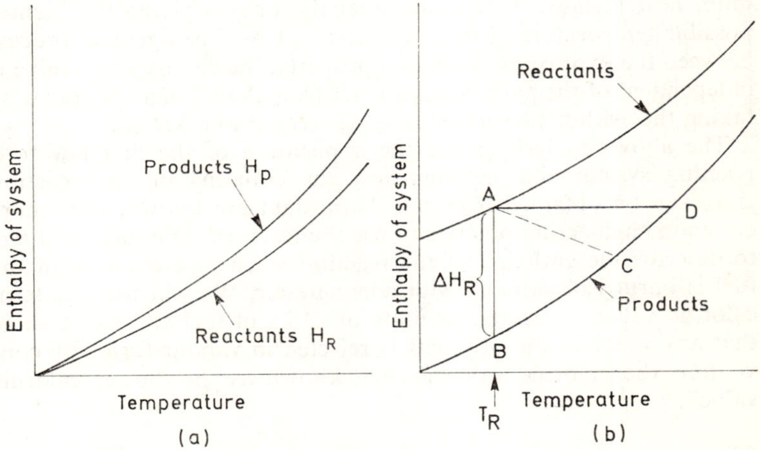

FIGURE 1.6 *Application of the first law to a combustion process*

reactants, referred to that at some datum temperature, against absolute temperature. The curve shown in Figure 1.6(a) is calculated from

$$H = \Sigma n K_p \Delta T$$

where n is the number of mols, K_p is the molal specific heat and ΔT is the temperature above the datum value. On the same graph is plotted H_p the total enthalpy of the products measured above the same datum value. Although Figure 1.6(a) shows a true picture of the enthalpies of reactants and products referred to some datum temperatures it does not take account of the energy *available in the reactants* which might be released by combustion. It is known, however, that if the mixture is ignited it will be necessary to remove a large amount of heat from the system if the products are to be kept at the same temperature as the reactants. An experiment might be carried out to measure this quantity of heat ΔH_R at some reference temperature T_R and a new graph Figure 1.6(b) could be drawn to show the more complete picture. In Figure 1.6(b) the H_R versus temperature curve of Figure 1.6(a) has been shifted upwards by the amount ΔH_R at T_R.

A graph similar to that of Figure 1.6(b) can be drawn for any reacting system as ΔH_R will be available for all common fuels in the appropriate reference books. The graph shows how much heat is rejected by the system for any temperatures of reactants and products. For example process A–B is an isothermal with the maximum heat release; A–D is an adiabatic one so giving the highest possible temperature of the products; and A–C is a general process between the two. As enthalpy is a property, the change in its value is independent of the path taken, so ΔH from A to C may be found by taking the arithmetic sum of $\Delta H_{\mathrm{A-E}}$, $\Delta H_{\mathrm{E-F}}$ and $\Delta H_{\mathrm{F-C}}$.

The above analysis shows the application of the first law to a reacting system and explains how the enthalpy of any reaction ΔH_R can be obtained if certain basic data are known. It is more common engineering practice to use the term *calorific value of a fuel* to describe the enthalpy of the reaction when unit quantity of the fuel is burnt isothermally with adequate air. We will use this term calorific value, measured in units of kJ/kg of fuel and will assume that any water in the products is rejected in vapour form. By convention the calorific value is then known as the "lower calorific value".

Dissociation and Chemical Equilibrium

A chemical equation for a combustion process such as

$$H_2 + \tfrac{1}{2}O_2 \rightarrow H_2O + 242\ \mathrm{MJ/mol\ of\ } H_2$$

is a convenient shorthand way of describing a complex series of events. It says that molecules of hydrogen and oxygen may be made to combine and that, when they do, the difference between the binding energy of the molecules in the initial and final states is such that a considerable amount of energy is released to the surroundings. But the equation tells nothing of *how* the reaction can be made to occur—it merely gives an overall picture of what happens if a reaction *does* take place. It is well known, though, that a mixture of oxygen and hydrogen can be kept safely in a container without reacting until some spark ignites the mixture and triggers off the reaction.

Changes in the molecular structure of substances such as occur in chemical reactions will take place only if a certain minimum amount of energy is supplied to the molecules and this energy will normally come from collisions between molecules. The *activation energy* required to start a reaction will depend on the particular circumstances but the important point to note is that, given sufficient energy, almost any reaction can take place. To take the example above, it does not take much excess energy to cause hydrogen and oxygen to combine but, if sufficient energy is supplied, a reverse reaction can take place and the complex H_2O molecule will break down. This process is known as *dissociation.* Any compound may dissociate given the right conditions into simpler atoms or groups of atoms.

Temperature is always the most important parameter in dissociation effects because increasing temperature steps up the general level of the kinetic energy of the molecules in a gas. In any quantity of gas, the molecules in their random motion will have a very wide distribution of velocities. The vast majority will be at one preferred velocity but there will always be appreciable numbers moving slowly and appreciable numbers moving at high speed. These latter energetic high-speed molecules are the ones of greatest interest to us as they are liable to collide with other molecules and break them down into their atoms. At normal atmospheric temperatures the high-speed gas molecules ("high-speed" here applies to the molecules, not the gas!) are so scarce that there is little dissociation. On the other hand, as temperature increases, the number of energetic molecules goes up and the degree of dissociation can increase rapidly. In fact the distribution of molecular energies follows a logarithmic law—the Maxwell–Boltzmann distribution. The number of molecules n with energy E in a total of n_0 molecules at temperature T is given by

$$n = n_0 e^{-E/RT}$$

It should now be clear that a chemical equation should really be written with arrows going each way, as for example

$$H_2 + \tfrac{1}{2}O_2 \rightleftharpoons H_2O$$

because both the forward and reverse reactions will be taking place at the same time. For any temperature there will be an *equilibrium composition* of the constituents (H_2, O_2 and H_2O in the above case) and the forward and reverse reactions will continue until this equilibrium composition is reached. The composition is defined by an equilibrium constant K_p for the reaction, which is a function of temperature and which gives a measure of the mol fractions of the various constituents in the mixture. As shown in thermodynamics textbooks, if a reaction is of the form

$$a \text{ mols of A} + b \text{ mols of B} \rightleftharpoons c \text{ mols of C} + d \text{ mols of D}$$

the equilibrium constant will be given by

$$K_p = \frac{(p_C)^c(p_D)^d}{(p_A)^a(p_B)^b} \tag{1.20}$$

where p_A, p_B, p_C, p_D are the partial pressures (equivalent to mol fractions) of each of the constituents.

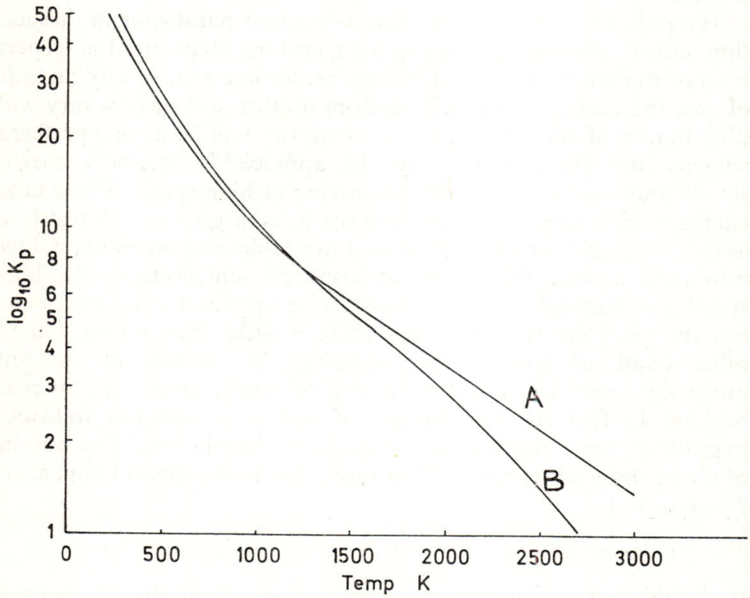

FIGURE 1.7 *Equilibrium constants for typical reactions*

Curve A: $H_2 + \frac{1}{2}O_2 \rightleftharpoons H_2O$
Curve B: $CO + \frac{1}{2}O_2 \rightleftharpoons CO_2$
(Partial pressures in atmospheres)

Figure 1.7 shows K_p versus temperature for two typical reactions and the following example illustrates how this data may be used.

Example 1.3 A container holds equal mol fractions of oxygen and hydrogen which are caused to react. After certain heat transfers have taken place the container is found to contain an equilibrium mixture of oxygen, hydrogen and water vapour, 5% of the total water vapour having dissociated to form oxygen and hydrogen. At these conditions the total pressure of the gases in the vessel is 1 atmosphere. Find the temperature of the mixture at the second condition.

Solution: As equal mol fractions of oxygen and hydrogen are stipulated this implies that equal numbers of molecules are present in the original mixture. The reaction may therefore be written

$$H_2 + O_2 \rightarrow H_2O + \tfrac{1}{2}O_2$$

Heat is now transferred from or to the system until 5% of the H_2O has dissociated. The equilibrium mixture will then consist of

$$0{\cdot}95\ H_2O + 0{\cdot}05\ H_2 + 0{\cdot}05 \times \tfrac{1}{2}O_2$$

If the amount of excess oxygen present is added to this, the total mixture will be

$$0{\cdot}95\ H_2O + 0{\cdot}05\ H_2 + (0{\cdot}5 + 0{\cdot}025)O_2$$

Adding up the total mols present as 1·525 mols we can then write the mol fractions of each constituent as

H_2O	0·95/1·525 = 0·623
H_2	0·05/1·525 = 0·033
O_2	0·525/1·525 = 0·344
	1·000

But with a total pressure of 1 atmosphere the mol fractions will be equal to the partial pressures in atmospheres. The equilibrium constant K_p for the reaction is

$$K_p = \frac{p_{H_2O}}{p_{H_2}(p_{O_2})^{1/2}}$$

Putting in the values gives

$$K_p = \frac{0{\cdot}623}{0{\cdot}033(0{\cdot}344)^{1/2}} = 32{\cdot}8$$

$$\log_{10} K_p = 1{\cdot}516$$

The scale of Figure 1.7 is not sufficiently large to allow an exact determination of the temperature corresponding to this K_p but reference to more detailed data shows that the temperature is actually 2 900 K.

(Insufficient information has been given to allow calculation of the amount and direction of heat transfer from the container. If the appropriate books are consulted it will be found that a mixture of hydrogen with twice the stoichiometric amount of oxygen will give temperatures very much higher than the 2 900 K found above, so the heat transfer must be a heat *rejection*. Calculation of the actual amount would be quite a difficult process as it would depend on the initial temperature and would need a knowledge of the properties of water vapour, oxygen and hydrogen at high temperatures.)

What then is the significance of dissociation? The most important point to note is that when dissociation occurs it generally implies the *absorption* of energy as opposed to its release in a chemical reaction. This means that the peak temperature attained in a combustion process is reduced. In addition, as the temperature of a mixture falls, the dissociation products will recombine and release energy but at a lower temperature. From the entropy change equation 1.15 the lower temperature will mean a larger entropy increase for a given heat transfer. Entropy increases, as has been seen, should always be kept to a minimum. Finally, as will appear later in the chapter on rockets, dissociation leads to some uncertainty as to the true composition of the exhaust gases resulting from a high temperature reaction.

1.4 PRINCIPLES OF FLUID MECHANICS

Our interest in fluid mechanics will be in those parts related particularly to gases and especially to gases in motion—the gas dynamics aspects of the complete field of fluid flow. At times we will find that it will be an acceptable approximation to treat the fluid as if it were incompressible but this will always be a simplifying assumption which must be treated with care. It is assumed also that the gases are not at such a low pressure that the mean free path of the molecules is a significant dimension compared with the size of an object we are considering—this assumption only breaks down at the low pressures encountered at extreme altitudes on the edge of space. The aim throughout this section again, as with the preceding ones, is to provide in a convenient form a statement of the main principles of the subject with a minimum of proofs and once more to emphasize the physical picture of processes. For a rigorous treatment of the science of fluid mechanics with full proofs the reader must consult standard textbooks such as refs. 5 and 6.

Standard Atmospheres

Aero-engines, being aerodynamic machines, are very sensitive to changes in the atmosphere which surrounds them. Variations in atmospheric conditions over the world from day to day are so random that it is necessary to specify some standard atmospheric conditions to which aircraft and engine performance can be referred. In this book we will refer conditions in general to the International Standard Atmosphere (ISA) conditions which are tabulated in Table 1.2 for the range sea-level to 20 km. (The reference quoted also gives proposed standard conditions for atmospheres over a much larger range of altitudes, but 20 km will be adequate here.) Table 1.2

shows that at ISA sea-level the pressure is 101·33 kN/m² and the temperature is 288 K (15°C) and that the temperature falls by about 13 degC for each 2 000 m increase in altitude to a height of 11 km. Above this height the temperature remains constant over the range we are considering but starts to rise again at about 24 km. Atmospheric pressure falls continuously but non-linearly over the whole range of altitudes. The region below 11 km is known as the *troposphere* and above as the *stratosphere*, while the separating surface is known as the *tropopause*. This tropopause, at the discontinuity in the temperature versus height relationship, is a very real entity and can be used, for example, to reflect radio waves. The actual height of the tropopause will vary from time to time but will generally lie in the range of heights from 10 to 13 km. Also in Table 1.2 is shown the air density which is calculated from the characteristic equation $pv = RT$ and the viscosity of the air. Viscosity is discussed in greater detail on p. 54.

TABLE 1.2 *The International Standard Atmosphere*

Altitude km	Pressure kN/m²	Temperature K	Density kg/m³	Kinematic viscosity m²/s × 10⁵
0	101·33	288	1·23	1·46
2	79·5	275	1·07	1·72
4	61·7	262	0·823	2·03
6	47·2	249	0·663	2·42
8	35·7	236	0·528	2·90
10	26·5	223	0·415	3·53
11	22·7	217	0·366	3·90
12	19·4	217	0·313	4·56
14	14·2	,,	0·229	6·24
16	10·4	,,	0·167	8·54
18	7·57	,,	0·122	11·69
20	5·53	,,	0·089	15·99
22	4·05	,,	0·065	22·20

Based on Data Sheet *01*, by kind permission of The Royal Aeronautical Society, London

It should be emphasized once again that the ISA climatic conditions are a model only and approximate with varying degrees of accuracy to the real atmosphere at any given time and place. With a real atmosphere there may even at times be what is known as a temperature inversion wherein the air temperature will actually rise with increase in altitude. For reference purposes, though, it is essential to have some standard such as ISA climate to which to refer.

Properties of a Moving Gas Stream

In measuring the properties of a moving gas stream it is found that the readings of the instruments depend very much on the relative velocity between instrument and gas. It is possible to distinguish in particular between two extreme sets of conditions: (i) when the measuring instrument is moving with the gas stream and so is static with reference to the gas, and (ii) when the instrument is itself stationary so that the gas impinges on it with the full gas stream velocity. The properties measured under the former conditions are known as the *static* properties and are written without a subscript (p, T, h, etc.) while those recorded by ideal stationary instruments are known as *stagnation* or *total* properties, being written with subscript 0 (p_0, T_0, h_0, etc.).

Consider first a gas flowing in a horizontal duct with velocity V (Figure 1.8) and let there be two thermometers, A moving with the gas stream and B mounted in the walls of the duct. With perfect

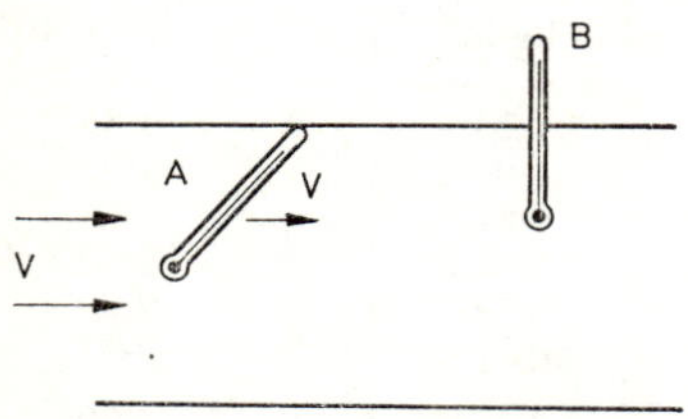

FIGURE 1.8 *Temperature measurement in a moving gas stream*

thermometers, A would read the static temperature and B would read the stagnation temperature. To find the relationship between the two temperatures, write the steady flow energy equation (equation 1.12) for a particle of air initially moving freely in the gas stream but then brought to rest by impinging on thermometer B. No heat is transferred, no external work is done so the energy equation reduces to

$$h_1 + \frac{V_1^2}{2 \times 1\,000} = h_2 + \frac{V_2^2}{2 \times 1\,000}$$

(The constant of 1 000 makes sure that both units are in kJ/kg.) But if the gas comes to rest relative to B, $V_2 = 0$ and we get the stagnation enthalpy h_0 as

$$h_0 = h + \frac{V^2}{2 \times 1\,000} \tag{1.21}$$

or

$$h_0 - h = \frac{V^2}{2 \times 1\,000} \tag{1.21a}$$

The relationship between stagnation and static temperatures can be found by combining equations 1.14 and 1.21a so that

$$T_0 - T = \frac{h_0 - h}{c_p} = \frac{V^2}{2 \times 1\,000 \times c_p} \tag{1.22}$$

This difference between stagnation and static temperatures is often called the *ram temperature rise*. Most of the serious problems associated with high-speed flight arise because of this ram temperature rise. It means that all air needed by the engines, by the passengers and for cooling purposes, will be subjected to the ram temperature rise.

Deceleration of the air from the full free stream velocity to stagnation conditions is a reversible adiabatic process and so the isentropic relations will apply. The temperature rise gives a pressure rise which can be calculated from

$$\frac{p_0}{p} = \left(\frac{T_0}{T}\right)^{\gamma/\gamma-1} \tag{1.23}$$

This *ram pressure ratio* is a useful bonus for the propulsion engineer in high-speed flight. Taking a value of $\gamma = 1{\cdot}4$ we find for example that in the stratosphere the isentropic ram pressure ratio for a flight speed of 3 200 km/h (2 000 miles/h) is 39. The difference between stagnation pressure p_0 and static pressure p of a gas stream is known as the *dynamic head*.

It must be emphasized that it is the *static* properties of a gas which determine such parameters as the gas density and the speed of sound in the gas. That this must be so is obvious if one thinks that the molecules of a gas have no "knowledge" of their velocity with respect to a randomly chosen reference point. The air in a room does not know that it is moving through space at very great speed and this motion does not alter the fact that 1 kg of the air occupies about 0·8 m^3, and that the speed of sound is about 330 m/s.

The Basic Equations for Steady Flow

Three fundamental conditions of conservation must be observed by the gas in any steady flow process. These are

(i) *Conservation of Mass.* Matter can be neither created nor destroyed in the systems we are dealing with.

(*ii*) *Conservation of Energy.* The first law of thermodynamics must be obeyed.

(iii) *Conservation of Momentum.* The net force acting on a system must be equal to the overall rate of change of momentum.

Equations can be written to express each of these conditions and these equations can take a number of different forms. For the present the law of conservation of mass for flow in a duct will be stated as

$$\dot{m} = \rho A V = \text{constant} \tag{1.24}$$

at all sections of the duct, ρ being the gas density.

The equation for conservation of energy can most usefully be used in the form of the steady flow energy equation.

In order to derive the equation for the conservation of momentum it is important to look carefully at the gas flow and to bring in all the forces which are acting. Figure 1.9 represents a system where gas is flowing in a duct. Forces are acting on an element of gas due to velocity variations, to gravitation effects and to friction

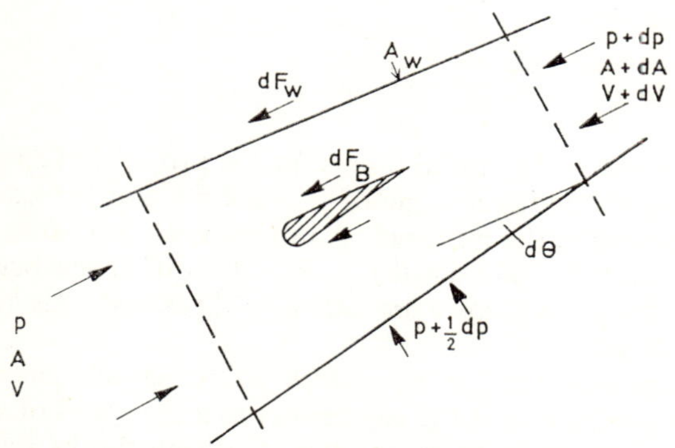

FIGURE 1.9 *Forces on an element of gas flowing through a duct*

at the walls and around bodies immersed in the stream. The net value of all these forces must be balanced by the rate of change of momentum of the gas stream resolved in the direction of motion. Neglecting gravitational terms the equation can be written

$$pdA - (p + dp)(A + dA) + (p + \tfrac{1}{2}dp)\, A_W \sin\theta - dF_W \cos\theta - dF_B$$
$$= \rho A V\,[(V + dV) - V]$$

But $A_w \sin \theta = dA$ and if θ is small $\cos \theta$ is approximately unity. Neglecting products of differentials this equation can be expanded and simplified, as is done in standard fluids textbooks to give

$$dp + \rho V \, . \, dV - \frac{dF_W}{A} - \frac{dF_B}{A} = 0 \tag{1.25}$$

If there are no friction forces equation 1.25 reduces to

$$dp + \rho V \, . \, dV = 0 \tag{1.26}$$

which is known as *Euler's Equation.*

This in turn can be integrated to give the well-known *Bernoulli Equation*

$$p + \tfrac{1}{2}\rho V^2 = \text{constant} \tag{1.27}$$

For low-speed flow the term $\frac{1}{2}\rho V^2$ is the dynamic pressure of the gas stream.

The Acoustic Velocity and Mach Number

One of the most important parameters for a gas which must be known in the study of gas dynamics is the local velocity of sound in the gas. Sound consists of pressure waves of very small amplitude passing through the gas. The compression wave is adiabatic and reversible and it can be shown that the speed of propagation of such a wave through the medium is given by

$$a = \sqrt{\left(\frac{dp}{d\rho}\right)_s} \tag{1.28}$$

Now $p/\rho = RT$ and for an isentropic process p/ρ^γ = constant, so

$$a = \sqrt{(\gamma \, RT)} \tag{1.29}$$

In this equation if the units are to be consistent, the characteristic constant must be measured in J/kg-K and so for air R will have a value of 0·2871 × 1 000. Adding further the value of $\gamma = 1{\cdot}4$ for air at normal atmosphere temperature we get the sonic velocity in metres per second,

$$a = 20{\cdot}0\sqrt{T} \tag{1.29a}$$

where T is the *static* temperature in kelvins.

Consider now the more general case for any gas of molecular weight M. The characteristic constant is found from the Universal Gas Constant and the molecular weight by the formula

$$R = \bar{R}/M$$

Looking at this in conjunction with equation 1.29 shows that, if variations in γ are neglected to a first order, gases with a low molecular weight will have a high acoustic velocity and vice versa. This fact will arise again in discussion of rocket propulsion in Chapter 10.

Much important early work on supersonic flow was done by the Austrian physicist Ernst Mach (1838–1916) and his name is commemorated in the *Mach Number M* which is the ratio of the velocity of a gas stream to the local acoustic velocity, i.e.

$$M = V/a$$

For subsonic conditions $M < 1$, for supersonic $M > 1$, while if $M > 5$ the flow is said to be hypersonic.

One of the fundamental differences between subsonic and supersonic conditions can be illustrated by Figure 1.10, which shows a

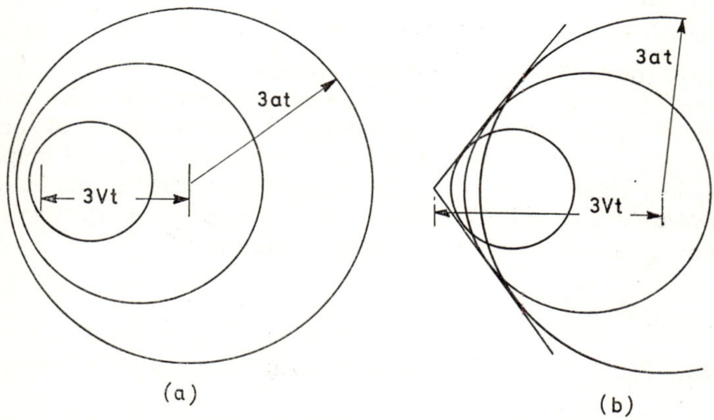

FIGURE 1.10 *Disturbance produced by a point source moving through a gas*
(a) Subsonic (b) Supersonic

particle moving through a gas with velocity V with the pressure waves "rippling" out from the particle at the speed of sound a. In Figure 1.10(a) the flow is subsonic so that V is less than a, and the pressure waves giving notice of the movement of the particle through the gas form a number of circles each lying completely within the original circle—the locus of the centre of the circles is moving with velocity V and the original circle is moving outward with velocity a. The picture in Figure 1.10(b) showing the supersonic case is very different. Here the centres of disturbance no longer lie within the circles of influence and the envelope for the sound waves is in the form of a conical sheet with its apex at the disturbance and with a half cone angle of $\sin^{-1}a/V$. Outside this cone is a region known as the "zone of silence" where the gas is undisturbed and knows nothing of the passage of the particle. For the case of a point source of disturbance where the disturbance is infinitesimally small, the envelope lines are known as *Mach Lines*, and the angle made between Mach lines and the direction of motion is the *Mach Angle*.

A useful physical picture of the meaning of the acoustic velocity follows from consideration of the kinetic theory of gases. Fundamentally, the speed of passing a message through a medium depends on the speed of the messengers. When the mails were carried by relays of horsemen who could average about 12 miles per hour it would take 30 hours to get a message from Edinburgh to London. Now the messengers which pass an acoustic signal through a gas are the molecules of the gas. One would expect, therefore, that the speed of sound in a gas would be closely related to the mean speed of the molecules and this is in fact the case. Additionally, we have seen before that there is a tie-up between the internal energy of a gas, the total kinetic energy of the molecules and the temperature, the relationship being such that the temperature varied as the square of the molecular velocity. In confirmation of this we have equation 1.29a, which shows that the speed of sound in a gas depends only on the square root of the absolute temperature.

Isentropic Flow Relationships

Very often in propulsion work the case arises of the flow of gases through passages or ducts where there is negligible heat transfer and no external work. In these cases the stagnation enthalpy and temperature of the gas stream are constant and, in the ideal case, there will be no change in stagnation pressure—in practice, though, all real flow will have *some* loss in stagnation pressure however slight. The ideal flow is, therefore, adiabatic and reversible and a number of very useful isentropic flow relationships can be derived. This work with the resulting design curves, was first presented in a paper by Jamison and Mordell (ref. 7).

Consider functions which can be expressed in terms of the Mach number of the gas stream, starting with the stagnation to static temperature and pressure ratios.

$$T_0 - T = \frac{V^2}{2c_p}$$

$$\frac{T_0}{T} - 1 = \frac{V^2}{2Tc_p}$$

But $V = Ma = M\sqrt{(\gamma RT)}$ and therefore

$$\frac{T_0}{T} - 1 = \frac{M^2\gamma RT}{2Tc_p} = \frac{\gamma RM^2}{2c_p} \qquad (1.30)$$

From equation 1.16

$$\frac{R}{c_p} = \frac{\gamma - 1}{\gamma}$$

Writing this in equation 1.30a and rearranging gives

$$\frac{T_0}{T} = [1 + \tfrac{1}{2}(\gamma - 1)\, M^2] \qquad (1.31)$$

Also, from equation 1.16 we can write

$$\frac{p_0}{p} = \left(\frac{T_0}{T}\right)^{\gamma/\gamma-1} = [1 + \tfrac{1}{2}(\gamma - 1)M^2]^{\gamma/\gamma-1} \qquad (1.32)$$

These two equations show, therefore, that the stagnation to static temperature and pressure ratios are each a unique function of Mach number for a given value of γ.

The pressure ratio needed to give a Mach number of unity is known as the *critical pressure ratio* for the gas. Putting $M = 1$ in equation 1.32, shows that the critical pressure ratio for a gas with $\gamma = 1{\cdot}4$ is

$$p_0/p = [1 + \tfrac{1}{2}(\gamma - 1)]^{\gamma/\gamma-1} = (1{\cdot}2)^{3{\cdot}5} = 1{\cdot}893$$

Supersonic flow is not possible unless the stagnation pressure available is at least this value times the outlet static pressure.

Coming now to the *velocity* of a gas stream a simple velocity functions can be derived as follows:

$$V = Ma = M\sqrt{(\gamma RT)}$$

$$\frac{V}{\sqrt{T}} = M\sqrt{(\gamma R)} \qquad (1.33)$$

and

$$\frac{V}{\sqrt{T_0}} = M\sqrt{(\gamma R)}\sqrt{\frac{T}{T_0}} = \frac{M\sqrt{(\gamma R)}}{\sqrt{[1 + \tfrac{1}{2}(\gamma - 1)M^2]}} \qquad (1.33a)$$

The latter equation 1.33a is generally more useful as the stagnation temperature will usually be known rather than the static temperature.

Mass flow parameters can be derived from the continuity equation as follows:

$$\dot{m} = \rho A V$$

But $\rho = p/RT$ and $V = aM = M\sqrt{(\gamma RT)}$. Therefore

$$\dot{m} = \frac{pAM\sqrt{(\gamma RT)}}{RT}$$

$$\frac{\dot{m}\sqrt{T}}{Ap} = M\sqrt{\frac{\gamma}{R}} \tag{1.34}$$

Further relationships in terms of stagnation conditions are obtained as follows

$$\frac{\dot{m}\sqrt{T_0}}{Ap} = \frac{\dot{m}\sqrt{T}}{Ap}\sqrt{\frac{T_0}{T}} = M[1 + \tfrac{1}{2}(\gamma - 1)M^2]^{1/2}\sqrt{\frac{\gamma}{R}} \tag{1.35}$$

and

$$\frac{\dot{m}\sqrt{T_0}}{Ap_0} = \frac{\dot{m}\sqrt{T_0}}{Ap}\frac{p}{p_0} = \frac{M[1 + \frac{1}{2}(\gamma - 1)M^2]^{1/2}\sqrt{(\gamma/R)}}{[1 + \frac{1}{2}(\gamma - 1)M^2]^{\gamma/\gamma-1}}$$

$$= \frac{M\sqrt{(\gamma/R)}}{[1 + \frac{1}{2}(\gamma - 1)M^2]^{(\gamma+1)/2(\gamma-1)}} \tag{1.36}$$

Graphs for these relationships have been prepared for a gas approximating to air at normal temperatures with $\gamma = 1{\cdot}4$ and $R = 287{\cdot}1$ J/kg-K and are presented in Figure 1.11; these graphs are drawn for use with the SI systems of units. The following example shows how the curves may be used to aid the solution of a gas flow problem.

Example 1.4 Air flows from the atmosphere through a nozzle into a chamber which is maintained at sub-atmospheric pressure. At entry to the nozzle the stagnation pressure and temperature are 101·35 kN/m² and 15°C and the nozzle outlet velocity is to be 250 m/s. Assuming isentropic flow through the nozzle use the curves of Figure 1.11 to find what pressure must be maintained in the chamber. Calculate also the nozzle outlet static temperature and Mach number and the mass flow rate of air for a nozzle outlet area of 0·01 m².

Solution: Through the nozzle the stagnation temperature must remain constant at 288 K. At outlet therefore

$$V/\sqrt{T_0} = 250/\sqrt{288} = 14{\cdot}8$$

From the figures, the stagnation to static pressure ratio to give this is $p_0/p = 1{\cdot}50$. If the flow is isentropic then the stagnation pressure must be constant through the nozzle at 101·35 kN/m² and hence the nozzle outlet static pressure is

$$p = 101{\cdot}35/1{\cdot}50 = 67{\cdot}6 \text{ kN/m}^2$$

Assuming that the chamber is large so that gas velocities in the chamber are negligible this pressure must equal the stagnation pressure in the chamber. Also from the figures we see that $M = 0{\cdot}79$ and $T_0/T = 1{\cdot}125$ at this pressure ratio. Hence the nozzle outlet static temperature is 288/1·125, i.e. 256 K.

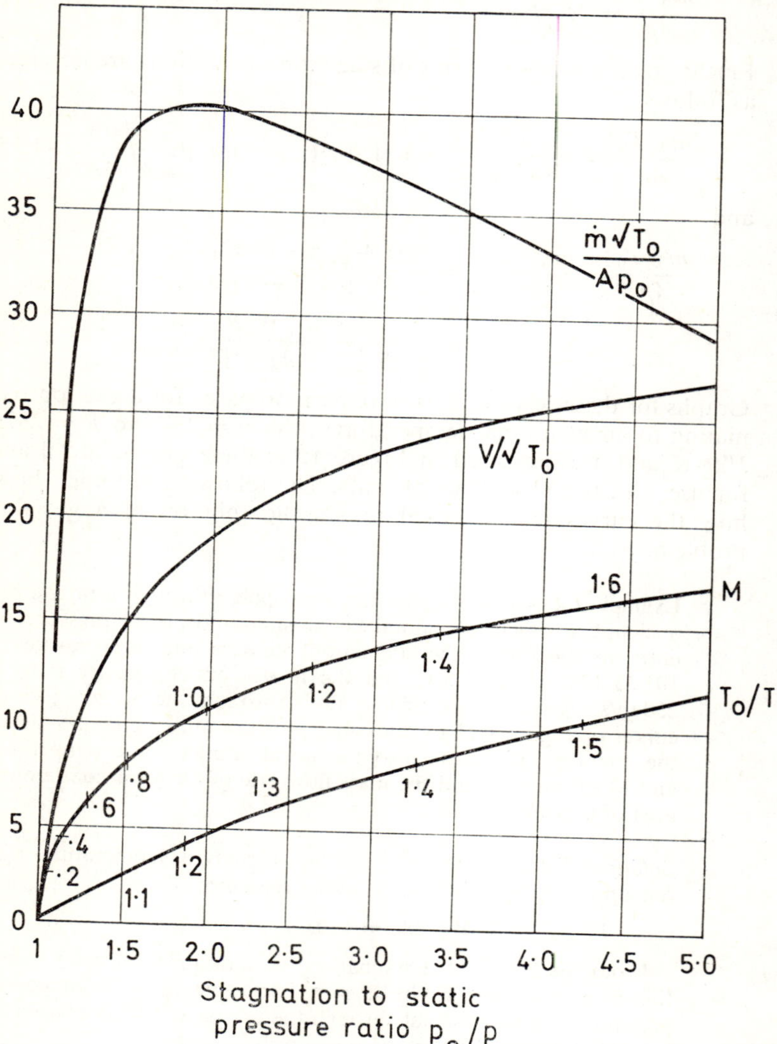

FIGURE 1.11 *Isentropic flow relationships*
$\dot{m}$ in kg/s; Ap_0 in kN; V in m/s; T_0 in K

The most convenient curve to use in finding the mass flow rate is that of $\dot{m}\sqrt{T_0}/Ap_0$, this parameter having a value of 38·8. Hence

$$m = (38{\cdot}8 \times 0{\cdot}01 \times 101{\cdot}35)/\sqrt{288} = 2{\cdot}88 \text{ kg/s}$$

Examination of the curves of Figure 1.11 reveals one very interesting point. All the curves, with one exception, increase continuously with increasing Mach numbers. The one which does not conform, that of $\dot{m}\sqrt{T_0}/Ap_0$, reaches a maximum at a Mach number of unity. (That this must be so can be demonstrated analytically by differentiating equation 1.35 and equating to zero. If this is done it will be found that γ cancels out from the equation and the curve passes through a maximum at $M = 1$ for all gases.) Now, for isentropic flow in a duct, the stagnation temperature and pressure are constant along the duct and the mass flow rate will also be constant, so the area required for a given flow must vary inversely with $\dot{m}\sqrt{T_0}/Ap_0$. As this passes through a *maximum* at $M = 1$, then the flow per unit area must pass through a *minimum* at the same condition.

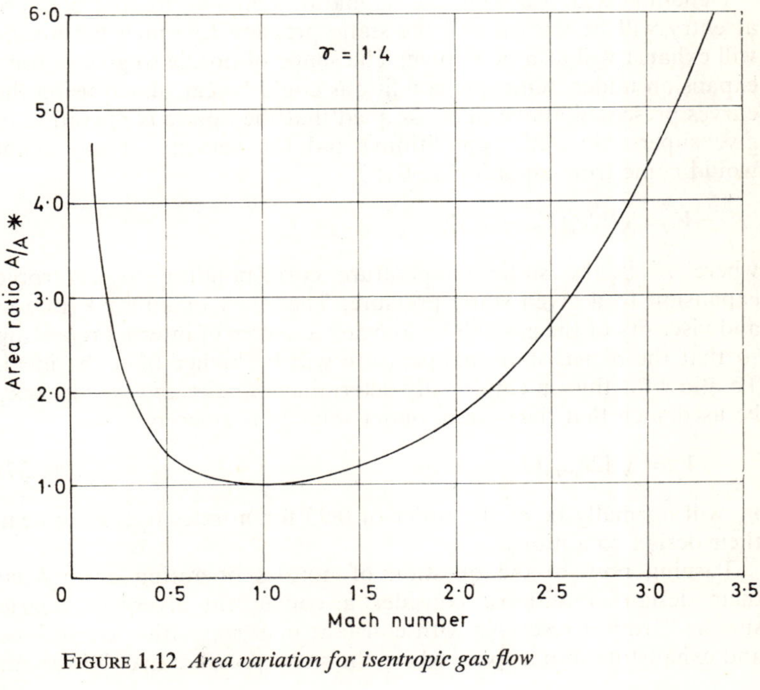

FIGURE 1.12 *Area variation for isentropic gas flow*

From the above, it will clearly be possible to calculate the area variation required in a duct which is to accelerate a flow up to any velocity. It is convenient to express the areas in terms of an area ratio A/A^* where A^* is the minimum area of the flow, that corresponding to a Mach number of unity. Figure 1.12 shows A/A^* versus Mach number for a gas with the same values of γ and R as for the previous curves. The graph shows that the distinctive feature of any duct which is to accelerate gases to supersonic conditions is that it must be of a convergent–divergent shape. If the outlet Mach number is to be very high, as occurs for example in rockets for space flight, the exit area will be very large compared with the throat area.

Non-ideal Flow through Nozzles

The most interesting cases for variable area flow are those for flow through nozzles which are used to accelerate a gas to high velocity. In these cases allowances must be made for the irreversibilities that always occur under real operating conditions and it is important to know what happens when a convergent–divergent nozzle is operated away from its design conditions.

Generally with a nozzle, the stagnation temperature and pressure at entry will be known and the static pressure to which the nozzle will exhaust will also be known. The shape of nozzle to give smooth expansion under isentropic conditions could be calculated using the curves presented above (it is assumed that the nozzle is operating to give supersonic outlet conditions) and the velocity at any point would come from equation 1.30a:

$$V = \sqrt{[2c_p(T_0 - T')]}$$

where T' is the static temperature corresponding to isentropic expansion to a given static pressure. The effect of internal friction and viscosity of the gas will be to cause a degree of internal reheating so that the *actual* static temperature will be higher than the ideal. To allow for this an empirically determined *nozzle efficiency* η_N may be used such that the correct outlet velocity is given by

$$V = \sqrt{[2c_p\eta_N(T_0 - T')]} \tag{1.37}$$

η_N will normally be of the order of 0·95 for nozzles operating near their design conditions.

Turning now to the question of nozzles operating *away from* their design conditions, consider a convergent–divergent nozzle supplied from a reservoir with constant inlet stagnation conditions and exhausting into a space where the static pressure may be varied.

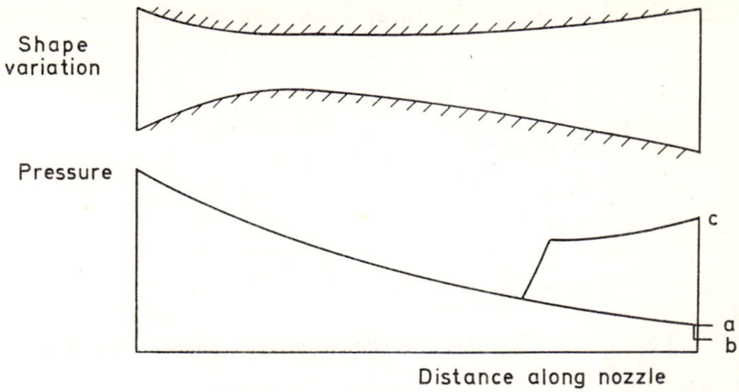

FIGURE 1.13 *Pressure variation through a nozzle at off-design conditions*

Figure 1.13 shows the arrangement and below is plotted the pressure variation along the nozzle for varying back pressures. Looking at the lines labelled (a), (b) and (c), the following points can be deduced:

Line (a)—the design case: At design conditions there is a smooth variation in pressure throughout the nozzle.

Line (b)—back pressure below design value: The gas flow downstream of the throat is supersonic and it has been noted before that signals cannot be transmitted upstream in a supersonic gas stream. A reduction in back pressure below the design value can, therefore, have no effect on the flow inside the nozzle so the pressure–distance variation will be as before. Once it is outside the nozzle abrupt changes in pressure will take place.

Line (c)—back pressure above design value: When the back pressure starts to rise *above* the design value, something very drastic must happen to the gas flow. A smooth variation of conditions along the entire length of the nozzle is no longer possible and a discontinuity must occur at some station in the diverging portion of the nozzle. If measurements are made it is found that at this discontinuity the flow changes from supersonic to subsonic and that there is a rise in static pressure, in static temperature and in gas density. The change in conditions is very abrupt and the flow is said to pass through a "normal shock".

Normal Shock

The normal shock phenomenon is of very great importance as such a shock will occur wherever flow decelerates from supersonic to subsonic conditions. The three basic conservation equations must

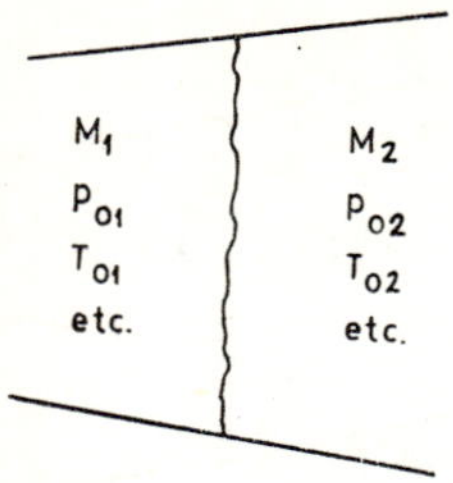

FIGURE 1.14 *Variation in conditions across a plane normal shock*

hold across the shock, and the shock thickness is so small that the cross-sectional areas for flow are virtually the same immediately before and immediately after the shock. A qualitative description of the change in thermodynamic conditions across the shock may be stated as follows (nomenclature as in Figure 1.14):

(i) The Mach number must decrease as $M_1 > 1$ and $M_2 < 1$.

(ii) No external work is transferred to the gas and the process is adiabatic so the stagnation enthalpy and temperature must be unchanged, i.e.

$$h_{01} = h_{02} \qquad T_{01} = T_{02}$$

(iii) As the gas is decelerating with constant stagnation temperature, the static temperature must rise, i.e.

$$T_2 > T_1$$

(iv) The static pressure and density will both rise across the shock, i.e.

$$p_2 > p_1 \qquad \rho_2 > \rho_1$$

(v) The very abruptness of the change of conditions in a shock suggests that turbulence must be present and this would imply a loss in stagnation pressure. For the same reasons the process is irreversible and the entropy will increase, i.e.

$$p_{02} < p_{01} \qquad s_2 > s_1$$

For a full derivation of the equations defining the change in properties across a normal shock, the reader is referred to standard texts on gas dynamics such as refs. 5 or 6. Some of the most important relationships are presented in Table 1.3 for a gas with $\gamma = 1{\cdot}4$ and the following example shows the use of these tables.

TABLE 1.3 *One-Dimensional Normal Shock Functions* ($\gamma = 1{\cdot}4$)

M_1	M_2	$\frac{p_2}{p_1}$	$\frac{\rho_2}{\rho_1}$	$\frac{T_2}{T_1}$	$\frac{p_{02}}{p_{01}}$
1·0	1·000 00	1·000 00	1·000 00	1·000 00	1·000 00
1·1	0·911 77	1·245 0	1·169 1	1·064 94	0·998 92
1·2	0·842 17	1·513 3	1·341 6	1·128 0	0·992 80
1·3	0·785 96	1·805 0	1·515 7	1·190 9	0·979 35
1·4	0·739 71	2·120 0	1·689 6	1·254 7	0·958 19
1·5	0·701 09	2·458 3	1·862 1	1·320 2	0·929 78
1·6	0·668 44	2·820 1	2·013 7	1·388 0	0·895 20
1·7	0·640 55	3·205 0	2·197 7	1·458 3	0·855 73
1·8	0·616 50	3·613 3	2·359 2	1·531 6	0·812 68
1·9	0·595 62	4·045 0	2·515 7	1·607 9	0·767 35
2·0	0·577 35	4·500 0	2·666 6	1·687 5	0·720 88
2·1	0·561 28	4·978 4	2·811 9	1·770 4	0·674 22
2·2	0·547 06	5·480 0	2·951 2	1·856 9	0·628 12
2·3	0·534 41	6·005 0	3·084 6	1·946 8	0·583 31
2·4	0·523 12	6·553 3	3·211 9	2·040 3	0·540 15
2·5	0·512 99	7·125 0	3·333 3	2·137 5	0·499 02
2·6	0·503 87	7·720 0	3·448 9	2·238 3	0·460 12
2·7	0·495 63	8·338 3	3·559 0	2·342 9	0·423 59
2·8	0·488 17	8·980 0	3·663 5	2·451 2	0·389 46
2·9	0·481 38	9·645 0	3·762 9	2·563 2	0·357 73
3·0	0·475 19	10·333	3·857 1	2·679 0	0·328 34
4·0	0·434 96	18·500	4·571 4	4·046 9	0·138 76
5·0	0·415 23	29·000	5·000 0	5·800 0	0·061 72
∞	0·377 96	∞	6·000 0	∞	0·000 00

Based on Keenan and Kaye, *Gas Tables*, by kind permission of the publishers, J. Wiley & Co.

Example 1.5 A gas with specific heat ratio $\gamma = 1{\cdot}4$ passes through a plane normal shock. Immediately before the shock the Mach number is 2·0, the stagnation pressure is 600 kN/m^2 and the stagnation temperature is 400 K. Find the static pressure and static temperature immediately downstream of the shock.

Solution: From the normal shock functions of Table 1.3 we see that for $M_1 = 2{\cdot}0$, $M_2 = 0{\cdot}577$ and $p_{02}/p_{01} = 0{\cdot}721$.

Therefore stagnation pressure downstream of shock is $0{\cdot}721 \times 600 = 433$ kN/m^2 and the stagnation temperature is unchanged at 400 K.

With a Mach number of 0·577,

$$T_0/T = [1 + \tfrac{1}{2}(\gamma - 1)M^2] = (1 + 0{\cdot}2 \times 0{\cdot}577^2) = 1{\cdot}0666$$

Static temperature = 400/1·0666 = 376 K

$$p_0/p = (T_0/T)^{\gamma/\gamma-1} = 1{\cdot}254$$

Static pressure = 433/1·254 = 345 kN/m²

(It would have been as convenient to read the stagnation to static temperature and pressure ratios from the isentropic flow curves in Figure 1.11.)

Before leaving the question of normal shock for the moment, two points should be emphasized. Firstly, it has been convenient to introduce the topic by dealing with the special case of off-design operation of a nozzle but this is only one case where normal shocks occur; as stated above, a flow decelerating from supersonic to subsonic conditions will *always* pass through a normal shock. Secondly, in the case of the flow through the nozzle, wall effects will inevitably introduce some "fuzziness" in the shock pattern and it will not form a true plane across the whole nozzle flow area.

Diabatic, Frictionless, Constant Area Flow

Usually in engineering work a deliberate effort is made to alter only one variable at a time when analysing complex processes. In this way things can be simplified and a final effect can be estimated by adding together a number of simpler cases. Care must be taken, of course, to see that important interactions are not ignored but in very many cases it is acceptable to make changes one at a time and sum the results to find an overall effect.

We want now to study the effect of heat addition to a moving gas stream as this kind of process often occurs in propulsion work. But in real plant the effect of heat addition will be complicated by such factors as the friction of the gas flow, variations in area, and even variations in the mass and composition of the working fluid undergoing the process; most heat addition is accomplished by adding fuel and burning it. In order to make the problem manageable attention will be confined to frictionless flow through a constant area duct with a constant mass flow of a gas whose chemical composition is invariant. The term *diabatic* used to describe the flow, means "with heat transfer". No external work will be done on or by the gas.

Consider a quantity of gas flowing through a pipe at subsonic velocity with a quantity of heat Q being added (Figure 1.15). What basically will be the effects of adding this heat? If the temperature of the gas rises, the density will fall, so, for the same mass flow rate

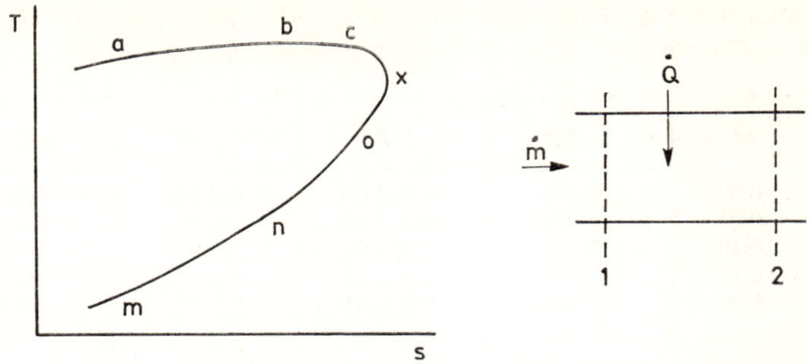

FIGURE 1.15 *Effect of heat transfer to a frictionless constant-area gas flow—the Rayleigh Flow curve*

and flow area, the velocity and Mach number must increase. An acceleration of the gas can only be accomplished (as we see from conservation of momentum) if there is a static pressure gradient with a fall in static pressure along the pipe. Considering then the entropy, heat is being transferred *to* the system so the entropy $[= \int (dQ/T)_{rev}]$ must be increasing and this means that the stagnation pressure must be decreasing. The combined effects of increasing temperature and increasing entropy must mean that the process can be illustrated on the *T–s* diagram by a line such as a–b–c in Figure 1.15. Finally, the conservation of energy equation tells us that heat addition will lead to a rise in stagnation enthalpy and temperature.

If the initial flow is supersonic, such as point m in Figure 1.15, the effects of heat addition are rather different in some cases. As for the subsonic case, there must be increases in the static and stagnation temperatures and in the entropy, so the condition line will be such as m–n–o. But the density–velocity relationship is reversed compared with the subsonic case and the flow will in fact *decelerate* with a decrease in Mach number.

The qualitative analyses above suggest that something special must happen if heat is supplied in sufficient quantities to raise the Mach number of the gas stream to unity—point x in Figure 1.15. Any attempt to pass this point, approaching from either subsonic or supersonic conditions, would imply a *decrease* in entropy of the fluid and so would contravene the Second Law of Thermodynamics. The limiting heat addition to a gas stream is, therefore, that which would result in the gas attaining sonic velocity from either initially subsonic or initially supersonic conditions.

This type of flow is known as *Rayleigh Flow* and a line such as

TABLE 1.4 *One-Dimensional Compressible Flow Functions. Diabatic Flow Constant Area, no Friction, Rayleigh Line* ($\gamma = 1{\cdot}4$)

M	T_0/T_0^*	T/T^*	p/p^*	p_0/p_0^*
0·00	0·000 0	0·000 00	2·400 0	1·267 9
0·05	0·011 92	0·014 30	2·391 6	1·265 7
0·10	0·046 78	0·056 02	2·366 9	1·259 1
0·15	0·101 96	0·121 81	2·326 7	1·248 6
0·20	0·173 55	0·206 61	2·272 7	1·234 6
0·25	0·256 84	0·304 40	2·206 9	1·217 7
0·30	0·346 86	0·408 87	2·131 4	1·198 5
0·35	0·438 94	0·514 13	2·048 7	1·177 9
0·40	0·529 03	0·615 15	1·960 8	1·156 6
0·45	0·613 93	0·708 03	1·869 9	1·135 1
0·50	0·691 36	0·790 12	1·777 8	1·114 0
0·55	0·759 91	0·859 87	1·686 0	1·093 97
0·60	0·818 92	0·916 70	1·595 7	1·075 25
0·65	0·868 83	0·960 81	1·508 0	1·058 20
0·70	0·908 50	0·992 89	1·423 5	1·043 10
0·75	0·940 09	1·014 03	1·342 7	1·030 10
0·80	0·963 94	1·025 48	1·265 8	1·019 34
0·85	0·980 97	1·028 54	1·193 1	1·010 91
0·90	0·992 07	1·024 51	1·124 6	1·004 85
0·95	0·998 14	1·014 63	1·060 30	1·001 21
1·00	1·000 00	1·000 00	1·000 00	1·000 00
1·20	0·978 72	0·911 85	0·795 76	1·019 41
1·30	0·957 98	0·859 17	0·713 01	1·043 65
1·40	0·934 25	0·805 40	0·641 02	1·077 65
1·50	0·909 28	0·752 50	0·578 31	1·121 5
1·60	0·884 19	0·701 73	0·523 56	1·175 6
1·70	0·859 70	0·653 77	0·475 63	1·240 2
1·80	0·836 28	0·608 94	0·433 53	1·315 9
1·90	0·814 14	0·567 34	0·396 43	1·403 3
2·00	0·793 39	0·528 93	0·363 64	1·503 1
2·50	0·710 05	0·378 70	0·246 16	2·221 8
3·00	0·653 98	0·280 28	0·176 47	3·424 4
6·00	0·536 33	0·078 49	0·046 69	38·946
10·00	0·507 02	0·028 97	0·017 02	381·62
∞	0·489 80	0·000 00	0·000 00	∞

Based on Keenan and Kaye, *Gas Tables*, by kind permission of the publishers, J. Wiley & Co.

a–b–c–x or m–n–o–x in the T–s diagram is known as the Rayleigh line. The position and shape of the Rayleigh line are uniquely determined by the initial state point of the gas flow. Although we have spoken throughout in terms of heat *addition* to the gas stream (as this is the more common occurrence in engines) exactly the same arguments can be applied in reverse for heat *rejection* and the process for cooling follows the same Rayleigh line but in the opposite direction.

A quantitative method for carrying out Rayleigh flow calculations was presented in a classic paper by Hawthorne and Shapiro (ref. 8) and the tables needed for complete calculations have been presented in such standard texts as ref. 5. In this method all conditions are referred as ratios to a reference condition, denoted by a star (e.g. p^*, T_0^*, etc.), where the Mach number is unity. Table 1.4 contains an extract from tables for a gas with $\gamma = 1{\cdot}4$ and the method of use can best be explained by taking a specific example as follows.

Example 1.6 The stagnation temperature of gas stream is to be raised from 800 K to 1 500 K while it flows through a constant area duct. Initially the stagnation pressure is 300 kN/m^2. Compare the losses in stagnation pressure resulting from the heat addition for initial Mach numbers of 0·20 and 0·40 assuming that the flow is frictionless in each case.

Solution: Use is to be made of the Rayleigh flow tables, Table 1.4. Taking first the flow with an initial Mach number of 0·20 we see that at this Mach number $T_{01}/T_0^* = 0{\cdot}173\ 55$ and $p_0/p_0^* = 1{\cdot}2346$. Therefore

$$T_0^* = 800/0{\cdot}173\ 55 = 4\ 610 \text{ K}$$

The physical meaning of this is that if heat were added so that the stagnation temperature reached 4 610 K, the Mach number in the duct would reach unity. However, heat is only added until a point is reached where the temperature is 1 500 K—we can call this T_{02}.

$$T_{02}/T_0^* = 1\ 500/4\ 610 = 0{\cdot}326$$

This corresponds to a Mach number (which can be found by interpolation in the tables) of 0·29. It does not matter from the point of view of the tables that heat addition did not *continue* until the flow reached unity Mach number. It is the *ratios* which are important.

To find the loss in stagnation pressure we can write

$$p_{02}/p_{01} = (p_{02}/p_0^*)(p_0^*/p_{01})$$

The tables show $p_0^*/p_{01} = 1/1{\cdot}2346$ and interpolation at a Mach number of 0·29 gives p_{02}/p_0^* of 1·203. Therefore

$$p_{02}/p_{01} = 1{\cdot}203/1{\cdot}2346 = 0{\cdot}975$$

The heat addition results therefore in a 2½% loss in stagnation pressure, that is a drop of 7·5 kN/m^2.

In the second case with $M_1 = 0{\cdot}40$ we can write directly

$$T_{02}/T_0^* = (T_{01}/T_0^*) \times (T_{02}/T_{01}) = 0{\cdot}529\,03 \times 1\,500/800$$
$$= 0{\cdot}991$$

$$M_2 = 0{\cdot}895$$

By interpolation in the tables we get $p_{02}/p_0^* = 1{\cdot}006$. Therefore

$$p_{02}/p_{01} = 1{\cdot}006/1{\cdot}1566 = 0{\cdot}87$$

The 13% loss in stagnation pressure (39 kN/m^2) is clearly much more important and this example shows that if heat is to be added to a gas stream the initial Mach number should be kept low in order to avoid large pressure losses.

Constant Area Adiabatic Flow with Friction

A further "standard" form of flow is that of adiabatic flow in a constant area duct making allowance for friction—what is known as *Fanno Flow*. The same approach as for the previous section will be used, that is, a qualitative description of the effects will be given, followed by a brief résumé of how frictional effects can be calculated.

Friction is one of the basic sources of irreversibilities and it would be expected, therefore, that a process involving friction would be one in which entropy would increase and stagnation pressure would decrease. With a fall in stagnation pressure there will result a fall in static pressure and so, neglecting temperature changes for the moment, a decrease in density. Taking first the subsonic case there must, therefore, be an increase in velocity due to friction. As the stagnation temperature will be constant and the velocity increases, the static temperature will fall. The process can, therefore, be illustrated on a *T–s* diagram by a line such as line a–b–c, on Figure 1.16. For the case with the flow initially supersonic, again there is the effect of the reversed density–velocity relationship compared with subsonic, so the curve for the process with friction is then of the form m–n–o with static temperature and entropy both increasing.

The second law of thermodynamics imposed a limit on the possible heat addition for Rayleigh flow by forbidding the Mach number to pass unity either while increasing or while decreasing, and the same condition will apply with Fanno flow. Point x on the Fanno Line of Figure 1.16 represents, therefore, the limit of either a–b–c–x or m–n–o–x. If friction were to be increased in a duct for which there was sonic flow at outlet (by, for example, reducing the opening of a valve in the line) then the initial state must change.

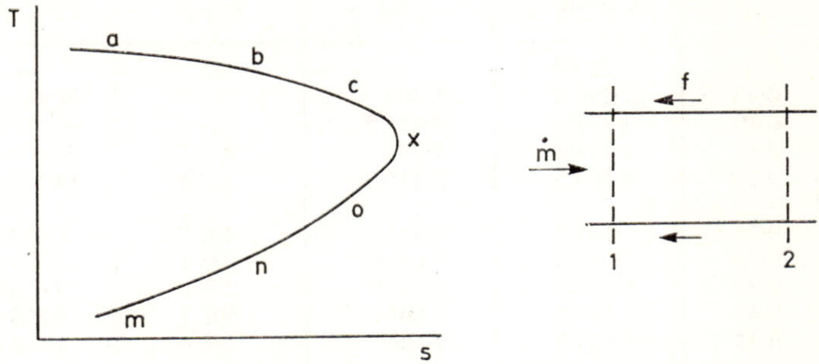

FIGURE 1.16 *Effect of friction on adiabatic constant-area gas flow—the Fanno Flow curve*

The loss in stagnation pressure in duct flow may be expressed by an equation such as the well-known *Darcy Formula*,

$$\Delta p_0 = 4f \frac{L}{D} \frac{V^2}{2} \tag{1.38}$$

where f is a friction factor, L/D is the duct length to diameter ratio, and V is the fluid velocity; f is a function primarily of Reynolds number and pipe roughness and a typical numerical value for turbulent flow through metal ducts would be $f = 0{\cdot}002$; with V in m/s this will give a pressure loss in N/m^2. Hawthorne and Shapiro (ref. 8) gave tables to allow the computation of flows with friction and again referred all conditions to the reference case where the outlet flow is sonic. In this case though the friction is expressed as a *distance* $(f \times L/D)^*$ to achieve sonic flow rather than a *ratio.* Table 1.5 gives an extract for the conditions for a gas $\gamma = 1{\cdot}4$ and Example 1.7 which follows illustrates the method of carrying out a calculation.

Example 1.7 Air flows adiabatically through a 15 m length of 50 mm bore pipe. At entry to the pipe the Mach number M_1 is 0·35 and the friction factor f, between air and wall has a value of 0·002. Find the percentage loss in stagnation pressure through the pipe and the percentage change in static temperature.

TABLE 1.5 *One-Dimensional Compressible Flow Functions. Adiabatic with Constant Area, Friction, Fanno Line* ($\gamma = 1{\cdot}4$)

M	T/T^*	p/p^*	p_0/p_0^*	$4fL_{max}/D$
0·0	1·200 0	∞	∞	∞
0·05	1·199 4	21·903	11·591 4	280·02
0·10	1·197 6	10·943 5	5·821 8	66·922
0·15	1·194 6	7·286 6	3·910 3	27·932
0·20	1·190 5	5·455 5	2·963 5	14·533
0·25	1·185 2	4·354 6	2·402 7	8·483 4
0·30	1·178 8	3·619 0	2·035 1	5·299 2
0·35	1·171 3	3·092 2	1·778 0	3·452 5
0·40	1·162 8	2·695 8	1·590 1	2·308 5
0·45	1·153 3	2·386 5	1·448 6	1·566 4
0·50	1·142 9	2·138 1	1·339 9	1·069 08
0·55	1·131 5	1·934 1	1·254 9	0·728 05
0·60	1·119 4	1·763 4	1·188 2	0·490 81
0·65	1·106 50	1·618 3	1·135 6	0·324 60
0·70	1·092 90	1·493 4	1·094 36	0·208 14
0·75	1·078 65	1·384 8	1·062 42	0·127 28
0·80	1·063 83	1·289 2	1·038 23	0·072 29
0·85	1·048 49	1·204 7	1·020 67	0·036 32
0·90	1·032 70	1·129 13	1·008 87	0·014 513
0·95	1·016 52	1·061 29	1·002 15	0·003 280
1·00	1·0	1·0	1·0	0·0
1·20	0·931 68	0·804 36	1·030 44	0·033 64
1·30	0·896 86	0·738 48	1·066 30	0·064 83
1·40	0·862 07	0·663 20	1·114 9	0·099 74
1·50	0·827 59	0·606 48	1·176 2	0·136 05
1·60	0·793 65	0·556 79	1·250 2	0·172 36
1·70	0·760 46	0·512 97	1·337 6	0·207 80
1·80	0·728 16	0·474 07	1·439 0	0·241 89
1·90	0·696 86	0·439 36	1·555 2	0·274 33
2·00	0·666 67	0·408 25	1·687 5	0·304 99
2·50	0·533 33	0·292 12	2·636 7	0·431 97
3·00	0·428 57	0·218 22	4·234 6	0·522 16
6·00	0·146 34	0·063 76	53·180	0·729 87
10·00	0·057 14	0·023 90	535·94	0·786 83
∞	0·0	0·0	∞	0·821 53

Based on Keenan and Kaye, *Gas Tables*, by kind permission of the publishers, J. Wiley & Co.

Solution: Looking at the Fanno flow tables we see that at $M_1 = 0{\cdot}35$ the parameter $4fL_{max}/D$ has a value of 3·4525. Putting in the appropriate figures gives

$$L_{max} = \frac{3{\cdot}4525D}{4f} = \frac{3{\cdot}4525 \times 0{\cdot}05}{4 \times 0{\cdot}002}$$

$$= 21{\cdot}4 \text{ m}$$

The physical meaning of this is that if the flow were to pass through 21·4 m of pipe the Mach number would rise from 0·35 to 1·00. In fact, the pipe is cut off after 15 m and the Mach number M_2 at *this* station can be obtained from the tables, thus.

The L_{max} for station 2 will be that for station 1 minus the actual distance between 1 and 2, i.e.

$$L_{max\ 2} = L_{max\ 1} - L$$

Therefore

$$\left(\frac{4fL_{max}}{D}\right)_2 = \left(\frac{4fL_{max}}{D}\right)_1 - \left(\frac{4fL}{D}\right)$$

$$= 3{\cdot}4525 - \frac{4 \times 0{\cdot}002 \times 15}{0{\cdot}05} = 1{\cdot}0525$$

By interpolation $M_2 = 0{\cdot}501$.
Also from the tables at $M_1 = 0{\cdot}35$,

$$p_{01}/p_0{}^* = 1{\cdot}7780 \text{ and } T_1/T^* = 1{\cdot}1713$$

while at $M_2 = 0{\cdot}501$,

$$p_{02}/p_0{}^* = 1{\cdot}338 \text{ and } T_2/T^* = 1{\cdot}1422$$
$$p_{02}/p_{01} = 1{\cdot}338/1{\cdot}778 = 0{\cdot}752, \text{ a loss of } 24{\cdot}8\%$$
$$T_2/T_1 = 1{\cdot}1422/1{\cdot}1713 = 0{\cdot}975, \text{ a loss of } 2{\cdot}5\%$$

Representation of Normal Shock on the T–s Diagram

It is of interest to look at the Fanno and Rayleigh lines plotted together on the *T–s* diagram (Figure 1.17). Both these flows are for constant area ducts and, for both, the three basic conservation equations must be obeyed. Looking at the diagram, it is seen that the lines cross twice, at a in the supersonic region and at b in the subsonic flow part. A line joining a to b will connect two points at the same stagnation temperature (Fanno condition), one supersonic and one subsonic. As all three conservation laws have been obeyed, a and b must represent the conditions immediately before and immediately after a normal shock. There is an increase in entropy through a normal shock and it is therefore an irreversible process—it would never be possible to set up flow conditions so that a flow went spontaneously from b to a.

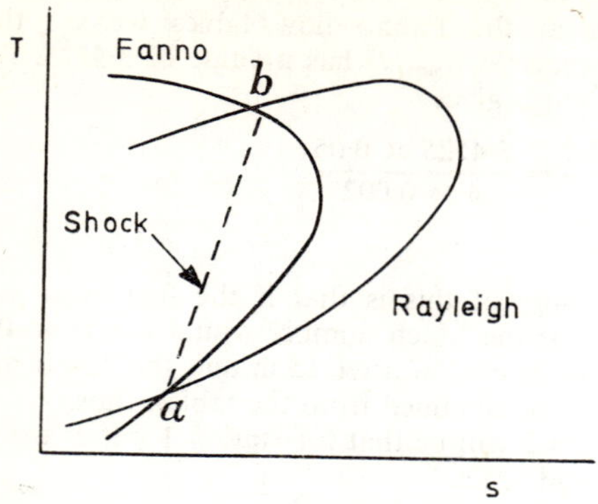

FIGURE 1.17 *Representation of normal shock on the temperature–entropy plane*

Oblique Shocks

Suppose that a gas moving at supersonic conditions meets some obstruction such as a wedge or a cone which causes the flow to be deflected. Such a situation is illustrated in Figure 1.18 which shows

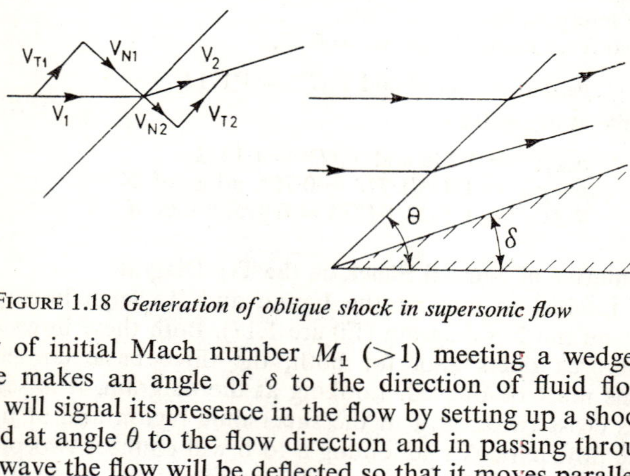

FIGURE 1.18 *Generation of oblique shock in supersonic flow*

a flow of initial Mach number M_1 (>1) meeting a wedge whose surface makes an angle of δ to the direction of fluid flow. The wedge will signal its presence in the flow by setting up a shock wave inclined at angle θ to the flow direction and in passing through this shock wave the flow will be deflected so that it moves parallel to the wedge surface.

To analyse the effect of the oblique shock on the gas, break down the velocity into two components V_N normal to the shock and V_T parallel to the shock, as shown in Figure 1.18. Fundamentally, the shock wave can have no effect on the flow component parallel to it, so this component will be unchanged through the shock. The normal component will, on the other hand, be decelerated and the ratio of Mach numbers before and after the shock will be the standard one for a plane normal shock as given in Table 1.3. In addition all the other relationships—static pressure ratio, loss in stagnation pressure, etc.—will be as for a normal shock of this intensity. One crucial difference between oblique and normal shocks however, is that the downstream Mach number will normally still be greater than unity. (Flow downstream of a normal shock must always be subsonic.)

The change in conditions across an oblique shock is determined uniquely by the upstream Mach number and the wedge angle. Figures 1.19(a) and (b) give characteristics of Mach numbers, deflection and static pressure ratio for oblique shocks in a gas with $\gamma = 1{\cdot}4$. Other relationships such as loss in stagnation pressure and static temperature ratio can be calculated from the standard formulae as illustrated in Example 1.8.

Example 1.8 A wedge of included angle 15° is immersed in an air-stream of Mach number 2·2 with its lower surface parallel to the flow direction, so that it sets up a plane oblique shock in the flow. The stagnation pressure and temperature upstream of the shock are 108 kN/m² and 400 K respectively. Calculate the static pressure and temperature downstream of the shock and the loss in stagnation pressure through the shock.

Solution: Figures 1.19(a) and (b) are used to solve this problem. From the Mach number relationships in (a) we see that the downstream Mach number may have the value of 1·63 or 0·60. In practice, the first solution will always be the one that appears under real flow conditions, despite the fact that either result will satisfy the equations.

The static pressure ratio across the shock can be read from Figure 1.19(b) as $p_2/p_1 = 2{\cdot}29$.

To find the actual value of the static pressure downstream, p_1 must first be calculated from

$$p_{01}/p = [1 + \tfrac{1}{2}(\gamma - 1)M_1{}^2]^{\gamma/\gamma-1} = (1 + 0{\cdot}2 \times 2{\cdot}2^2)^{3{\cdot}5}$$
$$= 10{\cdot}7$$
$$p_1 = 108/10{\cdot}7 = 10{\cdot}1 \text{ kN/m}^2$$
$$p_2 = 10{\cdot}1 \times 2{\cdot}29 = 23{\cdot}1 \text{ kN/m}^2$$

The stagnation temperature is constant through the shock at 400 K and

$$T_{02}/T_2 = [1 + \tfrac{1}{2}(\gamma - 1)M_2{}^2] = (1 + 0{\cdot}2 \times 1{\cdot}63^2) = 1{\cdot}53$$

Therefore, static temperature downstream of shock is

$$400/1{\cdot}53 = 262 \text{ K}$$

At this Mach number, $p_0/p = (T_0/T)^{\gamma/\gamma-1} = (1{\cdot}53)^{3{\cdot}5} = 4{\cdot}43$. Therefore

$$p_{02} = 4{\cdot}43 \times 23{\cdot}1 = 102 \text{ kN/m}^2$$

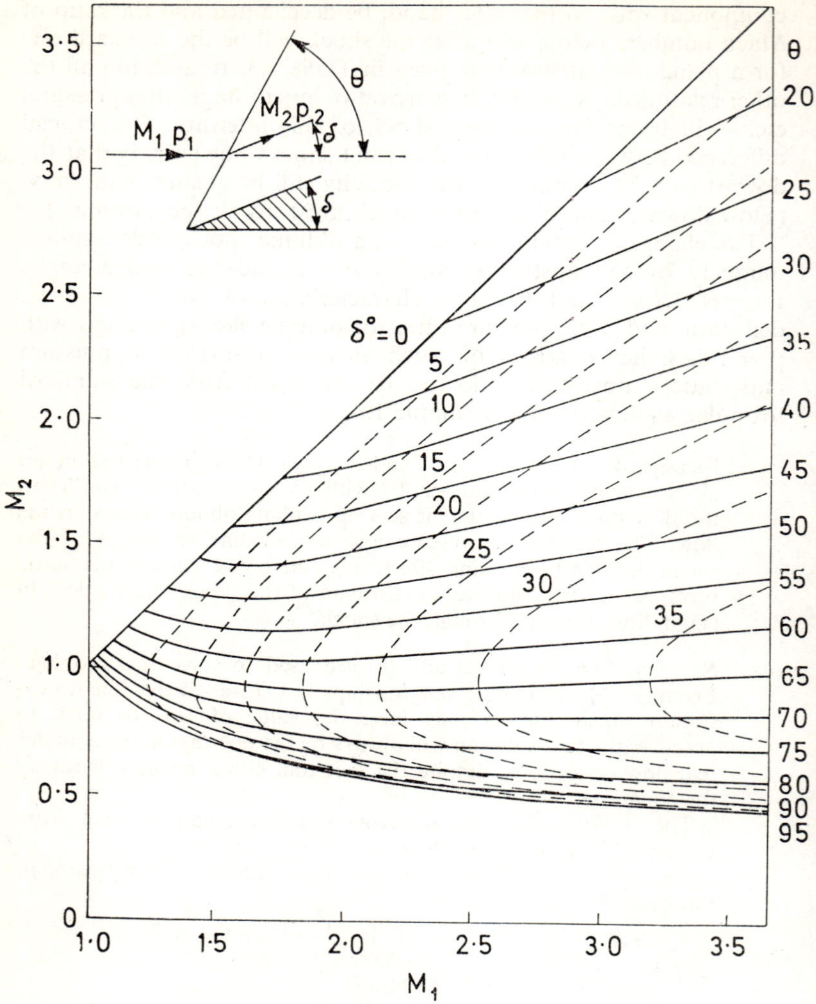

(a)

FIGURE 1.19 *Oblique shock characteristics*

(Reproduced by courtesy of the Royal Aeronautical Society)

Loss in stagnation pressure through shock = 6·0 kN/m²

Although the Oblique Shock Charts have been used to solve this particular example it will be seen that at certain other conditions they will not give a solution. For example, if the deflection angle

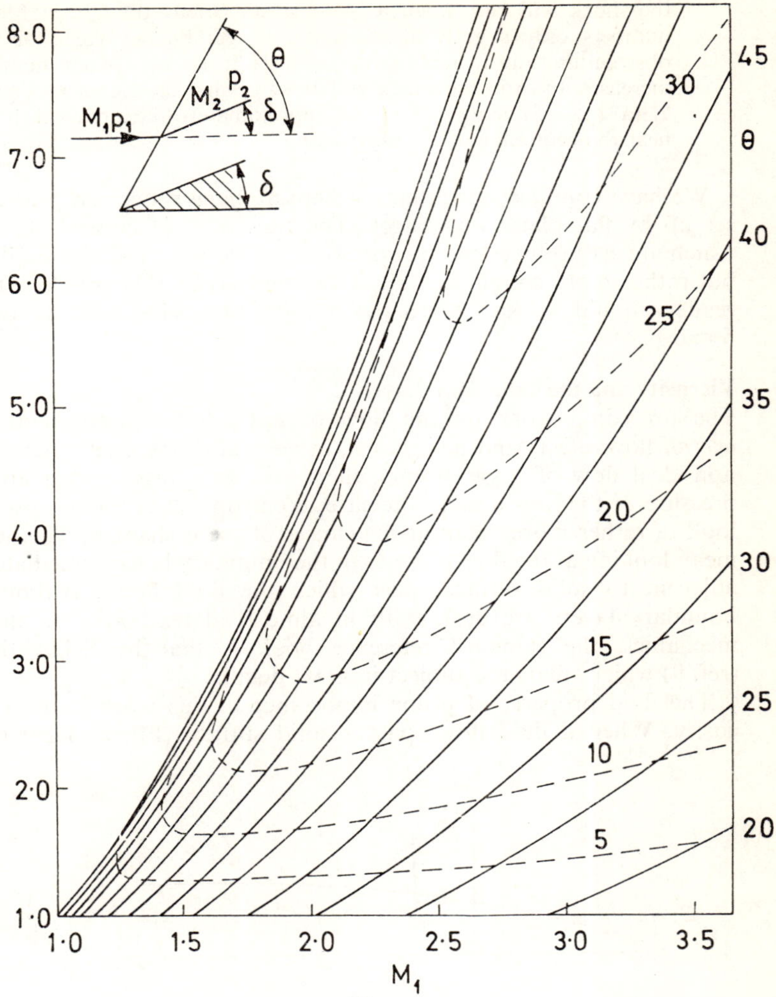

(b)

FIGURE 1.19

had been 15° but the upstream Mach number had been less than 1·6 no solution would have been possible. Under these circumstances a shock pattern is set up *ahead* of the obstruction in the stream and it is said to be a "detached shock". Reference should be made to such textbooks as Shapiro (ref. 8) for a full discussion of the form the shock will take when it is detached. Briefly though, at Mach numbers only slightly above unity, it will be far ahead of the obstruction but it will move towards it as the Mach number increases. In form the shock will be somewhat as shown in Figure 7.3(a) (p. 215) being curved and approximately like a normal shock near the centreline of the obstruction.

We have confined our attention here to *plane* shocks such as are set up by flat plates or wedges. The treatment of conical shocks which are set up by cones in supersonic gas flow is generally similar, but rather more complicated. For an analysis of this problem the reader should consult textbooks in gas dynamics such as refs. 5 and 6.

Viscosity and the Boundary Layer

The preceding work on the flow of gases has concentrated on *overall* flow effects and has shown, in general terms, that whenever non-ideal flow of a gas occurs, there will be a loss in stagnation pressure of the flow and an increase in entropy. It is useful now to look in rather more detail at the causes of these changes. This will mean looking at the flow of gases in the boundary layer immediately adjacent to solid surfaces past which the fluid flows. Although boundary layers are dealt with in almost all textbooks on fluid mechanics, the standard reference book is that by Schlichting (ref. 9) which treats the subject in great detail.

The fluid property of prime importance in this work is the viscosity. When a fluid flows past a solid surface (Figure 1.20) the

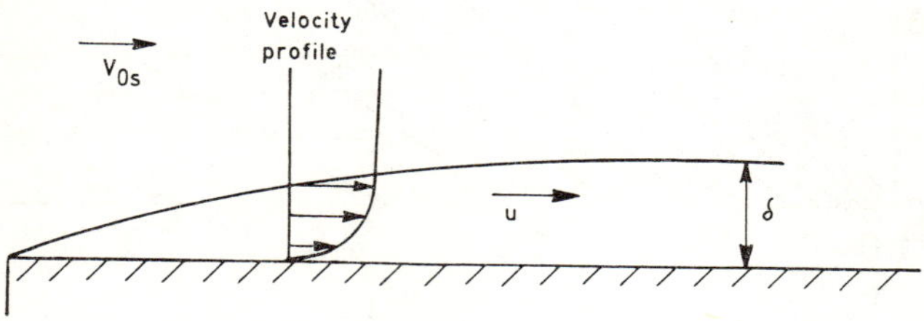

FIGURE 1.20 *Development of the boundary layer*
Free stream velocity is $V_{0_1 s}$ ahead of surface.

fluid which is immediately adjacent to the wall will be stationary with respect to the wall, so there will be some velocity variation through the fluid, with the velocity varying from this value of zero at the wall to the free stream value some distance away. The wall will exert a retarding force on the gas which will result in a shear stress

τ between layers of fluid lying parallel to the wall. From this, the coefficient of viscosity μ can be defined by the equation

$$\tau = \mu \frac{du}{dy} \tag{1.39}$$

where du/dy is the velocity gradient through the fluid perpendicular to the wall.

The coefficient defined by this equation is known as the coefficient of *absolute* viscosity. It is often more convenient to work in terms of the *kinematic* viscosity ν which is the abolute viscosity divided by the fluid density ρ, that is

$$\nu = \frac{\mu}{\rho} \tag{1.40}$$

A check on the dimensions shows that ν has the dimensions of L^2/T, so that the appropriate units in the SI system are m^2/s. It is those units which are used for the kinematic viscosity of air given in Table 1.2 above.

Once again the kinetic theory of gases gives a useful insight into the physical reason for the drag forces resulting from a velocity gradient in a gas. The motion of the molecules of gas are random in direction with a bulk gas velocity superimposed so, due to the non-flow-stream direction components, there will be continuous movements of low velocity molecules to higher velocity regions and vice-versa. Associated with each molecule there will be a momentum component in the free stream direction so there will be continuous momentum interchange between the layers; this appears as the shearing force opposing slippage between the layers. As the gas temperature rises there will be a higher random velocity of molecules so there will be a more vigorous transfer of momentum between layers of gas. This would lead us to expect an increase in gas viscosity with temperature—a result which measurements confirm.

When a body of given shape is immersed in a subsonic fluid stream, the characteristics of the flow—the shape of the streamlines, the drag, etc.—will depend partly on the fluid inertia and partly on its viscosity. The criterion that the flow around two similar bodies should be similar, is that the Reynolds number should be the same for both flows where Reynolds number, *Re*, is defined by the equation

$$Re = \frac{Vl}{\nu} \tag{1.41}$$

In this V is the fluid velocity, l is a characteristic length and ν is the kinematic viscosity. The physical interpretation of Reynolds number is that it gives a measure of the relative importance of the viscosity and inertia type forces in the fluid. At low values the viscosity effects predominate.

Mention has been made of the boundary layer which is the region close to a solid surface in a free stream where the velocity is less than the free stream value. The thickness of the boundary layer increases from zero at the leading edge, the rate of growth depending on such factors as the free stream Reynolds number and the overall characteristics of the flow such as whether it is accelerating or decelerating. In theory the disturbance caused by a solid object will spread throughout the fluid but in practice effective limits of boundary layer thickness can be defined. Various forms of thickness such as the displacement, momentum and energy thicknesses are used at different times. One of the most useful is the momentum thickness θ^* defined by the equation

$$\theta^* = \int_0^\delta \frac{u}{V_{0s}}\left(1 - \frac{u}{V_{0_1s}}\right) du$$

The symbols for this equation are as shown in Figure 1.20.

The main value of the momentum thickness lies in the fact that the loss in stagnation pressure of a fluid stream flowing past an obstacle (such as a wing or a compressor blade) is related directly to the momentum thickness of the wake. We will see later, when dealing with losses through an axial flow compressor, that separation of the boundary layer may sometimes occur and that this leads to a considerable thickening of the wake of disturbed air behind the blade and a subsequent drastic fall in efficiency. Separation is particularly liable to occur when flow is decelerating as it is then flowing into a region of increasing static pressure. Equation 1.27 shows that for subsonic flow, this deceleration (or diffusion) will occur in any flow passage where the area increases in the flow direction. This type of passage is known as a diffuser.

2 The Background to Aero-Engine Design

2.1 INTRODUCTION

By far the greatest part of this book is devoted to seeing how the principles of engineering science are applied to the design and performance of the components of an aero-engine. But this is not the complete story and there are other, less tangible, aspects of the design which must not be ignored. Ultimately the criterion of success in making a product will be measured by *economic* standards as well as by performance considerations; the designer must keep in mind *all* the various factors which affect the final product. An appreciation of the importance of the "background to design" should always be present and it is for this reason it is brought in at this early stage.

When dealing with problems in engineering science—in gas dynamics or applied mechanics for example—it is usually possible to state some logical rules which a process obeys. The design problem consists in recognizing the most promising rules, selecting best values for the appropriate constants, and carrying out the necessary calculations to find some optimum value. Topics which come under our background to design are less easy to handle. In many cases the logical rules may be so complicated as to defy analysis while in other cases only broad generalizations may be possible. For example, there is a general rule that well-established engines sell better than newer and technically better engines which have not yet built up a back-log of development experience. How does one quantify an obvious point like this? Again, certain engines get a head start by having a makers' name of high repute on the name plate. No one knows just how much this goodwill is worth when balanced against technical improvements which a competitor may offer.

Interesting though these points may be, they are not amenable to exact analysis, at least at the present state of knowledge, and we must confine ourselves to more tangible aspects of the work. In this chapter, therefore, we will look at four main topics of aero-engine design and we will attempt to formulate some principles to be followed. The propositions made are that for an aero-engine:

1. The overall weight of the power unit should be as low as possible. (Although a statement like that looks to be a glimpse of the obvious, it does merit discussion.)
2. The safety and reliability should be as high as is economically feasible.
3. The overall economics of the power unit should be as good as possible.
4. The complete vehicle must be socially acceptable.

Clearly, these four aspects of design will be interrelated to some degree but it is convenient to treat them in turn as distinct topics.

2.2 DESIGNING FOR LOW WEIGHT

Instinctively it is realized that any power units which are to be fitted to machines that fly should be of the lowest possible weight but it is useful to analyse the reasons why this should be so and to try to make a measure of the effect of excessive weight. Basically, the main reasons for wanting lightness are as follows:

(a) Extra Weight Means Extra Drag

The overall shape of an aircraft will decide its aerodynamic characteristics and a factor which will give a measure of the efficiency of the aircraft will be its lift/drag ratio. Normally for economy of flying, the aircraft should be operated at something close to its optimum lift/drag ratio when in cruising flight. In steady flight, the lift must equal the weight of the aircraft so, at a given lift/drag ratio, any increase in the aircraft weight will give a *pro rata* increase in drag.

This general statement oversimplifies matters as a number of practical problems may necessitate operation at other than the optimum lift/drag ratio. For example, for air traffic control purposes, a pilot may be required to use a flight plan which calls for special altitudes and flight speeds. The basic rule does hold though that excess weight in an aircraft or engine *must* introduce some drag penalty.

(b) Extra Weight Means Reduced Payload

Before an aircraft is granted a Certificate of Airworthiness an extensive series of flight trials is carried out and, as a result of these, limitations are placed on the maximum operating weight of the aircraft. Any subsequent increase in the basic aircraft structure or the engine must be compensated for by reducing either the fuel load or the payload. Reduction in fuel will mean loss in range, while loss in payload may mean leaving behind a fare-paying passenger.

Again many factors complicate this simple statement of the general rule. If aircraft always operated with their maximum payload it would be possible to make a reasonable financial estimate of the penalty but in fact aircraft operate full only on a relatively small percentage of the total number of flights.

(c) Extra Weight Means a Heavier Structure
It is not always appreciated that extra weight in an engine has a snowballing effect. It necessitates a stronger and therefore heavier engine mounting. This in turn requires a stronger and heavier airframe structure in the vicinity of the engine. In an extreme case this would mean more power being required from the engine with a further round of weight increases—but fortunately the process is seldom divergent! As a corollary to this point, the particular case of the Hawker-Siddeley Gnat fighter may be cited. The Gnat was designed to the philosophy that if weight could be *saved* on one part, further savings on adjacent parts would be possible. Everything hinged on being able to keep a tight control on weights and avoid any escalation. By applying this tight control the makers were able to produce an effective fighter of low weight and cost.

These, then, are the main reasons for the continual struggle to keep engine weights as low as possible. Before going on to look at some of the principles involved it is worth stressing that the price of low weight is eternal vigilance and that every proposed modification to an engine must be scrutinized to see if the weight increases are acceptable.

For many years it was not fully appreciated that the important weight to minimize was not that of the engine alone but that of engine plus fuel required to complete a given mission. In the days of the piston engine, high output for a given weight went hand in hand with improved fuel consumption as both were related to the engine compression ratio. With the arrival of the gas turbine though, it became much more feasible to design a really lightweight engine if one were prepared to forgo good efficiency. (The cycle analysis of the next chapter shows how output and efficiency can be varied as functions of compressor pressure ratio and turbine entry temperature.)

Consider three separate engines and let us plot for each the total of engine weight plus fuel weight, against running time at some required thrust. A picture of the form shown in Figure 2.1 will be obtained, engine A being light but inefficient, engine C heavy and efficient, and engine B intermediate in these characteristics. Taking the important parameter as being engine plus fuel weight we see that for short times of operation the "cheap and nasty" engine A is best but for long durations it is worth while to make the heavier, more efficient engine C.

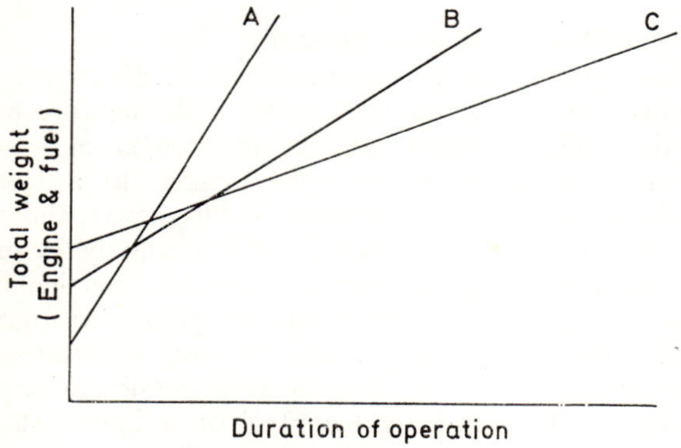

FIGURE 2.1 *Variation of (engine + fuel) weight with duration of operation for different engines*

One field of propulsion which has grown important in recent years and where this work is particularly relevant is in the case of turbojets designed as lift engines. Lift engines were originally developed by Rolls-Royce for their vertical take-off aircraft, the idea being that lift engines would be installed with sufficient thrust to overcome the all-up weight of the aircraft, but that once wing-born flight had been achieved, the lift engines would be shut down. When the aircraft was ready to land they would be re-started. At first sight it would seem to be hopelessly uneconomic to install engines which were to be used only for these two brief periods of flight but the question can then be asked, what is the alternative? An alternative *does* exist with the Bristol Engine Division system of vectored-thrust engines where one power unit is used for both lift and propulsion. When the calculations are carried out, however, it is found that in very many applications an engine which is big enough to lift the aircraft is far too big for propulsion. A second look may then be taken at the idea of the lift engines and the figures show that if a very light-weight lift engine can be built, it will give an acceptable solution. By very careful design Rolls-Royce have produced turbojets which can give about three times the thrust/weight ratio of a conventional turbojet. Not only are such engines suitable for lift purposes but, with their very low weight, they may also find applications as boost engines to assist the take-off performance of civil airliners.

Since the earliest days of heavier-than-air flight, there has been speculation as to the implications of the *Square–Cube Law* for aircraft. This law states the irrefutable fact that, for objects which are different sizes but dimensionally similar, the surface area varies with the square of the linear dimensions while the weight (for a given density of material) varies with the cube. If one were to build two aircraft, one exactly twice the size of the other, the larger would have four times the wing area but eight times the weight. As it is the wing loading which mainly determines the take-off speed of an aircraft, the larger one would have considerably more difficulty in getting airborne. For years the square–cube law has been held to be an ultimate barrier to building large aircraft but for years the engineers have found ways round it by using thinner sections, better structures, and stronger materials.

Coming now to the case of the turbine engine, the thrust or power will vary mainly with flow areas, that is with the square of a linear dimension, while the weight will vary with the cube. Big engines will therefore have a worse specific weight (weight per pound of thrust). As in the case of aircraft, the picture is complicated by the fact that not all parts scale fully—there are lower practicable limits for sheet metal thicknesses and fastenings for example. Given all these factors there *does* seem to be a minimum specific weight for turbojets somewhere in the range of 20 000–60 000 N design thrust. It is a very flat minimum, however, and square–cube law effects will seldom be the dominant considerations in selecting the size of an engine. If large amounts of thrust are needed, there are practical advantages from the installation point of view in providing them in large single units.

On the question of detail design for lightness a great deal must depend on the experience of the designer but two general principles are worth stating. Firstly, throughout the aircraft and aero-engine industries there is always a great emphasis on specific development testing to determine the life and strength of components. With a land-based system it is acceptable to use large "factors of safety" to cover unknown quantities in stressing and it is more economical to put in extra material than to find the ultimate limit of what a structure can stand. For an aero-engine on the other hand, special tests are worth while to ensure that parts are not too strong. In general an engine will be deficient in strength as first designed and the whole object of development testing is to find the weakest spots and strengthen them particularly—taking care not to add more extra material than is necessary. By this means, the engine can be proved and improved at one and the same time.

The second general principle on detail design is concerned with what is known as *Kinematic Design*. A proposition may be stated that, if a structure is to be subject to varying stresses due to varying loads or thermal expansion, two courses are open to the designer: either he may make a rigid structure so strong that all the stresses can be withstood by the members or he may design a structure such that relative movement between the members is possible. Needless to say the direction and amount of movement in the latter case must be very carefully controlled and sufficient restraints must be kept to ensure that *necessary* forces are transmitted. It is in these considerations that the principles of kinematic design came into use.

For a fuller discussion on kinematic design reference should be made to Braddick (ref. 1). This work is primarily concerned with the design of measuring instruments, and it explains how, in order to achieve complete location between parts, six restraints, *and not more than six restraints*, are needed. (A proviso is made that they are all independent of each other.) If some controlled relative movement is needed, then one of the six must be removed. Besides helping in designing for lightness, kinematic design principles can be usefully applied to give improvements in both ease and repeatability of assembly. For example, a four-legged stool can be stood firmly on a level surface only if all legs are made exactly the same length and the surface is truly level; by removing one of the redundant restraints and making a three-legged stool a firm location in any reasonable plane is possible. (Complete location in three dimensions will need a further three restraints.)

2.3 DESIGNING FOR SAFETY

For aircraft, more than for any other form of transport machine, *the* paramount requirement is to achieve a high degree of safety. In many ways this is incompatible with the requirements of getting high power from small, light-weight engines but ways must be found of achieving the safety by means, for example, of duplicating highly loaded components. Failure in *parts* are acceptable provided the vehicle as a whole can still be operated safely.

Each country which operates aircraft has its own authority to control the relevant aspects of design and performance. In Britain, the Air Registration Board (ARB) is responsible for maintaining airworthiness standards in civil aircraft and when the Board is satisfied with a design, and has carried out a series of flight trials, a Certificate of Airworthiness will be issued for the aircraft. Any further modifications to the design of the aircraft, engines, or components must be cleared by the ARB. In addition, on the basis of the proved flight performance, a Flight Manual is issued laying down the prescribed operating conditions for the aircraft. The Flight Manual specifies such things as the take-off distances required and the limiting operational weights at varying atmospheric pressures and temperatures.

Authorities such as the ARB and its American equivalent the Civil Airworthiness Authority (CAA) make a very deep study of the whole question of reliability and keep under continuous assessment all developments in new materials and manufacturing techniques. In a way, the ARB in Britain operates almost as a form of supra-company quality control for the aircraft industry, as the company designers must first convince the Board's assessors of the acceptability of any new practices. The interest in the reliability aspects has lead to a study of the mathematics of reliability.

A more general paper on designing for safety is that by Hafner on *Safe Mechanisms* (ref. 2). In this paper the author first reviewed the background to safety, dealing with such matters as the reliability of complex systems and the cost of achieving safety. He then went on to write about methods of detecting failure and discussed some means of designing safety into systems. Raoul Hafner has devoted his life to the world of helicopters and vertical take-off aircraft so most of his examples in the paper are taken from his experience in helicopter engineering. This does not in any way detract from the usefulness of the paper as the lessons pointed out have very wide applications.

To examine some of Hafner's points in greater detail, it should be appreciated from the start that safety and reliability cost money and the cost can, to a certain extent, be calculated. An airliner operator will know just how much it will cost him if he loses an aircraft with all its occupants and this price will be reflected in his insurance charges which are part of the direct operating costs of the aircraft. (Other factors, such as the loss in traffic and goodwill resulting from an accident are not as easy to calculate.) The operator must, therefore, make a calculated decision on what is an "acceptable risk" and plan his aircraft operations accordingly. It will always be possible to improve safety by accepting some performance or commercial penalty—for example by building an aircraft which is grossly overpowered—but in practice the "acceptable risk" used in civil air transport is so low that reasonable and safe commercial operation is possible. Occasionally controversies do arise, such as that concerning the use of wide-cut aviation fuel in place of the safer kerosene or on the provision of rearward-facing seats for the passengers. If a real increased danger is seen to be present, public opinion or the airworthiness authorities may force a change.

If one is confronted with a system made up of a number of components and is asked to improve the overall reliability, a first instinctive step might be to duplicate some of the more important components. In fact, very great care is needed in any multiplication of components as the overall reliability will only be improved in this way if the system can operate quite safely with one component out of action—otherwise the effect of duplication is to *reduce* the system reliability. To take a specific example, a twin-engined aircraft is safer than a single-engined machine only if it can operate safely under all conditions with one engine shut down. Often it will not be able to fly safely on half-power, especially if there are out-of-balance forces due to an engine being out of action, hence the hazard has in fact been doubled. A basic rule for duplication can therefore be stated:

"Duplication of components will improve the overall reliability of a system only if the system can function fully with one component inoperative. Otherwise the overall reliability is impaired."

The most important lesson to be learnt from this rule is that the best way to improve reliability is to keep systems as simple as possible. Each time an extra link or an extra unit is added the cumulative hazard of *some* failure is increased and the overall hazard will only be reduced if some of the links are redundant under normal operating conditions. Especial care must be taken in adding "safety devices" to a system as they may in themselves introduce greater risks than they cover. An unreliable warning light which comes on when it should not might cause a pilot to be distracted or to take risks in flying which would not normally be justified.

An example of a system where a very high degree of reliability is needed is the Blind Landing Unit of the auto-pilot made by the Aviation Division of Smiths Industries for such aircraft as the Hawker Siddeley Trident. Three separate input units are used, each of high reliability, and the auto-pilot believes two out of the three input signals. Hence, if one unit should malfunction, the sensing unit will ignore its signals. The chance of two channels going wrong at the same time is so low as to be an acceptable risk.

Turning now to the problems of failures, the development of any failure in a component may be divided into a number of stages as shown in Figure 2.2. Generally there will be a preliminary stage

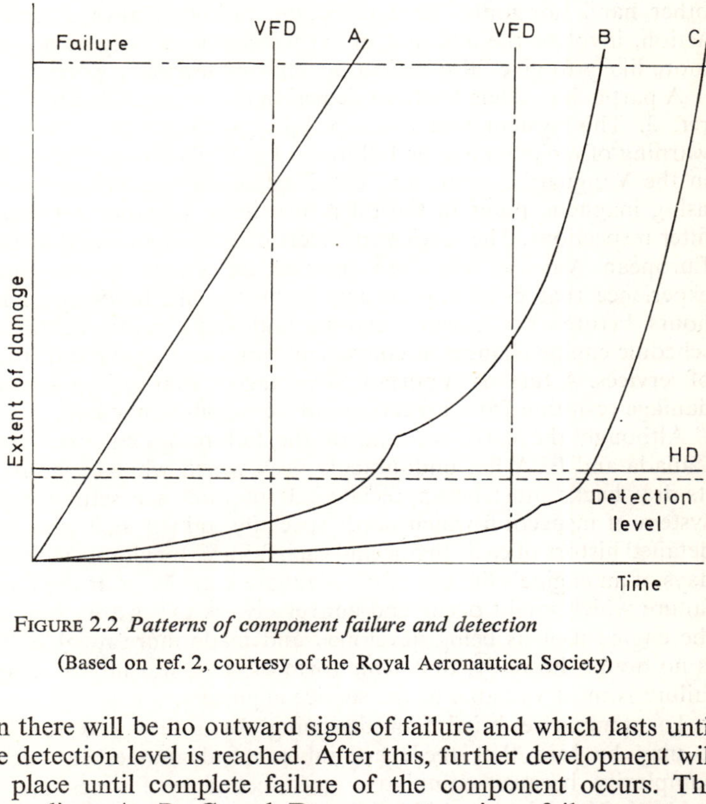

FIGURE 2.2 *Patterns of component failure and detection*

(Based on ref. 2, courtesy of the Royal Aeronautical Society)

when there will be no outward signs of failure and which lasts until some detection level is reached. After this, further development will take place until complete failure of the component occurs. The various lines A, B, C and D represent various failure patterns which may be encountered. Figure 2.2 is based on graphs from Hafner's paper (ref. 2) and in the paper Hafner distinguishes between "vertical fault detection" and "horizontal fault detection". With vertical fault detection, inspections are carried out at discrete time intervals and if a failure has proceeded beyond the detection level at the time of an inspection it will be picked up safely. Horizontal fault detection on the other hand takes some signal from a component and if the signal reaches a certain value, it gives warning to the pilot.

A typical horizontal fault detector is the oil pressure warning light which comes on when the oil is getting low.

Clearly each type of detection will have its advantages and disadvantages. The vertical fault detection is liable to miss certain failures (such as type C on Figure 2.2), it is time-consuming, and it involves, in general, disturbing the cowlings of an engine. On the other hand horizontal fault detection, although always ready for action, involves the addition of extra items to an engine and is far more likely to give false indications and unnecessary worry.

A particular system for fault detection is that described by Eden in ref. 3. This system was developed by Air Canada to give early warning of incipient engine failure on the Rolls-Royce Tyne engines in the Vanguard aircraft and consisted of an inspection procedure using magnetic plugs in the oil return lines, together with routine filter inspections. The service engineers of Air Canada and of British European Airways who also operate the system found through experience that potential bearing failures could be detected many hours before they became serious and steps could be taken to schedule engine changes at convenient times and so avoid disruption of services. A further important advantage is that serious secondary damage resulting from a bearing failure might be avoided.

Although the early warning of the failure system used by Air Canada and BEA has undoubtedly been worth while economically, its drawbacks must be appreciated. It requires the setting up of a system of inspection which needs specially trained men who keep a detailed history of each engine and each failure. Inevitably in the early days of an engine's life there is inadequate knowledge of the types of failure which might occur and conversely, as the history builds up, the engine itself is being developed and made more reliable. There is no doubt, however, that *some* form of early warning of incipient failure is most valuable to the service engineers.

To summarize briefly on designing for safety and reliability, it may be said that there are inherent advantages in avoiding complexity because cumulative and secondary failures are then reduced. There is a considerable amount of truth in the American saying that an engineer is a man who can make for one dollar something that any fool could make for two. Also, in the early stages of an engine's life, the designer must make provision for both vertical and horizontal fault detection. By making parts readily accessible, servicing as well as inspection can be made more easy.

2.4 DESIGNING FOR ECONOMY OF OPERATION

It is only in comparatively recent years that air transportation has ceased to be somewhat of a technical miracle catering for a chosen few, and has become a means of mass transportation. In this new status it must compete with surface transport on an economic basis, accepting, of course, that people will be prepared to pay extra for additional speed and convenience. The days when the aircraft and engine manufacturers considered cost more or less as an afterthought have now gone and firms *must* be cost conscious.

When dealing with economics, it is important to consider the *overall* economy of operation rather than just the running costs of fuel and maintenance. An airliner is an expensive piece of equipment which has a fairly limited life and so will depreciate at a high rate, whether it is used intensively or not. In addition, with its spares holding, it represents a large investment of money which is tied up and so cannot be used for other purposes.

With this in mind, it is possible to apportion the overall costs of operating an aircraft under two main headings, the "capital charge" and the "running charge". Capital charges will cover the items which depend on the value of the aircraft and will include depreciation, insurance, and an interest-on-investment charge which represents the amount paid for borrowing the cost of the aircraft (or alternatively what could be obtained by investing its value on the money market). Running charges include such items as fuel and oil consumed, maintenance costs, and the wages and salaries of the crew operating the aircraft. The crucial difference between the two types of charge is that the capital charge per annum will have to be paid no matter how little the aircraft is used whereas the total running costs per annum will vary directly with the number of hours flown. Graphs can be drawn for different aircraft showing total cost per annum against hours flown and they will be ot the form shown in Figure 2.3.

Looking at Figure 2.3 it is seen that the factor determining which is to be the most economical aircraft is the utilization—the hours flown per year. As was seen before in dealing with the engine weight plus fuel weight parameter, for short duration work the "cheap and nasty" aircraft comes out best but for high hours per year it is worth while to go to considerable expense to improve the running costs. The lesson is also that very expensive aircraft must be flown intensively if they are to be economic. Reliability comes in here because unserviceability resulting from aircraft and engine snags can have a catastrophic effect on the utilization.

(Although our arguments on costing are stated specifically for civil aircraft operations, they have wide applications in all fields of life. One specific example where the picture is particularly clear is in the economics of electrical power generation. At one end of the spectrum of power stations there are the nuclear stations which, at least for the first generation, have exceedingly high capital costs but low fuel costs. The bulk of the electricity in Britain is generated in conventional steam stations but a small percentage is now supplied by gas turbine sets with low capital costs but high fuel costs. A graph such as Figure 2.3 shows that, to be economic, nuclear stations must be on base load operation with high utilization, while the gas turbines must be used only at times of peak demand.)

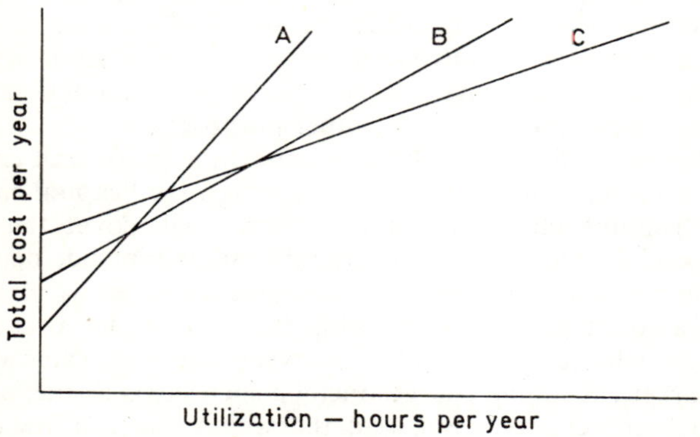

FIGURE 2.3 *Variation of total annual costs for various aircraft*

At one time each company producing aeroplanes would make estimates of how much it thought its planes would cost to operate and, not unnaturally, the figures produced by different companies would be inconsistent with each other and some would be wildly optimistic. In an effort to introduce some rationality into the business, a number of standard procedures for cost estimation have been developed. Among the best known of these are the British S.B.A.C. method, issued by the Society of British Aero-Space Companies and the American A.T.A. method. The S.B.A.C. method of estimating Direct Operating Costs (D.O.C.) brings in all the various factors which affect the costs and says what values are to be used for the different constants which appear in the calculation. For example, the aircraft is assumed to fly 3 000 hours in the year, the insurance rate is 4%, and each airman earns £3·1 per flying hour. Despite the fact that some of these values will vary from place to place and from time to time the method is invaluable for giving a standard of comparison between aircraft. It has the further advantage that it may be applied at the project stage before a new machine has flown, although naturally in these circumstances the figures obtained will be less reliable.

The final result for cost calculations by the S.B.A.C. method can be presented as a curve of direct operating cost against stage length (Figure 2.4). On the same graph has been plotted a typical curve of

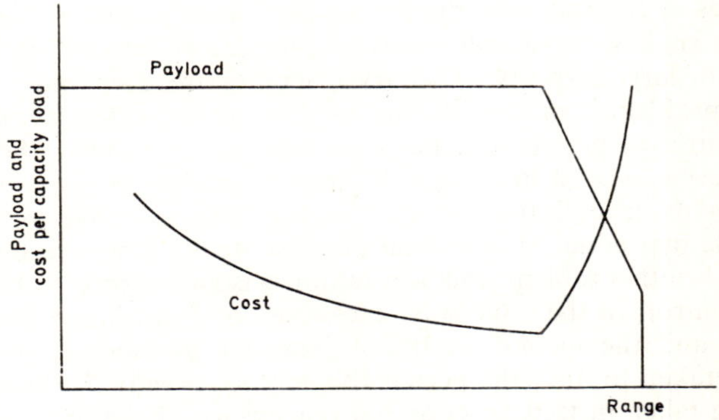

FIGURE 2.4 *Typical payload and cost curves for an aircraft*

payload against range. It will be seen that the payload is constant at its limiting design value up to a certain range and that to get longer ranges it is then necessary to reduce the payload and make-up to the limiting all-up weight by carrying more fuel. Finally the aircraft will reach its limiting range when its tanks are full. Regarding the cost curve, as the cost is expressed per *capacity* ton mile, an initial fall-off in cost will occur due to the improved block speed and the reduced landing fees per flight. Naturally when the point is reached where the payload starts to fall the cost per capacity ton mile will rise.

Although direct operating costs of an airliner are important, it must be appreciated that the total costs for an operator will be very much higher. For example, the usual load factor achieved in scheduled services is of the order of 60%, whereas the standard methods of calculating costs are on a *capacity* basis, that is 100% load factor. In addition, an airliner has high overheads to cover advertising, sales, crew training, catering and such-like items. It is for these reasons that the actual fare per mile is several times the calculated direct operating costs.

For anyone interested in the operation and economics of airlines the annual reports of BEA and BOAC make fascinating reading. These reports contain operational statistics showing the load factors and utilizations achieved by the different aircraft in the fleets and the financial returns show the operating costs of the different types.

2.5 DESIGNING FOR SOCIAL ACCEPTABILITY

Of the factors which go to influence the social acceptability of aircraft, by far the most important is the noise. Without doubt one of the curses of civilization is the uproar produced by modern airliners which is so bad as to make normal life almost impossible in the vicinity of large airports. (Military machines are even more noisy but the problem is rather different as they are not often based near large centres of population. For their neighbours though, their din can be even more distressing.) A parallel may be drawn with the problem of air pollution which was for long accepted to be an inevitable byproduct of industrial life. Eventually it was recognized to be a health problem and legislation was introduced to bring it under control. In the same way, excessive noise can injure both the physical and the mental health of innocent bystanders, so steps must be taken to limit the permissible levels. Already the first steps are being taken in this direction but the surface of the problem has only been scratched so far, and the technical problems ahead are very severe.

The whole question of aircraft noise has attracted a good deal of attention in recent years and a large number of papers have been published. Those listed as refs. 4 and 5 give a useful introduction to the problem.

One of the main troubles when considering the noise nuisance is that the human ear is a very sensitive device so that, even if only a small fraction of the energy emanating from a machine appears as sound energy, it can be very troublesome. To give an idea of the order of magnitudes, the loudspeaker in a radio will have a power rating of the order of watts while the energy in the jet efflux from the engines of a VC-10 at take-off conditions will be of the order of 200 megawatts. If even 0·1% of this energy is dissipated as sound it will represent an intense noise source. Also, a person's reaction to noise is very subjective and is difficult to quantify. Each individual has his own threshold of nuisance and this will vary from day to day and place to place. A noise which would be most disturbing at dead of night can be lost in the background in the daytime. As a general rule high frequency noises are more objectionable than low frequency and a noise source which is approaching is more disturbing that one which is going away.

When jet engines first came into use their distinctive noise characteristics compared with a piston engine were soon apparent.

On starting, there was the high-pitched compressor whine which was drowned completely at the take-off conditions by the roar of the propulsion jet. Sound is generated by the jet through the shearing action between jet and surroundings and the subsequent turbulent mixing which takes place downstream of the final nozzle. The intensity of the sound varies as some function of the mass flow, so that bigger engines are noisier than smaller, and as the seventh power of the relative velocity between jet and surroundings. Clearly this velocity term will be of overriding importance and advanced-design turbojets with intrinsically high jet velocity due to their high temperatures and pressure ratios will be exceptionally troublesome.

One of the most important steps towards a general reduction in engine jet noise has been the introduction of the turbofan-type engine. This type, as shown in Chapter 3, produces its thrust by accelerating a larger mass flow rate of air through a smaller velocity increment and so the noise is bound to be less. In addition, for both turbojet and turbofan engines silencers have been designed which aim to reduce the intensity of the mixing between jet and surroundings. These are discussed in greater detail in Chapter 5. (A cartoon in *The Aeroplane* many years ago commented on the efficacy of these sound suppressors. Two men were standing beside an engine running on the test bed and one was yelling into the ear of the other, "Now let's try it without the silencer!")

Although the turbofan makes a considerable improvement in the noise from the jet efflux it does bring in an important new problem because the noise from the front end becomes highly significant. Sound generation by an axial flow fan or compressor stage is due to siren-like effects as the blades pass through the disturbed wakes from previous rows of blades (see ref. 4). It is far more concentrated at distinct frequencies than is the noise from the turbulent jet and, as the frequency is often high, the nuisance value is worse. In addition, there is the psychological effect mentioned above as most of the noise radiates forward and is heard as the aircraft approaches the observer. This is more disturbing than to hear an intense but receding noise source.

What can be done about intake noise? Once the geometric characteristics have been determined it is possible to suggest solutions but decisions must then be made on whether the solutions are technically acceptable. For example, it is found that the spacing between blade rows has an important effect on the amount of sound being generated but any increase in clearances will affect engine performance and size. Another suggestion that has been made is to manufacture the intake in the form of a venturi with sonic conditions at the throat but, although this would block the sound, it would give pressure losses and would need to be of variable geometry to cover different operating conditions—the engine aerodynamicists are trying all the time to keep intakes short and as open as possible, so fancy sound suppression designs will be unpopular.

It is relevant at this stage to ask how it came about that the Britannia was able to earn the name of the Whispering Giant. At some conditions, such as when running the engines at full power on the ground there was a large amount of propeller noise but when flying past the Britannia was notable for its quietness. A number of factors contributed to this, the most important probably being that the engine intake was buried deep inside a cowling and the gas velocity leaving the final nozzle was only of the order of 200 m/s. Other turboprops such as the Viscount, Friendship and Vanguard have intakes which have a much clearer "sight" of the cowlings of the intake so there is far more opportunity for sound to come out.

From the point of view of noise nuisance around airports the one other main variable, apart from the weather, is the distance between sound source and observer. As the noise intensity decreases inversely with the square of this distance, any improvements in rate of climb out of an airfield will help those living around, provided of course that the improved climb is not won by increasing the thrust greatly. In the same way a steeper descent will mean that fewer people living near the approved route would be disturbed. Unfortunately these alterations in flight plan—the steeper climb and descent—could only be brought in at the risk of compromising flight safety to some extent. Clearly very careful thought is needed before safety is reduced in any way.

On the other noise problem, that of the sonic boom, very little can be said at the moment. Although the reasons for the boom are now quite well understood no information has yet been published on just how intense the boom will be from a civil transport flying at 20 000 m or more in altitude. Again the reaction of people depends very much on the individual and while some will be unconcerned, other may object to it very strongly. There is an element of resentment also in a lot of the reaction and a common expression of opinion is that of "why should we be inconvenienced so that 100 people can get to New York three hours sooner?" It is certain that every effort will be made to find ways of reducing the intensity of the boom so that there is no fatalistic acceptance of it being inevitable and always with us.

Turning now to the other problems of designing for social acceptability the main things to ensure are that the integrity of an engine should be as high as possible and that it should not emit noxious fumes. Regarding the integrity, this means firstly that the engines should be reliable. An aircraft which was always becoming unserviceable so that schedules were disrupted and passengers were stranded far from home would not meet the criterion of social acceptability. But integrity means more than that—it means that the engine should not drip oil on the airfield and that no parts should fall off in flight. This may seem to be a ridiculous thought, but in the past design faults have often resulted in bits of cowling, large lumps of ice, and other pieces of hardware falling from aircraft and damaging property on the ground. On the question of noxious fumes, if the combustion system of an engine is not designed properly, various products of incomplete combustion may be emitted which might cause staining or smarting of the eyes. The quantities of fuel burnt by a modern large engine are so high that even small losses in combustion efficiency result in large quantities of unpleasant products.

2.6 CONCLUDING REMARKS

In this chapter an attempt has been made to show that there are always a number of factors to be taken taken into account in producing a successful design. There is often a tendency, particularly in a highly technical field such as that of aircraft propulsion, for the specialists in different fields to attract the greatest glory. Although credit should be given where credit is due, the contribution of the designer must not be neglected. A good designer will not necessarily be a specialist in any particular topic, but he will appreciate the significance of advances in a number of fields and must follow up these advances as closely as possible while at all times striking a balance between conflicting requirements. Above all, he must be able to translate *ideas* into practical *hardware* and tell the manufacturing people what form things should take.

The work of this chapter has shown that many factors other than purely technical ones must be taken into account when preparing a design. For example, a new improved compressor might involve some novel manufacturing techniques which were more expensive than existing ones. Thought would be needed as to whether the extra cost was justified by the improved performance. All too often in the past, improved performance has been sought regardless of cost and to the detriment of overall economics.

Above all it must be stressed that design work involves compromise at every stage and that there must be a continuous series of optimization studies. It should be appreciated though that design time costs money and a halt must be called at some stage in the proceedings. This is why the exercise of criticizing a design and showing how it can be improved is often comparatively easy; a designer can always look back on what he has done and decide that, if he had had a little more time, he would have done things rather differently. The aim must always be to produce a design which is the technical and economic best within limits of the time allowed.

3 Propulsion Cycles

3.1 INTRODUCTION

The design work for making an aero-engine is partly concerned with such matters as achieving reliability and light weight as discussed in Chapter 2, and is partly concerned with choosing the best thermodynamic cycle. It has been assumed that an engine to meet certain requirements *could* be designed but in fact the problem of deciding what will be the optimum cycle is a very complex one. Not only is there a variety of different types of reaction propulsion engine but, with each type, there is a wide range of possible operating conditions. In aircraft work the economic penalties of choosing the wrong cycle are generally high, so the initial choice of cycle is very important.

In this chapter the basic mechanics of reaction propulsion are first reviewed briefly and then the question of engine efficiencies is discussed. Two main efficiencies can be recognized, a *thermodynamic efficiency* of an engine producing a propulsive jet and a *propulsion efficiency* which shows how well the jet is used. As a general rule, the aim will be to optimize the product of these two separate terms and find how much useful power can be produced per unit quantity of fuel. Having covered this basic work, the cycle of the turbojet and its related engines are considered in more detail. The work shows how generalized plots can be constructed which can be used at the preliminary design stage for helping to choose the main operating parameters of an engine.

Besides the turbine engines for aircraft propulsion, this chapter also mentions the ramjet and rocket cycles, although each of these engines has a special chapter to itself later in the book. Some more unusual engine cycles are also covered, such as the turborocket and turboramjet. No practical examples of these have yet been made but their characteristics are such that they may find a place in the future of high-speed propulsion.

3.2 THE MECHANICS OF REACTION PROPULSION

Newton's laws of motion were discussed briefly in dealing with the fundamentals of applied mechanics in Chapter 1. These laws form the basis for the design of reaction propulsion engines. The first and second laws describe how a force must be applied if a mass is to be accelerated, while Newton's third law say that to very action there must be an equal and opposite reaction. A jet propulsion engine applies a force to a stream of working fluid in order to accelerate it backwards and uses the forward reaction force to act as the useful engine thrust.

Consider the engine shown in Figure 3.1 which is fitted to an

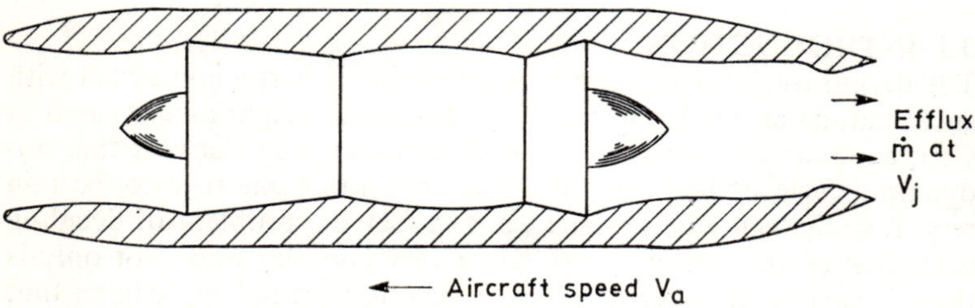

FIGURE 3.1

aircraft flying at a speed V_a. The engine has a mass flow of $\dot{m}$ kg/s and the jet efflux is expelled backwards at a velocity of V_J m/s relative to the final nozzle. It will be recalled from Chapter 1 (equations 1.2 and 1.3) that the thrust needed to accelerate a mass can be calculated from the rate of change of momentum, that is from the mass flow rate times the change in velocity.

$$\text{Thrust exerted by engine} = \dot{m}V_J \tag{3.1}$$

This thrust is defined as the *Gross Thrust* of the engine and we will denote it by X_G. In the case of a rocket vehicle, where the working fluid is carried by the vehicle, the gross thrust is the most important figure affecting the performance of the aircraft but for an air breathing engine, account must be taken of the fact that the working fluid is picked up from rest. The act of accelerating the air to flight speed before it enters the engine means that a force must be applied to the air and there will be a corresponding negative thrust or drag. This will be defined as the *Intake Momentum Drag* with symbol ΔX_I such that

$$\Delta X_I = \dot{m}V_a \tag{3.2}$$

The useful thrust of the engine is the difference between gross thrust and intake momentum drag and is known as the *Net Thrust*, X_N. Combining equations 3.1 and 3.2 gives

$$X_N = X_G - \Delta X_I = \dot{m}(V_J - V_a) \tag{3.3}$$

In all three of the above equations the thrust will be found in newtons (N), if the mass flow rate $\dot{m}$ is in kg/s and the velocities are measured in m/s. It is assumed that complete expansion to atmospheric pressure takes place in the final nozzle. Actually, as is discussed in greater detail while dealing with nozzles in Chapter 5 (p. 177), it is very seldom that the ideal expansion is achieved. However, the errors arising from making the simplifying assumption of complete expansion are not great in many practical cases.

(No account has been taken of any difference between inlet and exhaust mass flow rates, though, strictly speaking, allowance should be made for the effects of fuel addition. In the case of a turbojet, the fuel/air ratio is low and there are usually air bleeds for cooling purposes within the engine which are of the same order of magnitude as the fuel flows. For ramjet engines and for turbojets with reheat in operation, the fuel flow may be around 5% of the intake airflow and this should be allowed for in calculating the rate of momentum change.)

The thrust most easily measured for a turbine engine is that given on the test-bed and is quoted as the "static thrust". In this case, as the flight speed is zero, the gross thrust and net thrust are the same.

3.3 THE EFFICIENCIES OF REACTION PROPULSION

The prime purpose of an aero-space engine is to produce a useful thrust at a required flight speed. The product of thrust and flight speed will be the propulsive power and the most efficient machine will be that which provides this power with the lowest possible fuel consumption. Although the highest possible thermal efficiency may not be the overriding requirement in the design—Chapter 2 showed how important such things as weight and cost may be—it will almost always be important to try to get a high efficiency.

If an engineer were to look at the performance of a diesel-electric locomotive he would be interested in the overall performance of the unit but he would also be interested in finding where losses occurred. He would like to know how efficient the diesel engine was and also what efficiency was being attained by the electric drive. In a similar way we are interested in the *thermodynamic efficiency* of propulsion units and in the *propulsive efficiency* which shows how well the output is used to give a useful power.

The Thermodynamic Efficiency

The reaction propulsion engines described in section 3.2 above produce their thrust by accelerating backwards a working fluid. The efficiency of this process can be defined in terms of the ratio of the increase of kinetic energy of the fluid stream passing through the engine and of the energy content of the fluid. The thermodynamic efficiency η_T is defined as

$$\eta_T = \frac{\text{Increase in KE of fluid stream}}{\text{Energy input from fuel}} \text{ in unit time}$$

Using the same nomenclature as before (Figure 3.1) and with a fuel flow rate $\dot{F}$ and fuel calorific value C gives

$$\eta_T = \frac{\frac{1}{2}\dot{m}(V_J^2 - V_a^2)}{\dot{F}C} \tag{3.4}$$

The major part of this chapter is concerned with the performance of engine cycles and especially of how high thermodynamic efficiency can be achieved. For the moment it is sufficient to say that in choosing a cycle to give high thermodynamic efficiency some factors such as pressure ratio and temperatures are within the control of the designer. Other, such as compressor efficiency and combustion pressure losses, cannot yet be designed for with full confidence.

The Propulsive Efficiency

The normal transmission efficiency of a power drive is expressed as the ratio of power output to power input. Such a definition is not feasible for a jet engine as the engine produces *thrust* rather than power. To define the propulsive efficiency therefore, it is necessary to go back to first principles and write it as the ratio of useful work output to energy input. The energy input will be the same as that used in defining the thermodynamic efficiency, namely the increase in kinetic energy of the gas stream.

Consider the case of the engine shown in Figure 3.1 again. By equation 3.3 the net thrust is

$$X_N = \dot{m}(V_J - V_a)$$

The thrust power is the product of force and speed of the aircraft, i.e.

$$\text{Thrust power} = \dot{m}(V_J - V_a)V_a$$

The rate of energy addition to the gas stream is equal to the KE increase, i.e.

$$\text{Rate of energy addition} = \tfrac{1}{2}\dot{m}(V_J^2 - V_a^2)$$

$$\text{Propulsion efficiency} = \frac{\text{Useful output}}{\text{Rate of energy input}} = \frac{2(V_J - V_a)V_a}{(V_J^2 - V_a^2)}$$

Therefore

$$\eta_P = \frac{2}{1 + V_J/V_a} \tag{3.5}$$

This efficiency is often named the *Froude Efficiency* in commemoration of William Froude (1810–1879) who first defined it for marine propulsion.

Examination of equation 3.5 shows that high propulsive efficiency can be attained by making the jet velocity V_J close to the aircraft velocity V_a. Unfortunately, this is not practicable as the net thrust then tends to zero and the engine becomes excessively large. A compromise must be struck between the two conflicting requirements of high efficiency and a compact engine. The equation also shows why jet engines would never be practicable for surface use—except for breaking world speed records! At a typical jet efflux velocity of 500 m/s the propulsive efficiency at 120 km/h would be $12\frac{1}{2}\%$ and would fall to less than 3% at 15 km/h.

The propulsive efficiency may be used to show the best fields of use of varying types of aero-engine. The type of picture produced is that of Figure 3.2 which shows the efficiency achieved against

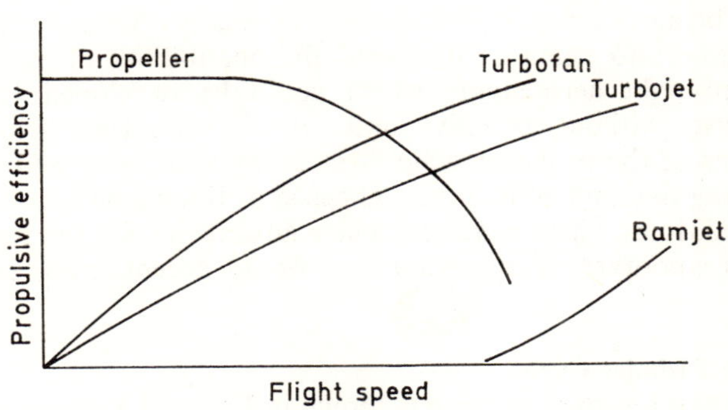

FIGURE 3.2 *Variation of propulsive efficiency with flight speed for different engines*

flight speed for the different types of engine. At low speeds the propeller-type engines give by far the best efficiency but propellers reach their useful limit at something under 750 km/h. At the high speed end, turbojets come into their own with good propulsion efficiency, while there is a small (but very significant) speed range in between where the turbofan engine is best. To achieve flight at very high speeds ramjet and rocket engines are needed.

The Overall Efficiency

Combining the thermodynamic and propulsion efficiencies gives the overall efficiency of the reaction propulsion engine. It can be shown easily that the expression is

$$\eta_O = \frac{\dot{m}(V_J - V_a)V_a}{\dot{F}C} \qquad (3.6)$$

where the symbols have their meaning as before.

From the point of view of running costs, it is this overall efficiency which is of the greatest importance—the two other efficiency measures are only steps on the way to finding the overall.

3.4 THE TURBOJET ENGINE

In starting the detailed study of propulsion cycles, pride of place is given to the turbojet engine. Partly this is because the turbojet was the first successful gas turbine aero-engine and partly it is because the main variants, the turbofan and turboprop engines, can be considered as adaptations of the basic turbojet theme. Although the turboprop and turbofan are used in considerable numbers, the turbojet is still the type in most common use throughout the world.

The turbojet story started seriously in the 1930's with extensive research being pursued in Britain and Germany. Basically, the jet engine came into being as it offered the possibilities of making a very lightweight aero-engine which could be developed to give high thrusts without excessive bulk. It offered release from the restrictions of the propeller which was bulky and heavy and which was nearing the limit of its practical speed as tip speeds approached sonic conditions. There were secondary advantages for the turbojet as it was smoother in operation and would accept lower quality fuels.

The Ideal Turbojet Cycle

The layout of a turbojet engine is shown in Figure 3.3. Reference to this figure shows that, with the subscripts as shown, the components are as follows:

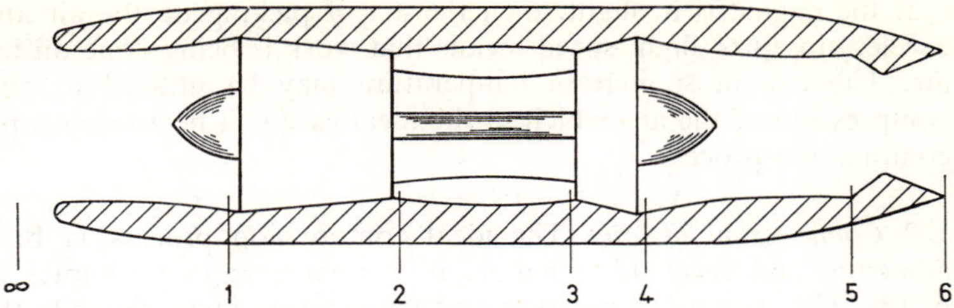

FIGURE 3.3 *Reference stations for turbojet cycle analysis*

Stations ∞–1 The Intake
Stations 1–2 The Compressor
Stations 2–3 The Combustion System
Stations 3–4 The Turbine
Stations 4–5 The Jet Pipe
Stations 5–6 The Final Nozzle

The operation of the engine can best be illustrated with reference to pressure–volume and temperature–entropy diagrams for 1 kg of working fluid. The diagrams of Figure 3.4 show the operation of

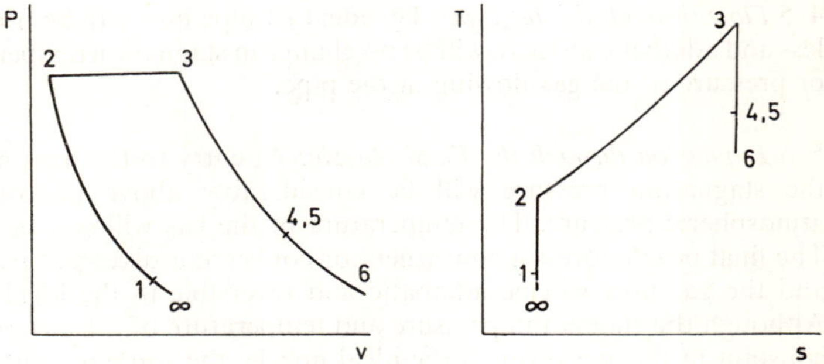

FIGURE 3.4 *The ideal turbojet cycle*

an ideal engine with the numbering of the points corresponding to that in Figure 3.3. The steps in the cycle are as follows:

∞–1 *The Intake Section:* Two cases are of interest, namely when the engine is at rest and when it is moving at some flight speed V_a. For the engine at rest the ideal intake is adiabatic and without friction, so process ∞–1 is at constant stagnation temperature and pressure.

If the engine is in flight, then the act of picking up the air and accelerating it to flight speed means that work is being done on the air. The rise in stagnation temperature may be utilized to give compression of the air, which in the ideal case will be an isentropic compression process.

1–2 *Compression of Air:* The ideal compression process is both adiabatic and reversible, that is, it is isentropic (see Chapter 1, p. 15). The stagnation pressure and temperature increase while the specific volume decreases.

2–3 *Heat Addition:* The heat transfer to the air in the combustion system is ideally at constant stagnation pressure. The entropy of the gas will increase considerably during the burning.

3–4 *Expansion in the Turbine:* The power required to drive the compressor is obtained by expanding the gases in the turbine. As in the case of the compression, the ideal expansion is adiabatic and reversible, i.e. isentropic. In going through the turbine the stagnation temperature and pressure of the gas will fall.

4–5 *Flow through the Jet Pipe:* The ideal jet pipe flow will be frictionless and adiabatic so there will be no change in stagnation temperature or pressure of the gas flowing in the pipe.

5–6 *Expansion through the Final Nozzle:* At entry to the final nozzle the stagnation pressure will be considerably above the outside atmospheric pressure. The temperature of the gas will still be high. The final nozzle forms a convergent or convergent/divergent passage and the gas flow will be adiabatic and reversible in the ideal case. Although the stagnation pressure and temperature of the gas remain constant in flowing through the ideal nozzle, the *static* pressure and temperature will fall so there will be a large increase in gas velocity.

At station 6 the static pressure of the gas stream should be equal in value to the static pressure of the surrounding atmosphere. In fact, a nozzle can only be designed to be of the correct shape at one overall pressure ratio, and when operating away from this design point the expansion may be incomplete. What happens under these circumstances was touched on in Chapter 1 and is discussed in greater detail in Chapter 5 when dealing with the design of final nozzles.

It is necessary now to calculate the work and heat transfers for each of the processes in the cycle. To do this, the steady flow energy equation, equation 1.12, is applied to each stage. It is convenient to work in terms of stagnation conditions—pressure, temperature and enthalpy—round the cycle. Considering each step in turn we get for the ideal cycle:

∞–1 *Intake of Air:* For the flight case the air is initially moving at the flight speed V_a relative to the engine. It is the *relative* motion which is important and we get the same result by flying the engine at speed V_a or blowing a wind of velocity V_a into the intake. The steady flow energy equation for this stream can be written as

$$T_\infty + \frac{V_a^2}{2c_p} = T_{01} \tag{3.7}$$

$(T_{01} - T_\infty)$ is the ram temperature rise. As the ideal flow is frictionless and adiabatic the isentropic relationship can be applied so that

$$p_{01}/p_\infty = (T_{01}/T_\infty)^{\gamma/\gamma-1} \tag{3.8}$$

The static case reduces to $T_{01} = T_\infty$ so that no work is done, no heat is transferred and the stagnation temperature and pressure are unchanged in the intake duct.

1–2 *Compression:* The ideal compression is adiabatic so

$$Q_{1-2} = 0 \tag{3.9}$$

Assuming the compressor is horizontal the energy equation reduces to

$$\begin{aligned} \dot{W}_{1-2} &= \dot{m}(h_{01} - h_{02}) \\ &= \dot{m}c_p(T_{01} - T_{02}) \end{aligned} \tag{3.10}$$

(This is, of course, negative implying a power input.)

Also as the ideal compression is isentropic equation 1.17 applies:

$$T_{02}/T_{01} = (p_{02}/p_{01})^{(\gamma-1)/\gamma} \tag{3.11}$$

$$\dot{W}_{1-2} = \dot{m}c_p T_{01}[(p_{02}/p_{01})^{(\gamma-1)/\gamma} - 1] \tag{3.12}$$

2–3 *Combustion:* No work is done in the combustion chamber so

$$\dot{W}_{2-3} = 0 \tag{3.13}$$

If the fuel flow is $\dot{F}$ and the velocities into and out of the combustion systems are about the same we have

$$\dot{F}h_{fL} = \dot{m}(h_{03} - h_{02}) + \dot{F}h_3$$

h_{fL} is the enthalpy of the liquid fuel entering the chamber and will be neglected.

$$\dot{Q}_{2-3} = \dot{m}(h_{03} - h_2) + \dot{F}h_3 \tag{3.14}$$

It is convenient to express the fuel flow in terms of a fuel/air ratio f and the air mass flow so that $\dot{F} = \dot{m}f$.

Equation 3.14 then reduces to

$$\dot{Q}_{2-3} = \dot{m}[h_{03}(1 + f) - h_{02}] \tag{3.14a}$$

3–4 *Turbine Expansion:* As with the compressor, the ideal flow through the turbine is adiabatic, so

$$Q_{3-4} = 0 \tag{3.15}$$

Making the same assumptions as for the compressor, that is, that the turbine is horizontal, the power equation becomes

$$\begin{aligned} \dot{W}_{3-4} &= \dot{m}(h_{03} - h_{04}) \\ &= \dot{m}c_p(T_{03} - T_{04}) \end{aligned} \tag{3.16}$$

In this case T_{03} is greater than T_{04} so the work output is positive. Again the isentropic relationship between temperature and pressure ratios can be written as

$$T_{03}/T_{04} = (p_{03}/p_{04})^{(\gamma-1)/\gamma} \tag{3.17}$$

and an equation similar to 3.12 can be written for the turbine power. One other very important stipulation must be made concerning the turbine process—a stipulation which is the key to the calculation of engine cycles. It is that the *turbine power output must be equal to the compressor power input* (plus, of course, any power losses in bearing friction, etc.). If there is an imbalance of power the engine will either accelerate or decelerate.

4–5 *Jet Pipe Flow:* The ideal jet pipe will be fully lagged to give adiabatic flow and there will be no work done. Therefore

$$Q_{4-5} = 0 \tag{3.18}$$

$$W_{4-5} = 0 \tag{3.19}$$

5–6 *Nozzle Flow:* Again ideally there will be no heat or work transfer, so

$$Q_{5-6} = 0 \tag{3.20}$$

$$W_{5-6} = 0 \tag{3.21}$$

To find the velocity change the energy equation is written in full (neglecting however the potential energy terms):

$$h_5 + \tfrac{1}{2}V_5^2 + \frac{Q_{5-6}}{m} = h_6 + \tfrac{1}{2}V_6^2 + \frac{W_{5-6}}{m}$$

As Q and W are both zero this becomes

$$h_6 + \tfrac{1}{2}V_6^2 = h_5 + \tfrac{1}{2}V_5^2 = h_{05}$$

$$V_6^2 = 2(h_{05} - h_6) \qquad (3.22)$$

$$= 2c_p(T_{05} - T_6)$$

The nozzle flow being isentropic we can write

$$T_{05}/T_6 = (p_{05}/p_6)^{(\gamma-1)/\gamma}$$

and hence find the nozzle temperature drop.

Component Efficiencies and the Real Cycle

Although much can be learnt from the study of *ideal* engine cycles, and various basic rules on performance may be discovered, a more realistic picture is obtained if allowance is made for the losses in an actual engine cycle. In any kind of thermodynamic machine such as a turbojet the losses can be considered as falling into three main categories. These are

(*a*) *Aerodynamic losses*, which show up as losses in stagnation pressure as fluids flow through ducts and past bodies. These losses are basically due to viscosity effects.

(*b*) *Thermodynamic losses* due to such things as incomplete energy release from fuels, and heat transfers from hot components to the surroundings.

(*c*) *Thermodynamic imperfections* which mean that compression and expansion processes depart from the ideal and are irreversible.

In Chapter 1 it was explained that for a process to be isentropic it had to be both adiabatic and reversible but that friction would inevitably introduce irreversibilities. A real gas with viscosity is bound to suffer frictional effects and so be non-isentropic in flow.

The main component efficiencies normally used in discussing real engine performances are based on these three types of loss. The efficiencies are defined as follows:

(a) Intake Section

The fact than an actual intake is non-isentropic and that there is, therefore, a smaller ram pressure ratio than in the ideal case can be expressed in a number of different ways. An *intake isentropic efficiency* η_I can be defined with reference to Figure 3.5. It is assumed that only a proportion of the ram temperature rise goes to give a useful pressure rise but that this proportion gives a stagnation

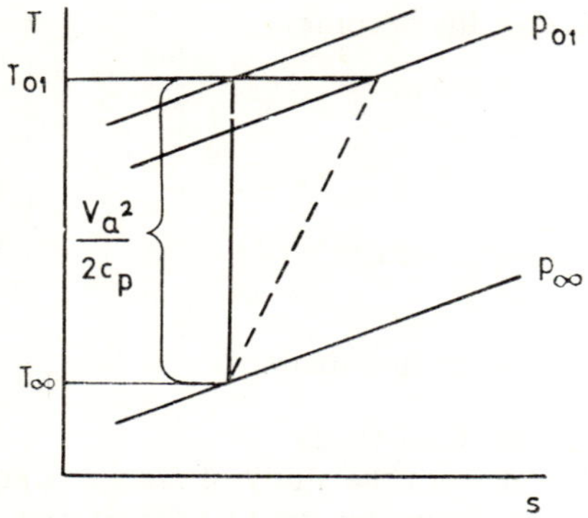

FIGURE 3.5 *The intake compression process*

pressure as if the compression had been isentropic. Using the symbols as in Figure 3.5 we write

$$p_{01}/p_\infty = \left(1 + \frac{\eta_I V_a^{\,2}}{2c_p T_\infty}\right)^{\gamma/(\gamma-1)} \tag{3.23}$$

(Later, in Chapter 7, while discussing supersonic intake, it will be seen that there are a number of alternative ways of defining intake losses. For reasons which are explained there, the efficiency defined by equation 3.23 is known as the "kinetic efficiency" and it is a particularly useful form.)

(b) The Compressor

A real compressor will require a larger work input for a given pressure ratio than will the ideal one. The efficiency is normally defined as the *compressor isentropic efficiency*, η_C, which is the ratio of the ideal specific enthalpy rise to the actual enthalpy rise for the same pressure ratio and inlet temperature.

$$\eta_C = \frac{c_p T_{01}[(p_{02}/p_{01})^{(\gamma-1)/\gamma} - 1]}{c_p(T_{02} - T_{01})}$$

To a close approximation the mean c_p for the actual compression will be the same as the mean c_p in the ideal case—it will depend on the mean temperature of the process—and so c_p can be cancelled out. The efficiency expression then becomes

$$\eta_C = \frac{T_{01}[(p_{02}/p_{01})^{(\gamma-1)/\gamma} - 1]}{(T_{02} - T_{01})} = \frac{(T_{02}' - T_{01})}{(T_{02} - T_{01})} \tag{3.24}$$

(A temperature with a prime, e.g. T_{02}' represents an ideal temperature.)

(c) The Combustion System

Imperfections in the combustion system appear both as a loss in stagnation pressure between stations 2 and 3 and as a reduction in the effective heating value of the fuel. The performance parameters corresponding to these losses are:

(i) A combustion chamber pressure loss coefficient

$$\lambda_C = \frac{p_{03} - p_{02}}{p_{02}} \tag{3.25}$$

(ii) A combustion efficiency which is the ratio of the actual combustion enthalpy rise to the calorific value of the fuel used. Writing the fuel calorific value as C gives

$$\eta_{COMB} = \frac{\dot{m}(h_{03} - h_{02})}{\dot{m}fC} = \frac{h_{03} - h_{02}}{fC} \tag{3.26}$$

(f, as above, is the fuel/air ratio.)

(d) The Turbine

Looking at the real expansion of gases through the turbine, the frictional effects in this case will cause the gases to leave the turbine hotter than they would do in the ideal case. This means that the temperature *drop* through the turbine will be less than the ideal one and so the power output will be reduced. The *turbine isentropic efficiency* η_T is the ratio of the *actual* power output to the *isentropic* power output for the same pressure ratio and inlet temperature—it is the inverse of the compressor efficiency expression.

$$\eta_T = \frac{c_p(T_{03} - T_{04})}{c_p T_{03}[1 - (p_{04}/p_{03})^{(\gamma-1)/\gamma}]}$$

As in the case of the compressor, the mean c_p will be very nearly the same for both processes and the expression reduces to

$$\eta_T = \frac{T_{03} - T_{04}}{T_{03}[1 - (p_{04}/p_{03})^{(\gamma-1)/\gamma}]} = \frac{T_{03} - T_{04}}{T_{03} - T_{04}'} \tag{3.27}$$

(e) The Jet Pipe

The loss in the real jet pipe appears as a fall in stagnation pressure so it can be expressed by means of a loss coefficient

$$\lambda_{JP} = \frac{p_{05} - p_{04}}{p_{04}} \tag{3.28}$$

Inevitably too, there must be some loss of heat through the jet pipe but it will normally be well lagged to keep heat losses small.

(f) The Final Nozzle

The expansion of gases through the final nozzle will again be accompanied by friction and so, as with the turbine, the gases will come out hotter than in the ideal case. A reduced enthalpy drop means of course a lower gas outlet velocity. Exactly as with the turbine, the *nozzle isentropic efficiency* η_N is defined as the ratio of actual to isentropic enthalpy drops. It is even more acceptable in this case to assume the same mean c_p for ideal and actual processes so the nozzle efficiency expression reduces to

$$\eta_N = \frac{T_{05} - T_6}{T_{05}[1 - (p_{05}/p_6)^{(\gamma-1)/\gamma}]} \tag{3.29}$$

Two other factors must be taken into account in analysing the final nozzle flow. Firstly, the effective area of flow for the nozzle will always be slightly less than the geometric area, the two being related by a discharge coefficient C_d where

$$C_d = \frac{\text{Effective nozzle area}}{\text{Actual nozzle area}} \tag{3.30}$$

Secondly, the flow streamlines leaving the nozzle will not all be parallel to the axis of the nozzle—Figure 3.6 shows a probable flow

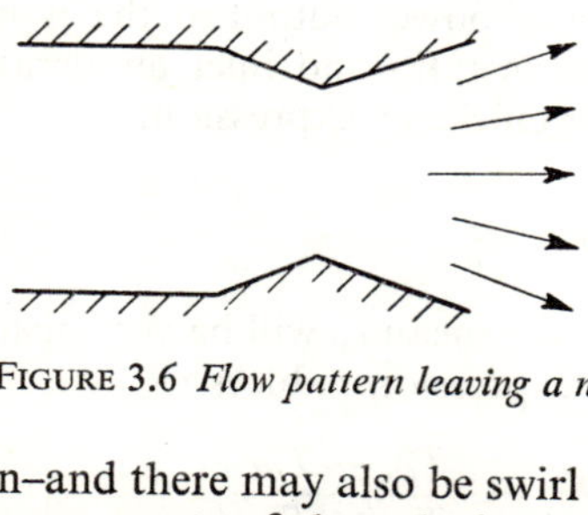

FIGURE 3.6 *Flow pattern leaving a nozzle*

pattern–and there may also be swirl in the turbine exhaust. Now it is only components of the velocity in the direction of the nozzle axis which give a useful thrust so the thrust of the nozzle will be slightly reduced. A thrust coefficient C_x can be defined as

$$C_x = \frac{\text{Actual nozzle thrust}}{\text{Thrust assuming all flow uniform and parallel to axis}} \tag{3.31}$$

When all these factors are taken into account, the complete engine cycle can again be illustrated on any convenient chart of the gas properties. This has been done in Figure 3.7 which shows the real

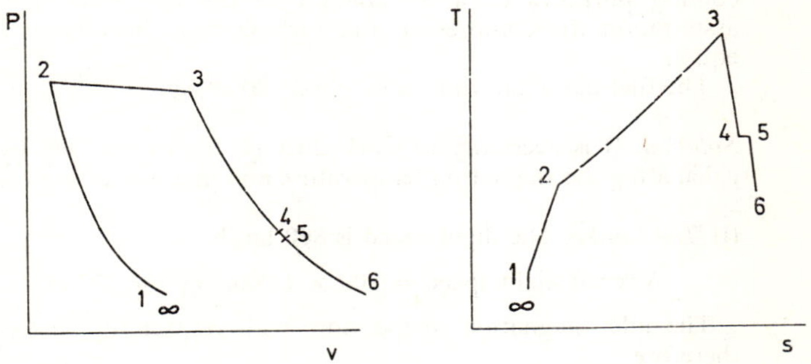

FIGURE 3.7 *The real turbojet cycle*

cycle in the p–v and T–s planes. It is worth noting that, as emphasized in Chapter 1, p. 16, all the real processes involve an increase in entropy of the working fluid.

Sample Turbojet Cycle Calculations

The past few pages which have explained in some detail the processes involved in gas turbine cycles have inevitably been somewhat lengthy and involved. Actually the work is very straightforward but it is important to follow it steadily step by step in order to build up a complete picture. To show the application of the equations which have been derived a sample cycle calculation is carried out in the following example. Realistic component efficiencies and losses have been used to give a typical turbojet performance.

Example 3.1 A jet aircraft is flying at 800 km/h true air speed at a height where the atmospheric pressure is 70 kN/m^2 and the temperature is 268 K. The efficiency of the intake defined by the kinetic efficiency (equation 3.23) is 0·90. Compression takes place through a stagnation pressure ratio of 10·0 in a compressor of isentropic efficiency 0·84. In the combustion chamber the stagnation pressure falls by 6% while the burning of the fuel raises the stagnation temperature to 1 200 K. The turbine has an efficiency of 0·90 and there is a loss of 2% in gas stagnation pressure as it passes through the jet pipe. Expansion to atmospheric pressure takes place isentropically in the final nozzle.

Calculate the net thrust produced by the engine for an air mass flow rate of 50 kg/s and the specific fuel consumption. To a first approximation it can be assumed that the gas passing through the engine is air with properties as given in Table 1.1 (p. 11). Assume also that there is an air bleed taken from compressor delivery for cooling purposes equal in amount to the fuel flow. With this assumption the compressor and turbine mass flow rates will be equal.

The fuel has a calorific value of 43 000 kJ/kg.

Solution. It is necessary to work through the engine step by step, calculating the stagnation temperature and pressure at each station.

(i) *The Intake.* The flight speed is 800 km/h.

$$\text{Aircraft flight speed} = 800 \times 1\,000/3\,600 = 222 \text{ m/s}$$

The inlet stagnation temperature with respect to the engine is therefore

$$T_{01} = T_{\infty} + \frac{V^2}{2 \times 1\,000 c_p} = 268 + \frac{222^2}{2 \times 1\,000 \times 1{\cdot}01}$$
$$= 268 + 24{\cdot}4$$
$$= 292 \text{ K}$$

From the definition of intake efficiency η_I

$$p_{01}/p_{\infty} = \left(1 + \eta_I \frac{(T_{01} - T_{\infty})}{T_{\infty}}\right)^{\gamma/\gamma-1}$$
$$= \left(1 + \frac{0{\cdot}9 \times 24{\cdot}4}{268}\right)^{1{\cdot}4/0{\cdot}4} = 1{\cdot}318$$

$$p_{01} = 70 \times 1{\cdot}318 = 92{\cdot}2 \text{ kN/m}^2$$

(ii) *The Compressor.* We are given that $p_{02}/p_{01} = 10{\cdot}0$. Therefore

$$p_{02} = 922 \text{ kN/m}^2$$

If the compression were isentropic giving a delivery temperature of T_{02}' then

$$T_{02}'/T_{01} = (p_{02}/p_{01})^{(\gamma-1)/\gamma}$$

Strictly speaking the true mean value of γ should be found by carrying out a trial and error process but we will assume $\gamma = 1{\cdot}4$.

Therefore

$$T_{02}'/T_{01} = (10)^{0\cdot286} = 1\cdot93$$
$$T_{02}' = 292 \times 1\cdot93 = 564 \text{ K and } T_{02}' - T_{01} = 272°\text{C}$$

But, by the definition of compressor efficiency,

$$\eta_C = \frac{T_{02}' - T_{01}}{T_{02} - T_{01}}$$

Therefore

$$T_{02} - T_{01} = 272/0\cdot84 = 323°\text{C}$$
$$T_{02} = 615 \text{ K}$$

To find the compressor power absorption use

$$\dot{W}_c = \dot{m}(h_{02} - h_{01})$$

From Table 1.1, $h_{02} = 621$ kJ/kg and $h_{01} = 295$ kJ/kg. Therefore

$$\dot{W}_c = 50\ (621 - 295) = 16\ 300 \text{ kW}$$

(iii) *Combustion.* At exit from the combustion chamber

$$T_{03} = 1\ 200 \text{ K and } p_{03} = 0\cdot94 \times p_{02} = 866 \text{ kN/m}^2$$

The heat added is given, to a close approximation, by

$$\dot{Q} = \dot{m}(h_{03} - h_{02})$$
$$h_{03} = 1\ 278 \text{ kJ/kg}$$

Therefore

$$\dot{Q} = 50\ (1\ 278 - 621) = 32\ 850 \text{ kW}$$

Fuel flow rate $= 32\ 850 \times 3\ 600/43\ 000 = 2\ 750$ kg/h

(iv) *The Turbine.* As was shown on p. 84 the crucial step is that turbine power must equal compressor power, i.e.

$$\dot{m}(h_{03} - h_{04}) = \dot{m}(h_{02} - h_{01})$$
$$h_{03} - h_{04} = 326 \text{ kJ/kg}$$

But $h_{03} = 1\ 278$ kJ/kg and therefore $h_{04} = 952$ kJ/kg.

This value of enthalpy shows that $T_{04} = 918$ K.

Equation 3.27 showed the turbine efficiency to be

$$\eta_T = \frac{T_{03} - T_{04}}{T_{03} - T_{04}'}$$

Therefore

$$T_{03} - T_{04}' = \frac{1\ 200 - 918}{0\cdot90} = 315°\text{C}$$

$$T_{04}' = 885 \text{ K}$$

The isentropic relationship may now be applied taking a mean value of $\gamma = 1{\cdot}333$ to give

$$p_{03}/p_{04} = (T_{03}/T_{04}')^{\gamma/\gamma-1} = (1\,200/885)^4 = 3{\cdot}42$$
$$p_{04} = 866/3{\cdot}42 = 254 \text{ kN/m}^2$$

(v) *The Jet Pipe*. Heat losses in the jet pipe will be negligible and no work will be done so

$$T_{05} = T_{04} = 918 \text{ K}$$

The 2% loss in stagnation pressure gives

$$p_{05} = 0{\cdot}98 \times 254 = 249 \text{ kN/m}^2$$

(vi) *The Final Nozzle*. Flow through the final nozzle is assumed isentropic and will be to an exit static pressure equal to atmospheric pressure, i.e. $p_6 = 70$ kN/m^2.

The nozzle exit static temperature T_6 comes then from

$$T_{05}/T_6 = (p_{05}/p_6)^{(\gamma-1)/\gamma}$$

with $\gamma = 1{\cdot}35$ approximately. Therefore

$$T_{05}/T_6 = (249/70)^{0{\cdot}259} = 1{\cdot}389$$
$$T_6 = 918/1{\cdot}389 = 661 \text{ K}$$

The jet exit velocity

$$V_6 = \sqrt{[2 \times 1\,000(h_{05} - h_6)]}$$
$$h_{05} = h_{04} = 949 \text{ kJ/kg} \qquad h_6 = 673 \text{ kJ/kg}$$

Therefore

$$V_6 = 743 \text{ m/s}$$

The overall turbojet performance can now be obtained using equation 3.3:

$$\text{Net thrust } X_N = \dot{m}\,(V_6 - V_a) = 50\,(743 - 222)$$
$$= 26\,050 \text{ N}$$

and

$$\text{Specific fuel consumption} = 2\,750/26\,050$$
$$= 0{\cdot}05 \text{ kg of fuel/N-h}$$

It is clear that calculation of the engine performance at a single flight condition involves a considerable amount of effort even though certain simplifying assumptions have been made. Luckily, digital computers are ideally suited to performing this type of calculation and they are capable of refining the assumptions made—for example on the gas properties—to any required degree of accuracy.

Generalized Turbojet Performance

The type of calculation carried out above could obviously be carried out for any engine operating conditions and for any level of efficiencies. If a whole series of calculations were carried out, it would be possible to prepare a generalized performance plot to show how the overall engine performance varied with design conditions. Clearly there are too many variables involved to allow all of them to be covered with one plot but, despite this restriction, many useful trends can be seen. Figure 3.8 shows a typical plot of

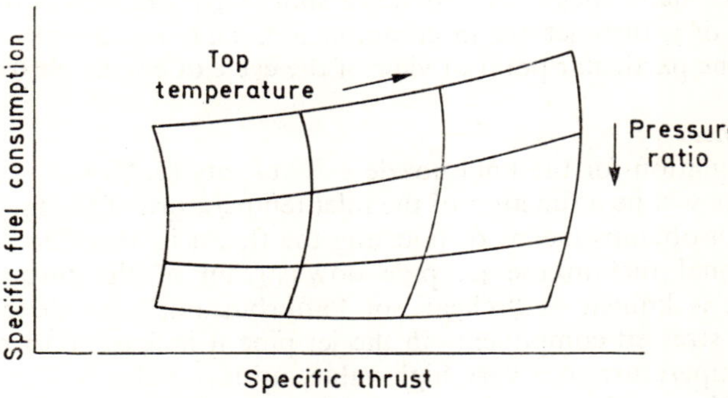

FIGURE 3.8 *Generalized turbojet performance*

specific thrust/specific fuel consumption (s.f.c.) for a turbojet at varying pressure ratio and turbine entry temperature. All the calculations are for a set flight condition and for set levels of component efficiencies and losses. Any alteration in these losses will affect the absolute value of engine output and efficiency without changing unduly the general shape of the curves.

Much can be learnt by looking at the generalized performance of the engine as shown in this diagram. One thing immediately apparent is the great significance of cycle pressure ratio in achieving a good s.f.c. This is one of the fundamental points of thermodynamics—to obtain high efficiency the heat should be added to the cycle at the highest possible temperature level. In recent years cycle pressure ratios have soared from 5–10:1 to 25:1 and more, with corresponding improvements in cycle efficiency. On the other hand, the main factor determining the specific *output* is the peak cycle temperature, but this is bought at the expense of s.f.c. Picking the best design for a given aircraft involves weighing up all these conflicting requirements.

Thrust Augmentation with the Turbojet

Certain types of aircraft, such as interceptor fighters, need the maximum power output from their engines for very limited times—of the order of a few minutes perhaps. It would be undesirable to fit an extra large size engine specially to cater for these limited occasions and so there are incentives to provide some form of thrust boosting above the normal maximum conditions of the engine. As the boost is wanted only for short times it does not matter overmuch if it is obtained inefficiently as the extra fuel used will probably not be a large amount compared with the total for a flight.

Three main means of providing short term augmentation of the thrust of a turbojet are in common use. They are described below from the particular point of view of the cycle of engine performance.

(*a*) *Reheat*

The equation for the final nozzle exit velocity (3.22) shows that the velocity will be a function of the inlet temperature of the final nozzle T_{05}. An obvious means of boosting the thrust is, therefore, to burn additional fuel in the jet pipe downstream of the turbine. This system is known as "reheat" or "afterburning." As there are no highly stressed components in the jet pipe it is feasible to raise the gas temperature to a very high value, up to a value, in fact, which nears the maximum attainable with normal fuels. With the high temperature rise there is, of course, a high fuel flow so the engine s.f.c. suffers, but for short times of operating this is acceptable.

There are drawbacks to the fitting of a reheat system besides the efficiency loss when it is in operation. For one thing, the mechanisms for stabilizing the flame in the jet pipe and for injecting the fuel introduce pressure losses which will reduce the basic engine performance. (This loss is known as the "cold loss" for the system and may amount to 3–4% of the basic engine thrust.) For another, it is generally necessary to fit a variable-area final nozzle to the jet pipe in order to accommodate the greatly increased volume flow of gases with reheat in operation. Such a variable nozzle is difficult to design and manufacture and introduces a weight penalty.

To give some idea of the thrust boost possible with reheat, Example 3.2 below shows how it would increase the thrust of the engine considered in Example 3.1.

Example 3.2 The turbojet of Example 3.1 is modified by adding a reheat system in which the stagnation temperature at entry to the final nozzle is increased to 1 800 K. Due to the extra mechanism in the jet pipe and to the effects of heat addition the jet pipe stagnation pressure loss increases from 2% to 8%. Calculate the net thrust and specific fuel consumption with reheat in operation assuming the reheat combustion efficiency is 0·90.

Solution. It can be assumed that the turbine outlet conditions will be unchanged with reheat in use so that $p_{04} = 254$ kN/m² and $T_{04} = 918$ K.

At entry to the final nozzle

$$p_{05} = 0{\cdot}92 \times 254 = 234 \text{ kN/m}^2 \text{ and } T_{05} = 1\,800 \text{ K}$$

First the fuel flow $\dot{m}_f$ must be calculated; it is necessary in this case to take account of the variation in mass flow rate of the gases through the combustion system. (In the main engine system the fuel flow was only $1\frac{1}{2}$% of the air flow so it could be neglected without bringing in too much of an error.) From the steady flow energy equation we get

$$\dot{Q} = \dot{m}_a\,(h_{05} - h_{04}) + \dot{m}_f(h_{05} - h_f)$$

where h_4 is the initial enthalpy of the fuel. This quantity will normally be small compared with the other terms.

Also

$$\dot{Q} = \dot{m}_f \times \text{calorific value of fuel} \times \text{combustion efficiency}$$

Therefore

$$\dot{m}_f \times 43\,000 \times 0{\cdot}90 = \dot{m}_a(h_{05} - h_{04}) + \dot{m}_f h_{05}$$

From the tables $h_{05} = 2\,003$ kJ/kg so this equation may be solved to give

$$\dot{m}_f = \frac{50\,(2\,003 - 949)}{(43\,000 \times 0{\cdot}9 - 2\,003)} = 1{\cdot}44 \text{ kg/s}$$

To find the nozzle outlet velocity, again assume isentropic expansion through the nozzle from a stagnation pressure of 234 kN/m² to an exit static pressure of 70 kN/m². Therefore

$$T_{05}/T_6 = (234/70)^{(\gamma-1)/\gamma} = 1{\cdot}33$$

(γ at this temperature is 1·31)

$$T_6 = 1\,800/1{\cdot}33 = 1\,353 \text{ K and } h_6 = 1\,457 \text{ kJ/kg}$$

Therefore

$$V_6 = \sqrt{[2 \times 1\,000\,(2\,003 - 1\,457)]} = 1\,050 \text{ m/s}$$

Because of the appreciable differences between inlet and exhaust mass flow rates, the gross and net thrusts should be calculated separately.

$$X_G = \dot{m}V_6 = (50 + 1{\cdot}44)1\,050 = 53\,900 \text{ N}$$

$$\Delta X_I = \dot{m}V_a = 50 \times 222 = 11\,100 \text{ N}$$

Therefore

Net thrust = 42 800 N, an increase of 64%

The total fuel flow is the sum of engine fuel plus reheat fuel, a total of 7 930 kg/h.

Specific fuel consumption = 7 930/42 800 = 0·185 kg/N-h

If these calculations are repeated for higher flight speeds and with reheat to a temperature of 2 000 K, it can be shown that two- or three-fold increases in net thrust are possible by the use of reheat, provided high specific fuel consumptions are acceptable.

(b) Liquid Injection at Compressor Inlet

A turbojet engine which is operating near its maximum temperature limitation may suffer severe performance penalties when operating at high atmospheric temperatures as it will then be necessary to reduce the rotational speed to avoid over-temperatures. These effects may be reduced by injecting liquid into the compressor inlet so that the liquid evaporates while passing through the compressor. Evaporation of the liquid entails heat extraction from the air which is being compressed, and this cooling leads to a reduction in the compressor power absorption. The power required from the turbine is reduced correspondingly and so, for given turbine entry conditions, a higher enthalpy drop is available for accelerating the gases in the final nozzle.

The properties required for the injectant are that the latent heat should be high and that it should not damage the engine. Distilled water is very suitable but there is a danger that any water not used on a take-off may freeze in the tank when the aircraft climbs to altitude. (Water injection is normally used to restore take-off power under tropical conditions.) To overcome this snag the water may be mixed with some alcohol to prevent freezing. Addition of alcohol will reduce the latent heat somewhat, but the alcohol can act as a fuel and burn in the combustion chamber. Whichever liquid is used it must be sprayed in as fine droplets to help it to evaporate readily—the water will pass through the complete compressor in a few milliseconds, so the time for evaporation is limited.

(c) Liquid Injection into the Combustion Chamber

A less common way of providing a thrust boost is to inject water or a water–alcohol mixture into the combustion chamber. There are two main effects in this case: a straightforward cooling of the gases and an increase in the turbine and jet pipe mass flow. Although larger amounts of water can be used than with compressor inlet water injection there are drawbacks, including the need to inject at a much higher pressure—of necessity well above combustion chamber pressure.

Of the three methods of thrust boosting described above, reheating is by far the most commonly used. For many years reheat has been used on engines for interceptor fighter and strike aircraft to give them the extra power needed for rapid take-off or closing to the attack. Water injection into compressor inlet on the other hand has found most of its applications for civil airliners which must carry high loads from tropical airfields. Very little has been done in Britain on the combustion chamber water injection system but it has been used on a number of American engines.

3.5 THE TURBOPROP AND TURBOSHAFT ENGINES

Long before jet propulsion for aircraft was first thought of, engineers had tried to make practical gas turbine power units. The step seemed to be a logical one and followed from comparison between reciprocating steam engines and steam turbines. Steam turbines had shown that smooth rotational power could be obtained without the need of converting reciprocating to rotary motion through the medium of connecting rods and cranks. They had shown too that very high volume flows of working fluid could be used, giving high powers in single prime movers. The advantages to be obtained of building a turbine which was independent of massive boilers and condensers—that is which used air as the working medium—were obviously very great.

Despite the prizes at stake, development of the gas turbine was slow. The main reason for this was the same as the reason for the relatively slow development of jet engines, namely that satisfactory materials were not available. However, once suitable materials had been developed there was little difficulty in making power-producing gas turbines. The efficiency of the plant was still low compared with that of well developed steam units or even of normal reciprocating i.c. engines, but the advantages of smoothness, and low weight and bulk for a given output, gave real advantages for aircraft work. In the aircraft field too, the gas turbine driving a propeller could be used at low to medium speeds where the propulsive efficiency of the turbojet was unacceptably low.

The sections that follow describe the cycles of operation of power producing engines. The two main classes are:

Turboprop Engines: Gas turbine power units, the *major part* of whose output goes to drive an aircraft propeller, but whose exhaust forms a propulsive jet.

Turboshaft Engines: Gas turbine power units, the *entire* useful output of which goes as a power output. (Such engines are used in helicopters and hovercraft and for surface applications.)

The differentiation between the two classes is one of degree; a turboprop always produces some useful output by ejecting the exhaust at an appreciable velocity but this "residual thrust" is virtually zero for a turboshaft engine.

The Turboprop Cycle

The basic layout of a turboprop engine is shown in Figure 3.9 and the similarity to the turbojet shown in Figure 3.3 will be obvious.

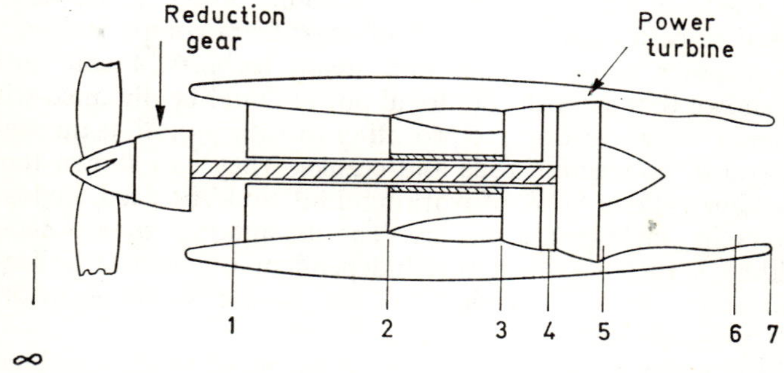

FIGURE 3.9 *Layout of a turboprop engine*

In fact, the turboprop uses the normal rotating assembly of a turbojet as a "gas generator" to form a source of high temperature gases which are above atmospheric pressure and so have available energy. For the turbojet, this energy is used to form a high velocity jet by expanding the gases through a nozzle. For the turboprop the gases expand first through a power turbine so extracting energy before they are passed on to the final nozzle. (Although Figure 3.9 shows a layout with a completely separate power turbine, it is just as practicable to design an engine in which the reduction gear and propeller are on the same shaft as the compressor. This "single shaft" arrangement is used on the Rolls-Royce Dart for example.

The "free turbine" arrangement represented in Figure 3.9 is used in a number of engines such as the Rolls-Royce Bristol Proteus.)

The cycle of opertiaon of a turboprop is shown on the temperature–entropy plot of Figure 3.10. It will be seen that up to point 4, the

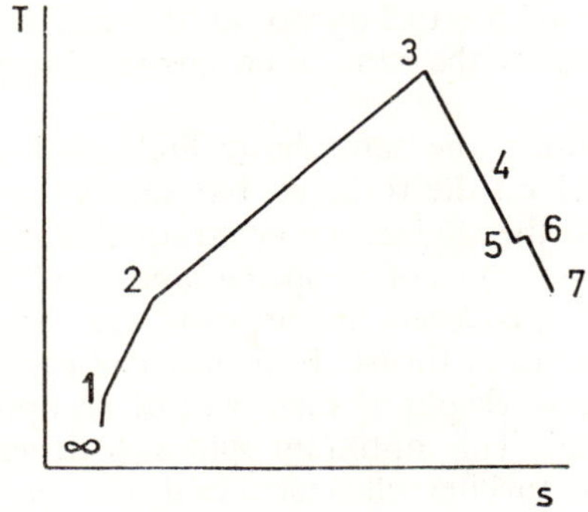

FIGURE 3.10 *The real turboprop cycle*

compressor turbine outlet, the diagrams are identical, but thereafter the single exhaust process 4–5 is replaced by separate power turbine expansion and final nozzle expansion. The power output of the power turbine is found by applying the steady flow energy equation to process 4–5. An equation similar to that for the compressor turbine can be derived namely:

$$\dot{W}_{4-5} = \dot{m}c_p(T_{04} - T_{05}) \tag{3.32}$$

Then, too, the power isentropic efficiency can be defined as before so that

$$\eta_{PT} = \frac{T_{04} - T_{05}}{T_{04}[1 - (p_{04}/p_{05})^{(\gamma-1)/\gamma}]} \tag{3.33}$$

The actual flow through the final nozzle can be treated exactly as before although this time the pressure ratio will be very much less and so the velocity will be lower. Nonetheless, the fact that the jet efflux will generally be at a higher velocity than the aircraft velocity means that there will be a useful momentum thrust which can be calculated from equation 3.3.

A power turbine is essentially a high tip speed device and so, if the diameter is to be kept low, the rotational speed must be high. On the other hand, propellers are large diameter and so the speed must be restricted so that excessive tip speeds do not result. It is always necessary, therefore, to interpose a reduction gear between the power turbine and the propeller. (The figures for the Rolls-Royce Bristol Proteus for example, show that at take-off conditions the power turbine runs at 11 111 rev/min while the propeller speed is 1 000 rev/min.) The necessity to fit a reduction gear has told heavily against the use of the turboprop, as the gearing adds weight and complexity, while at the same time introducing power losses and the risk of failures.

The following example takes the same basic flight and engine conditions as in Example 3.1 on the turbojet, but shows the effect of adding a power turbine to the engine. An arbitrary choice of the worksplit—the relative proportions of propeller and final nozzle thrust power components—has been made with the propeller producing about 80% of the total thrust. In a practical case great care is taken in deciding the worksplit as there will be an optimum value for a given flight speed. This optimum split will depend on such factors as the power turbine efficiency and the propeller efficiency.

Example 3.3 The turbojet of Example 3.1 is to be converted into a turboprop by adding a power turbine between the gas generator turbine outlet and the final nozzle. At the operating conditions of Example 3.1 the gas generator conditions are unchanged and expansion through the power turbine takes place to a stagnation pressure of 150 kN/m^2 with a turbine isentropic efficiency of 0·89. As before there is a 2% loss in stagnation pressure in the jet pipe and the expansion to atmosphere in the final nozzle is isentropic.

Calculate the total thrust power produced by the engine if the efficiency of the propeller is 75%.

Solution. We are told that the gas generator outlet conditions are unchanged with $p_{04} = 254$ kN/m^2 and $T_{04} = 918$ K. With expansion to 150 kN/m^2 the isentropic outlet temperature T_{05}' would be found from

$$T_{04}/T_{05}' = (p_{04}/p_{05})^{(\gamma-1)/\gamma}$$

Taking a mean value of $\gamma = 1{\cdot}35$ gives

$$T_{04}/T_{05}' = (254/150)^{0{\cdot}259} = 1{\cdot}146$$

$$T_{05}' = 800 \text{ K}$$

This gives an isentropic temperature drop of 116°C so the actual drop is 103°C. Therefore

$$T_{05} = 918 - 103 = 815 \text{ K and } h_{05} = 837 \text{ kJ/kg}$$

The power turbine output power is

$$\dot{W} = \dot{m}(h_{04} - h_{05}) = 50(949 - 837)$$
$$= 5\,600\ \text{kW}$$

To this power must be added that which comes from the residual thrust. At entry to the final nozzle $T_{06} = 815$ K and

$$p_{06} = 0{\cdot}98 \times p_{05} = 147\ \text{kN/m}^2$$

For isentropic expansion

$$T_{06}/T_7 = (p_{06}/p_7)^{(\gamma-1)/\gamma} = (147/70)^{0{\cdot}265} = 1{\cdot}217$$

(A suitable mean value for γ would be 1·36.)

$$T_7 = 815/1{\cdot}217 = 670\ \text{K} \qquad h_7 = 681\ \text{kJ/kg}$$
$$V_7 = \sqrt{[2 \times 1\,000(837 - 681)]} = 559\ \text{m/s}$$

Therefore

$$\text{Net thrust } X_N = 50(559 - 222) = 16\,850\ \text{N}$$

The total thrust horsepower will be that available at the propeller plus the net thrust times flight speed component.

$$\text{Total thrust horsepower} = \eta_{PROP}\dot{W} + \frac{X_N V_a}{550}$$
$$= 0{\cdot}75 \times 5\,600 + \frac{16\,850 \times 222}{1\,000}$$
$$= 7\,940\ \text{kW}$$

This can be compared with a thrust horsepower produced by the turbojet of

$$26\,050 \times 222/1\,000 = 5\,800\ \text{kW}$$

No attempt has been made in this case to optimize the work split between propeller and propulsive jet but even so this example does illustrate how useful the propeller is in producing propulsive power even up to flight speeds of 800 km/h.

Generalized Turboprop Performance

As in the case of the turbojet, calculations can be carried out for a turboprop engine cycle over a wide range of conditions and hence generalized performance curves can be plotted. Such a set of curves is shown in Figure 3.11 which illustrates the variation of s.f.c. with specific output (total power produced per unit mass flow rate through the engine), with compressor pressure ratio and compressor turbine inlet temperature as the design variables. The levels of efficiency and the worksplit between prop and jet are maintained constant for all points.

Although the general form of Figure 3.11 is similar to that for the turbojet in Figure 3.8, there is one particularly significant difference. This is that the thermal efficiency of the cycle may improve with increases in cycle maximum temperature. (For the turbojet the

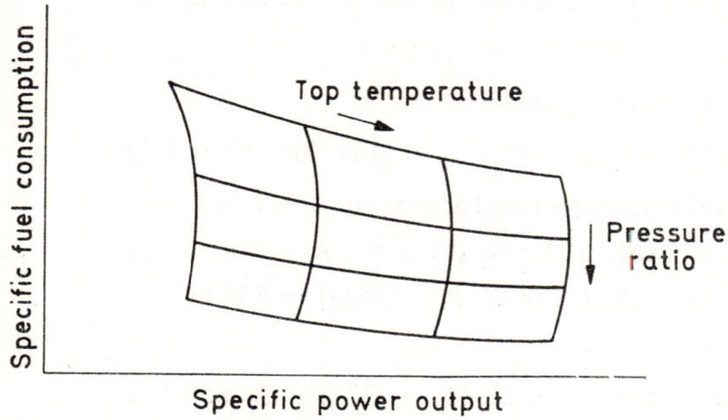

FIGURE 3.11 *Generalized turboprop performance*

opposite was the case and running the engine hotter always gave a worsening of the s.f.c.) There is, therefore, every incentive, from the performance point of view, to design a turboprop with the highest possible turbine inlet temperature.

Part Load Characteristics of Turboprop Engines

It has been said that two distinct types of turboprop are possible, namely the "free turbine" and "single shaft" varieties. Some of the biggest differences between the types arise in the field of control and part load performance. Before examining each type in more detail it may be said that part load control of a free turbine engine is generally easier in that the penalties of making a mistake are not so serious. Great care must be taken with the single shaft engine to ensure that dangerous operating regions are avoided.

With a free turbine engine, the operator must exercise control over two rotating systems, the gas generator system and the power-turbine/propeller system. The speed of the gas generator is controlled by varying the fuel flow and, at given flight conditions, there will be a unique relationship between compressor rotational speed and fuel flow. The speed of the propeller system, on the other hand, is fixed by varying the propeller blade angle and so the power absorption. Details of the types of control unit which may be used are discussed in Chapter 8; it is sufficient to say at this stage that a fairly simple control may suffice for the fuel flow and that the propeller speed control is a well understood problem with well tried equipment available to solve it.

Coming to the actual details of how the output of a free turbine engine varies with compressor speed, the characteristics are very similar to those of a boiler and steam turbine. In this type of plant, for given steam chest and exhaust conditions the power output of a turbine varies approximately parabolically with rotational speed. For a turboprop, the gas generator replaces the boiler and a set of parabolas of the form shown in Figure 3.12 can be drawn to illustrate the power variation with speed. The power output is zero at zero propeller rev/min (obviously!) and is also zero at some run-away speed where the torque is zero. A line passing through the peaks of parabolas shows the line of optimum power against optimum propeller rev/min. The equation of this line is of the general form

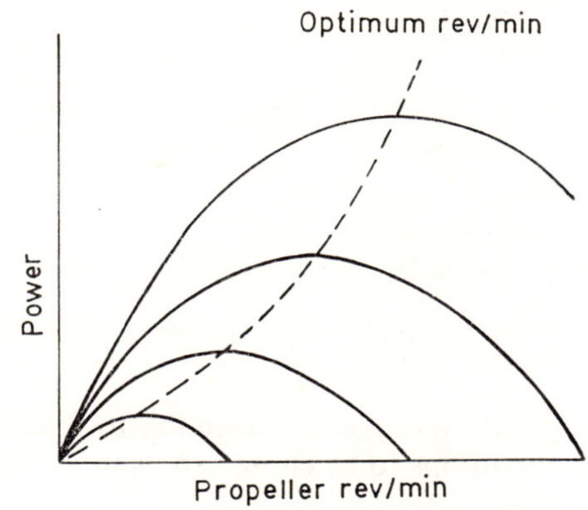

FIGURE 3.12 *Variation of output power with propeller speed for a free-turbine turboprop*

$$\dot{W}_{opt} = K(N_{opt})^n \qquad (3.34)$$

with K being a constant for the engine and the index n having a value of between 2·8 and 3.

The variation of propeller torque with rotational speed is also of interest. Figure 3.13 shows the general form of the torque versus speed lines at varying values of compressor speed. As a general rule the lines are approximately straight, passing through zero at the run-away speed. The torque with the propeller held stationary will, therefore, be approximately twice the torque at optimum rev/min.

It is clearly advisable from the performance point of view to operate as closely as possible to the optimum propeller rev/min for a given compressor speed. Provided, though, that the propeller speed does not depart too far from the optimum—about 10% variation is not excessive—the performance penalties are not great. This degree

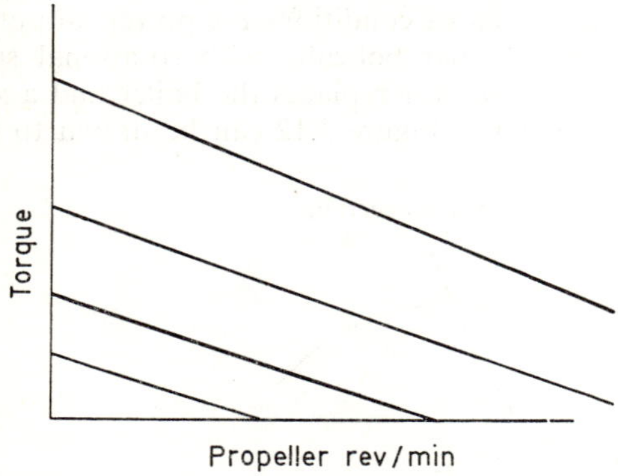

FIGURE 3.13 *Variation of output torque with propeller speed for a free-turbine turboprop*

of freedom is useful as it gives a measure of flexibility to the engine and allows the operating conditions to be chosen so as to match the best propeller characteristics.

The Single Shaft Engine

When it is required to change the power output on a single shaft engine, the operator has the choice of altering fuel flow, and thus the turbine entry temperature, or adjusting the air mass flow by varying the compressor rotational speed. Changes in compressor speed can be obtained by altering the propeller pitch and so varying the power absorption. From the performance point of view there is very little to choose between the two methods of control and the part-load efficiency is hardly affected by the choice. (Ref. 1 by Mallinson and Lewis gives a full discussion of the part load performance using the two methods.) In practice, though, there is every incentive to reduce cycle temperatures because in this way engine life is improved. On the other hand, the *noise* of a turboprop depends primarily on the tip speed of the propeller so any reduction in propeller rev/min will cut down the noise level. The usual practice is, therefore, to reduce the propeller speed slightly from the take off value and control power thereafter at constant speed.

Mention was made above that care must be taken with the control of single shaft engines to avoid running into dangerous conditions. It is necessary to interlock the fuel and speed controls so that rev/min is not reduced at constant fuel flow. If this should ever occur the fuel/air ratio would become excessive and there would be a risk of overtemperaturing the engine or causing the compressor to surge.

(One turboprop of special interest is the Rolls-Royce Tyne, a big engine which uses two compressors in series. The propeller is directly coupled through the reduction gear to the low-pressure system while the high-pressure compressor and turbine form a completely separate system. The Tyne is equivalent from the handling point of view to a single shaft engine.)

The Use of Heat Exchangers

Once the gases have left the power turbine of a turboprop or turboshaft engine they have given up most of their useful energy and are on their way to the exhaust. They are still at comparatively high temperature, however, and it is possible to use some of the energy to heat up the compressor delivery air before it goes to the combustion chamber. A diagrammatic layout of a turboshaft engine with heat exchanger is shown in Figure 3.14 and the corresponding temperature–entropy diagram is given in Figure 3.15.

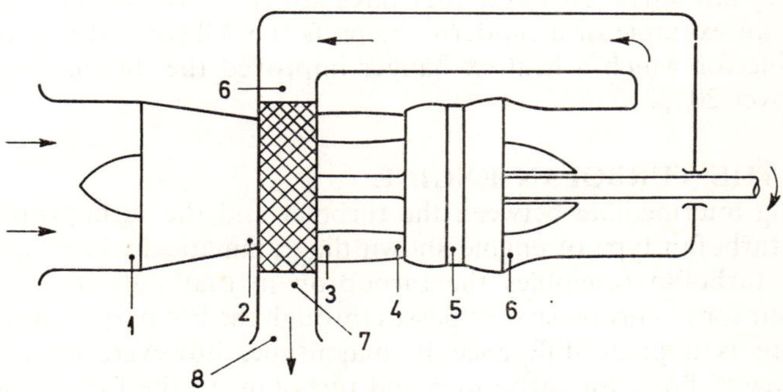

FIGURE 3.14 *Layout of turboshaft with heat exchanger*

Looked at from the cycle point of view the most important characteristic of the heat exchanger is that it improves the thermal efficiency of an engine *particularly at low pressure ratios* but at the same time it worsens the specific output. The marked improvement in efficiency at low pressure ratios arises because at these conditions there is the largest difference between the exhaust gas temperature (high) and the compressor delivery temperature (low). Fitting a heat exchanger is, therefore, especially useful in improving the *part load* performance of engines as then the compressor speed is reduced so reducing the pressure ratio. Losses in specific output arise from the inevitable pressure losses resulting from fitting the

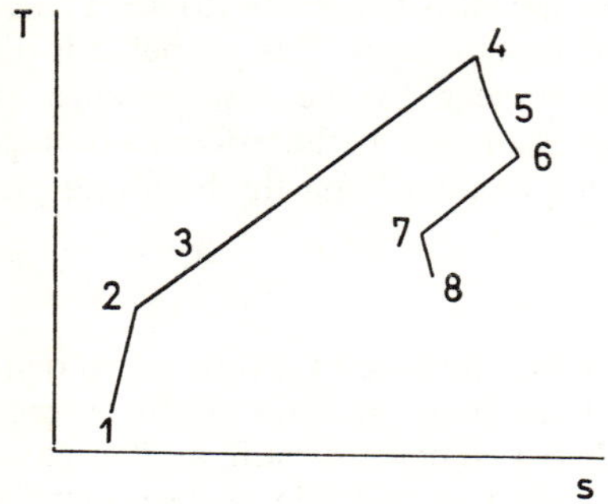

FIGURE 3.15 *The real turboshaft cycle with heat exchanger*

heat exchanger and there is, of course, a fairly severe weight penalty. For these reasons heat exchanger type engines have never been highly attractive. However they have always merited some attention and an example of a modern engine is the Allison T-68 turboshaft engine for which a heat exchanger improved the thermal efficiency by over 20%.

3.6 THE TURBOFAN ENGINE

Lying intermediate between the turbojet and the turboprop comes the turbofan type of engine shown diagrammatically in Figure 3.16. The turbofan resembles the turboprop in that only a portion of the air for the propulsive jet passes through the hot part of the engine. There is a great difference in magnitude, however, between the "bypass" flows for turboprop and turbofan. In the former case the air flow through the propeller may be fifty or more times the jet pipe mass flow; with the turbofan, the gas generator flow and the flow passing through the fan and cold nozzles are of the same general order of magnitude. The bypass ratio λ is defined as the ratio of cold duct flow to gas generator flow. Early turbofan engines such as the Rolls-Royce Conway were designed with the bypass ratio of about 0·5–1·0 but, with advances in technology, higher bypass ratios, of the order of 5–10 have become feasible and are now in common use. (It is in the application to the big subsonic jet transports that high bypass ratio turbofans show to greatest advantage.)

The location of the fan has considerable effect on the engine cycle of operation. Two main layouts are used, the "front fan" as in Figure 3.16(a) and the "aft fan" of Figure 3.16(b). The nomenclature

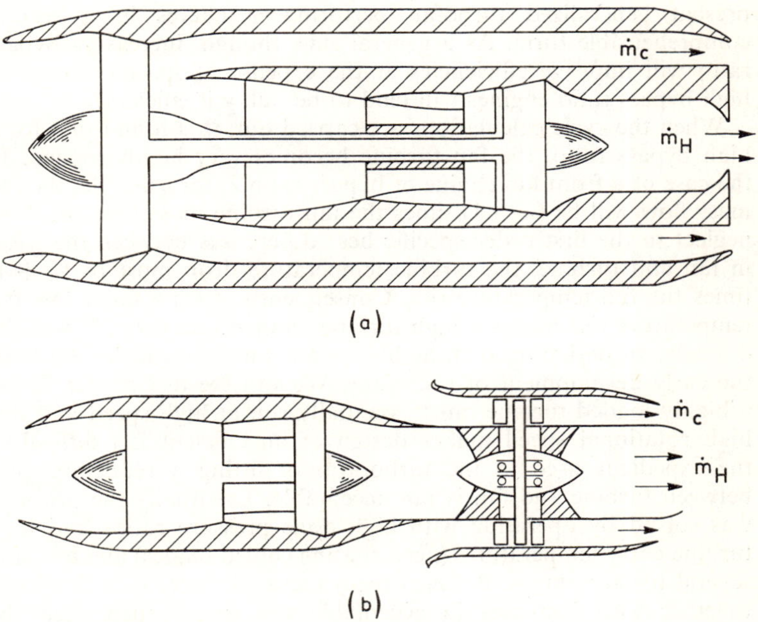

FIGURE 3.16 *Arrangements of turbofan engines*

is self-explanatory. For a front fan engine, the air passing through the fan is compressed before going to the high pressure system and so the overall pressure ratio is increased—a very important point. Although an aft fan does not offer this advantage, it is a much simpler unit and may be added easily to an existing design of engine as was done for example in the General Electric CJ-803. As the flow of a turbofan is divided between hot and cold systems it is not feasible to plot the cycle for a turbofan on a temperature–entropy chart as has been done for previous cycles. Performance calculations can be carried out as for the previous engines by applying the steady flow energy equation to each component of a turbofan.

When starting to choose a cycle, the designer has two further design variables to select besides those discussed previously. He will already have decided for the gas generator what pressure ratio and turbine entry temperature can be used and then, using the gas generator output, he will be able to select the *bypass ratio* and the *worksplit* or ratio between gas turbine and hot nozzle enthalpy drops. Because of these two additional variables—as well as the fact that front fan or aft layouts may be selected—it is not feasible to present generalized turbofan performance curves in a readily comprehensible form. As a general rule, though, increasing bypass ratio will improve efficiencies at the expense of specific thrust, so high bypass ratio engines will tend to be bulky if efficient.

When the cycle calculations are carried out, it is found that for a high bypass ratio, the fan turbine becomes very highly loaded. In the case of a front fan engine of bypass ratio λ, for example, the fan mass flow will be $(\lambda + 1)$ times the fan turbine mass flow so, if we neglect to the first order specific heat differences between the gases in fan and turbine, the turbine temperature drop must be $(\lambda + 1)$ times the fan temperature rise. Consequently, even a fairly low fan temperature rise means a high turbine temperature drop. It was the difficulty of designing such highly loaded turbines which held back the early development of turbofans. We will see in Chapter 5 that a highly loaded turbine must essentially run at high speed and this high rotational speed makes design of an efficient fan difficult—the expedient used in the turboprop of fitting a reduction gear between turbine and fan is not acceptable. Eventually the problem was solved by operating with high cycle pressure ratios and high turbine entry temperatures (necessitating cooled blades) and by using several turbine stages. Between them these advances in technology, together with increases in acceptable fan speeds, permitted the construction of the new generation of high bypass ratio engines.

One other important point to be decided in the design of a turbofan or bypass engine is whether to mix the hot and cold gas flows in the jet pipe or whether to expel them through separate nozzles. There are performance advantages in mixing the gas streams if it can be done without introducing too much turbulence. There are also important installation advantages as the single pipe is less complicated and the silencer and thrust reverser need only act in one gas stream. If mixing is to be used however, great care must be taken to see that the static pressure in the cold duct is above the turbine exhaust static pressure under all flight conditions. Any reverse flow of hot gases into the fan duct would be most undesirable.

Example 3.3 above illustrated the effect of converting a turbojet to a turboprop. To give some idea of the effect of adding an aft-fan to a turbojet Example 3.4 has been included. In order to permit comparison with the previous cases and same flight conditions and gas generator outlet conditions are taken as in the previous examples.

Example 3.4 The turbojet of Example 3.1 (p. 89) is to be converted into a turbofan of the aft fan variety by adding a free running turbine/fan unit. The fan has the same intake efficiency as the engine, a mass flow rate of 50 kg/s, an isentropic efficiency of 0·85 and imparts to the air a stagnation temperature rise of 40°C. For the fan turbine the isentropic efficiency is 0·89. There is a 2% loss in stagnation pressure in the jet pipe between fan turbine outlet and final nozzle. Flow through both hot and cold nozzles can be taken to be isentropic.

Calculate the net thrust and specific fuel consumption of this engine.

Solution. Let us define the stations of interest as

Station 4 Gas generator outlet*
Station 5 Fan turbine outlet*
Station 6 Hot nozzle inlet*
Station 7 Hot nozzle exit*
Station 8 Fan inlet
Station 9 Fan outlet or cold nozzle inlet
Station 10 Cold nozzle exit
(* as in previous examples)

Gas generator conditions are unchanged so $p_{04} = 254$ kN/m² and $T_{04} = 918$ K as before.

Consider the flow through the fan. As the fan intake has the same efficiency as the main engine we know $p_{08} = p_{01} = 92{\cdot}2$ kN/m² and $T_{08} = T_{01} = 292$ K. Therefore

$$T_{09} = T_{08} + 40 = 332 \text{ K} \qquad h_{09} = 333 \text{ kJ/kg}$$

The isentropic compression which gives the same pressure ratio would result in a temperature rise of $0{\cdot}85 \times 40 = 34$°C. Therefore

$$p_{09}/p_{08} = (T_{09}'/T_{08})^{\gamma/\gamma-1} = (326/292)^{1{\cdot}4/0{\cdot}4} = 1{\cdot}470$$

$$p_{09} = 135{\cdot}5 \text{ kN/m}^2$$

Expansion through the cold nozzle takes place isentropically so we can write

$$T_{09}/T_{10} = (p_{09}/p_{10})^{(\gamma-1)/\gamma} = (135{\cdot}5/70)^{0{\cdot}4/1{\cdot}4} = 1{\cdot}208$$

$$T_{10} = (332/1{\cdot}208) = 275 \text{ K} \qquad h_{10} = 275 \text{ kJ/kg}$$

$$V_{10} = \sqrt{[2 \times 1\,000(333 - 275)]} = 340 \text{ m/s}$$

Therefore

$$\text{Cold net thrust} = 50(340 - 222) = 5\,900 \text{ N}$$

Coming now to the hot stream, the turbine enthalpy drop must equal the fan enthalpy rise. Therefore

$$h_{04} - h_{05} = h_{09} - h_{08} = 333 - 293 = 40 \text{ kJ/kg}$$
$$h_{05} = 952 - 40 = 912 \text{ kJ/kg} \qquad T_{05} = 882 \text{ K}$$

Therefore

$$T_{04} - T_{05} = 918 - 882 = 36°\text{C}$$

and the equivalent isentropic drop is $T_{04} - T_{05}' = 36/0{\cdot}89 = 40{\cdot}5$. Therefore

$$T_{05}' = 918 - 40{\cdot}5 = 877{\cdot}5 \text{ K}$$

Taking a mean value of $\gamma = 1{\cdot}34$ at this temperature gives

$$(p_{04}/p_{05}) = (T_{04}/T_{05}')^{1{\cdot}34/0{\cdot}34} = 1{\cdot}248$$
$$p_{05} = 254/1{\cdot}248 = 204 \text{ kN/m}^2$$

With the 2% loss in stagnation pressure through the jet pipe we get

$$p_{06} = 0{\cdot}98 \times 204 = 200 \text{ kN/m}^2 \text{ and } T_{06} = 882 \text{ K}$$

As before

$$T_{06}/T_7 = (p_{06}/p_7)^{(\gamma-1)/\gamma} = (200/70)^{0{\cdot}35/1{\cdot}53} = 1{\cdot}313$$

Therefore

$$T_7 = 882/1{\cdot}313 = 670 \text{ K} \qquad h_7 = 681 \text{ kJ/kg}$$

Hence

$$V_7 = \sqrt{[2 \times 1\,000(912 - 681)]} = 680 \text{ m/s}$$

And the net thrust from the hot jet is

$$X_N = 50(680 - 222) = 22\,900 \text{ N}$$

The total engine thrust is therefore 28 800 N, an increase of 10% compared with the plain turbojet.

A similar comment may be made at this stage as was made at the end of Example 3.3—namely, that no attempt has been made to optimize the turbofan for this particular flight case and quite possibly some other worksplit or bypass ratio would give a better thrust. The example is designed to show how cycle calculations are carried out and it does illustrate how a useful thrust increase is obtained for the same fuel flow.

3.7 RAMJETS

Although very different mechanically from all the types of engine considered so far, the ramjet is closely related from a cycle point of view. It is, therefore, appropriate to consider the thermodynamics of this prime mover here while leaving a more detailed study of the design to Chapter 9.

The ramjet differs from turbine engines as the rotating assembly has been deleted—all the compression needed for it to operate as a heat engine comes from ram compression in the intake due to forward motion. Referring to the diagrammatic layout of Figure 3.17

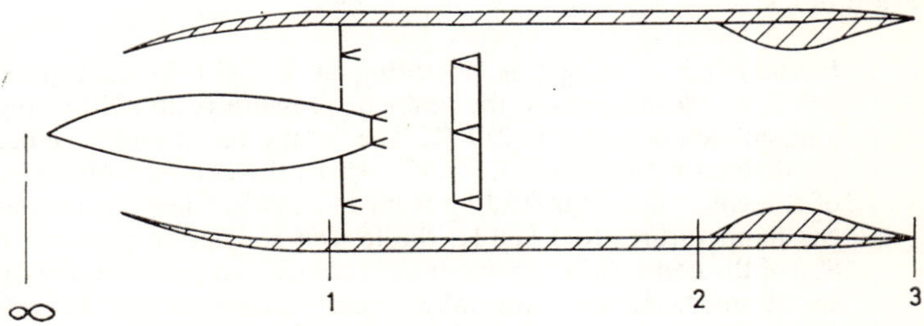

FIGURE 3.17 *Arrangement of ramjet engine*

it will be seen that the ramjet is basically a contoured duct with intake, combustion and expansion sections which together produce the high velocity gas stream. Until it is in motion, there can be no compression and no useful thrust can be produced. Initially therefore

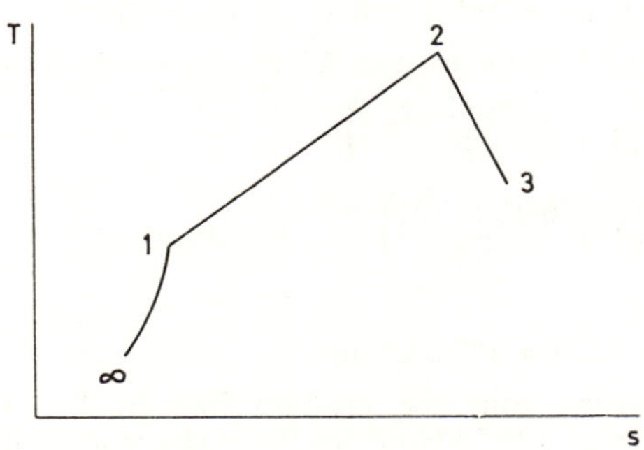

FIGURE 3.18 *The real ramjet cycle*

the ramjet must be accelerated to the light-up speed by some external means, such as rocket booster or air launch. In calculating the performances the steady flow energy equation can again be applied to each part of the cycle and, in fact, previously developed equations can be used. The real cycle shown in Figure 3.18 has been plotted on a temperature–entropy chart for air; the equations which are used for each process are

(a) *Intake Section* Equations 3.7, 3.8 and 3.23
(b) *Combustion* Equations 3.14, 3.26
(c) *Expansion* Equations 3.22, 3.29, 3.30 and 3.31

The use of these equations can best be seen from the following example.

Example 3.5 A ramjet is operating at a flight Mach number of 2·0 at an altitude where the ambient pressure is 20 kN/m² and the ambient temperature is 217 K. The intake has a kinetic efficiency (as defined by equation 3.23) of 0·92. In the combustion zone, fuel of calorific value 43 000 kJ/kg is burnt at 95% efficiency to raise the gas temperatures to 1 800 K, the loss in stagnation pressure being 8% of the combustion chamber inlet stagnation pressure. Expansion to atmospheric pressure takes place isentropically in a convergent/divergent nozzle with an exit diameter of 0·5 m. Calculate the net thrust produced by the ramjet and the corresponding specific fuel consumption.

Solution. As with all cycle questions, this one resolves to the calculation of stagnation pressures and temperatures through the engine. First find the intake delivery conditions

$$T_{01}/T_\infty = (1 + \tfrac{1}{2}(\gamma - 1)M^2) = (1 + 0{\cdot}2 \times 2^2) = 1{\cdot}8$$

$$T_{01} = 2{\cdot}7 \times 1{\cdot}8 = 391\ \text{K and } T_{01} - T_\infty = 174\ {}^\circ\text{C}$$

From equation 3.23 we know that for the intake efficiency η_I

$$p_{01}/p_\infty = \left(1 + \frac{\eta_I(T_{01} - T_\infty)}{T_\infty}\right)^{\gamma/\gamma-1}$$

$$= \left(1 + \frac{0{\cdot}92 \times 174}{217}\right)^{1{\cdot}4/0{\cdot}4} = 6{\cdot}88$$

Therefore

$$p_{01} = 6{\cdot}88 \times 20 = 137{\cdot}6\ \text{kN/m}^2$$

In the combustion zone the pressure falls by 8% so that $p_{02} = 0{\cdot}92 \times 137{\cdot}6 = 126{\cdot}5$ kN/m² while the temperature is raised to 1 800 K. These are therefore the conditions at entry to the final nozzle. If expansion takes place isentropically we can write

$$T_{02}/T_3 = (p_{02}/p_3)^{(\gamma-1)/\gamma}$$

Referring to the data of Table 1.1 a mean value of $\gamma = 1{\cdot}31$ would seem reasonable for this expansion. Therefore

$$T_{02}/T_3 = (126{\cdot}5/20)^{0{\cdot}31/1{\cdot}31} = 1{\cdot}548$$

$$T_3 = 1\,800/1{\cdot}548 = 1\,163\ \text{K}$$

Again from Table 1.1 by interpolation we get $h_{02} = 2\,003$ kJ/kg and $h_3 = 1\,235$ kJ/kg. Therefore

$$V_3 = \sqrt{[2 \times 1\,000(2\,003 - 1\,235)]} = 1\,238 \text{ m/s}$$

The nozzle outlet diameter is given as 0·5 m so the mass flow rate can be calculated from

$$\dot{m} = \rho AV = \frac{p_3 A V_3}{RT_3} = \frac{20 \times 0{\cdot}5^2 \times \pi \times 1\,238}{4 \times 0{\cdot}287 \times 1\,163} = 14{\cdot}5 \text{ kg/s}$$

The gross thrust of the ramjet is

$$X_G = 14{\cdot}5 \times 1\,238 = 18\,000 \text{ N}$$

As for the reheat example, Example 3.2 above, allowance must be made for the fuel addition. Using the same formula as there we have

$$\frac{\dot{m}_f}{\dot{m}_a} = \frac{h_{02} - h_{01}}{(\eta \times \text{C.V.} - h_{02})}$$

$$h_{02} = 2\,003 \text{ kJ/kg} \qquad \text{For } T_{01} = 391 \text{ K, } h_{01} = 392 \text{ kJ/kg}$$

η_c is given as 0·95 and calorific value as 43 000 kJ/kg. Therefore

$$\frac{\dot{m}_f}{\dot{m}_a} = \frac{2\,003 - 392}{6{\cdot}95 \times 43\,000 - 2\,003} = 0{\cdot}0416$$

(The air/fuel ratio $\dot{m}_a/\dot{m}_f$ is 24·3:1.)
The nozzle flow of 14·5 kg/s is the sum of $\dot{m}_f$ and $\dot{m}_a$, i.e.

$$1{\cdot}0414\,\dot{m}_a = 14{\cdot}5$$

$$\dot{m}_a = 13{\cdot}9 \text{ kg/s}$$

At a flight Mach number of 2·0 the aircraft flight speed is

$$2{\cdot}0\,\sqrt{(1{\cdot}4 \times 287 \times 217)} = 590 \text{ m/s}$$

Therefore

$$\text{Intake momentum drag} = 13{\cdot}9 \times 590 = 8\,200 \text{ N}$$

$$\text{Net thrust from engine } X_N = 18\,000 - 8\,200 = 9\,800 \text{ N}$$

$$\text{Fuel flow} = 0{\cdot}0416 \times 13{\cdot}9 \times 3\,600 = 2\,080 \text{ kg/h}$$

$$\text{Specific fuel consumption} = 0{\cdot}212 \text{ kg/N-h}$$

Due to the fact that the ramjet needs to be flying fast before it will operate, the intake momentum drag will always be large and the net thrust will, therefore, be the difference between two large quantities. This makes the engine performance particularly vulnerable to small losses in component efficiencies—a minor drop in efficiency will reduce the gross thrust by a small amount but the net thrust by a much larger percentage. The intake is a particularly difficult part to make efficient as will be seen later when supersonic intakes are considered. In the cycle calculations, it is important to allow for the increase in mass flow through the engine due to fuel addition as the fuel/air ratio is much larger than with a turbojet. Figure 3.19 shows the generalized performance curves for a family

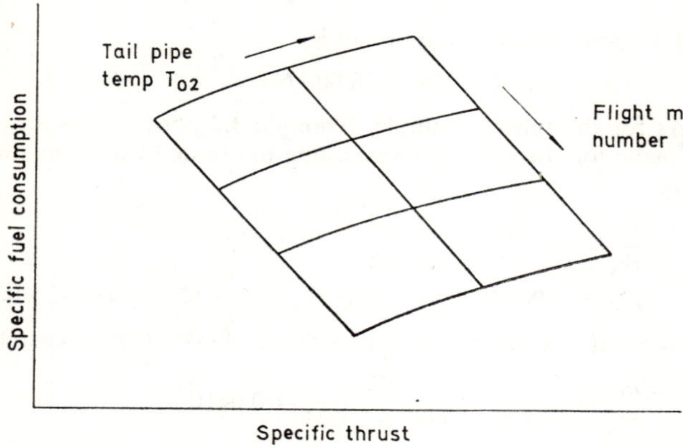

FIGURE 3.19 *Generalized ramjet performance*

of ramjets as a function of flight Mach number and maximum cycle temperature. Constant component efficiencies and losses must be assumed for all engines, as was done for the previous generalized plots.

3.8 THE ROCKET PROPULSION CYCLE

Historically the earliest of all reaction propulsion engines was the rocket, although it lagged behind air breathing engines in its practical application to large aircraft. This section considers the similarities from the cycle point of view between rocket engines and turbo and ramjet engines. Like them, the rocket is a form of heat engine so the basic laws of thermodynamics and fluid mechanics must hold true. The practical aspects of rocket propulsion are dealt within in greater detail in Chapter 10 but even a whole chapter can do little more than scratch the surface of the subject of rocket technology.

Although the solid propellant rocket is the simplest type from the points of view of construction and operation, the thermodynamic processes can be better understood by looking at a liquid fuelled engine. A diagrammatic layout is given in Figure 3.20 from which it will be seen that the two liquids, the fuel and the oxidizer, are pumped from tanks through an injector into the combustion chamber (process 1–2). When the liquids meet they are ignited and they react together to produce high temperature gases (process 2–3). The rates

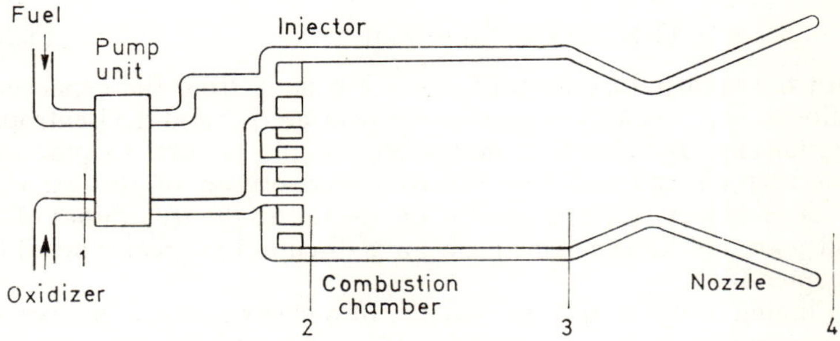

FIGURE 3.20 *Layout of a liquid fuel rocket*

of injection of the liquids are controlled so that the design pressure is achieved in the combustion chamber. Obviously if the flow rates are increased the chamber pressure would need to build up to eject the larger quantities of combustion products through the nozzle. In the same way as with all the cycles considered so far, the products of the gas generator (in this case the combustion chamber) will be accelerated to high velocity in the nozzle as they expand down to the ambient pressure (process 3–4).

Assuming that steady conditions have been attained, the steady flow energy equation can be applied as before to the different processes. Process 1–2, the pumping process, will ideally be adiabatic and reversible and there will be a work input equal to the enthalpy rise of the fluids. We can therefore write

$$\dot{W}_{1-2} = \dot{m}_f \Delta h_f + \dot{m}_o \Delta h_o \tag{3.35}$$

where suffices f and o refer to fuel and oxidizer respectively. In practice a pumping efficiency factor will be introduced to allow for pumping losses. The power required to drive the pumps will be only a small fraction of the total available energy from the combustion chamber reaction.

The actual reaction in the combustion chamber will ideally be adiabatic with no external heat losses and so, if the heat of reaction of fuel and oxidant is known, the methods of Chapter 1, page 19, can be used to find the temperature attained in the chamber. For the real case allowances would be needed for non-stoichiometric conditions and for incomplete mixing of the reactants in the space and time available. These factors, together with any heat losses, would reduce the temperatures at combustion outlet.

The steady flow energy equation can be written in its usual form to give the nozzle outlet velocity as

$$V_4 = \sqrt{[2 \times 1\,000 c_p (T_{03} - T_4)]} \tag{3.36}$$

with the temperature drop $(T_{03} - T_4)$ coming from the expansion ratio (p_{03}/p_4), the inlet stagnation temperature T_{05} and the isentropic relationship. But this is a deceptively simple picture. In practice, it is very difficult to know the true composition of the exhaust gases as they expand through the nozzle and so the true values of c_p and γ are not known. This problem is discussed in greater detail in Chapter 10.

Coming to the *overall* performance of rocket engines, a number of differences from the practice with the air-breathing engines are encountered. To start with, as a rocket carries all its propellants there is no intake momentum drag and so there is no need to distinguish between gross and net thrust—the "gross" thrust produced is *the* effective thrust on the vehicle. Also, it is not usual to refer to the specific fuel consumption of the engine but instead the specific impulse I is quoted. The specific impulse is the ratio of thrust to rate of consumption of propellants and is calculated from

$$I = X/\dot{m} \tag{3.37}$$

(In the foot-pound-second system, specific impulse has the units of lbf-sec/lb. For many years it was the practice to define the specific impulse as the thrust divided by the *weight* flow rate of propellants. Using this definition the units of specific impulse become "seconds". Although this practice now appears to be dying out, many older works will still be found quoting the specific impulse of propellants in the units of seconds.)

From equations 3.1 and 3.37 a simple relationship between specific impulse and the jet outlet velocity can be found. It is that

$$V_4 = I \tag{3.38}$$

(This assumes, of course, that all the thrust comes from acceleration of the gases, that is that the nozzle is working at its design pressure ratio.) A high specific impulse implies, therefore, a high outlet velocity.

The actual specific impulse achieved by a particular rocket engine will be rather less than the ideal values quoted for the fuels as combustion inefficiencies will inevitably reduce the combustion temperature. In addition, there will always be some pressure losses and thermodynamic irreversibilities which will reduce the ideal performance.

3.9 OTHER CYCLES

Looking around the world it is possible to find engines operating on each of the cycles so far described. In addition though, there are a number of minor cycle layouts available to the designer which either have found limited application in the past or give promise of fulfilling a particular need in the future. To complete the work of this chapter, a few of the better-known examples of these cycles will now be described with brief mention of their special attributes. The piston-type aero-engine driving propellers is not dealt with here, however, as it forms the subject of Chapter 11.

The Impulse Duct Cycle

The first jet propelled aircraft to be produced in large quantities was the German V-1 Flying Bomb of the 1940's. This weapon had

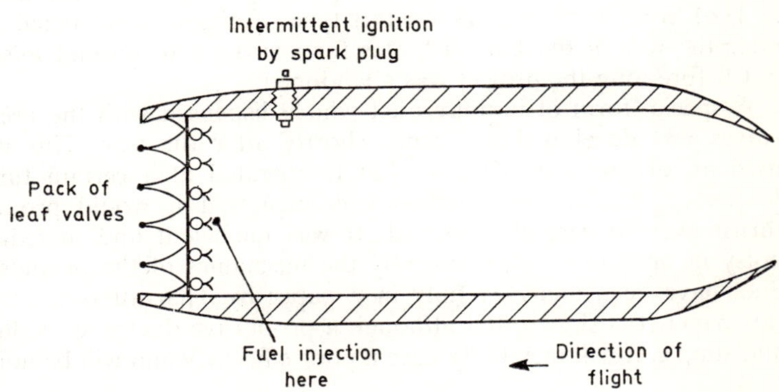

FIGURE 3.21 *The impulse duct engine*

a very simple and light power unit of the form sketched in Figure 3.21. Looking at this figure it will be seen that the engine is superficially similar to a ramjet, but that the intake area is covered by a valve grid. Operation of the engine was intermittent and could not start until a flight speed of about 250 km/h was achieved after which the cycle of operations was as follows:

(i) Owing to ram effects, the intake valves open and allow a supply of air to enter the combustion space. While passing through the intake grid fuel is sprayed in to mix with the air.

(ii) Ignition of the charge is achieved by means of a spark plug. As the combustion space pressure rises the air valves close (they were of a simple reed type). The combustion process is, therefore, at something between a constant pressure and a constant volume process. Because of the restriction effect of the outlet nozzle the pressure would build up above the inlet stagnation pressure.

(iii) The gases leave through the final nozzle. It will be clear from all the work we have done so far that the outlet velocity from the nozzle will be a function of the nozzle pressure ratio. The special non-constant pressure combustion process in the impulse duct will, therefore, give a higher outlet velocity and a better thrust per pound of air than would, say, a ramjet operating at the same flight speed.

Although the original V-1 engine carried out its job quite adequately and had the great advantages of light weight and low cost, its efficiency was low and the fact that it was intermittent in operation led to vibration problems. (The operating frequency was about 50 Hz.) Some further work on this type of engine was carried out after the war in the United States for a ground to ground missile but before long the project was abandoned.

A special form of impulse duct which dispensed with the use of valves was developed in France shortly after the war. This unit used an organ pipe effect so that it operated at a certain tuned frequency, and its construction was such that it would produce thrust even at zero flight speed. It was inefficient and incredibly noisy in operation. More recently the mechanics of the propulsive duct have been analysed fully in a paper by Tharratt (ref. 2). In this paper it is shown that although a propulsive duct may be light and simple, it will inevitably have a poor efficiency and will be noisy.

The Turboramjet

Comparison of the ramjet characteristics with those of the plain turbojet or turbofan shows that they complement each other if propulsion is wanted over a complete range from static conditions to a high supersonic flight speed. At first only the turbine engine will produce thrust but as supersonic flight is attained the ramjet begins to become competitive in efficiency. When very high speeds are reached the highly stressed rotating parts of the turbine engine are embarrassed by the high ram temperature and the turbojet efficiency will be little if anything better than that of the straight ramjet. There is clearly a case, therefore, for considering some form of composite engine combining the advantages of each main type.

The turboramjet results from an attempt to follow these arguments to their logical conclusion and so to devise a suitable composite cycle. There are a large number of possible variations on the theme because either a turbojet or a turbofan may be used as the turbine component and there are a number of ways in which the various parts may be assembled. A typical layout of plant is shown in Figure 3.22, from which it will be seen that the main feature is that both

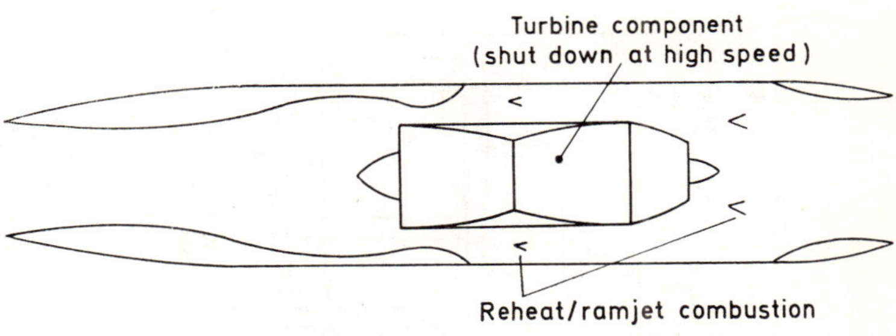

FIGURE 3.22 *The turboramjet engine*
(Courtesy Rolls-Royce Ltd., Bristol Engine Division)

ramjet and turbine engine share the same intake and exhaust ducting. At low speeds, only the turbine engine is operating but as flight speed is increased the ramjet is lighted up and at very high speeds the turbine engine is progressively reduced in rotational speed. If intake temperatures rise too high the turbine may be shut down completely.

Because of the wide variety of arrangements which can be used, it is difficult to give quantitative figures of the performance advantages that the turboramjet may offer. For any possible application it will be necessary to carry out fairly detailed optimization studies covering a range of configurations. Decisions will be needed on such matters as the type of turbine engine to be used, the relative sizes of turbine flow and duct flow and the relative thrusts of the components at different flight speeds. At the moment the most promising fields of use for the turboramjet would seem to be those in air-breathing boosters for the launching of space vehicles. The only supersonic aircraft which has flown to date with this type of composite engine is the French Griffon 02, the power unit of which is described by Gozlan in ref. 3. In this particular instance, the aircraft was accelerated to a flight Mach number of about 0·5 using the turbine engine only. Duct burning was initiated at this condition and with the full engine in operation the Griffon accelerated rapidly to supersonic conditions.

The Turborocket

Although it is aimed at the same general field of application as the turboramjet, that of providing power for high speed flight, the turborocket is very different in conception. A diagrammatic layout of this type of engine is given in Figure 3.23. This shows that it

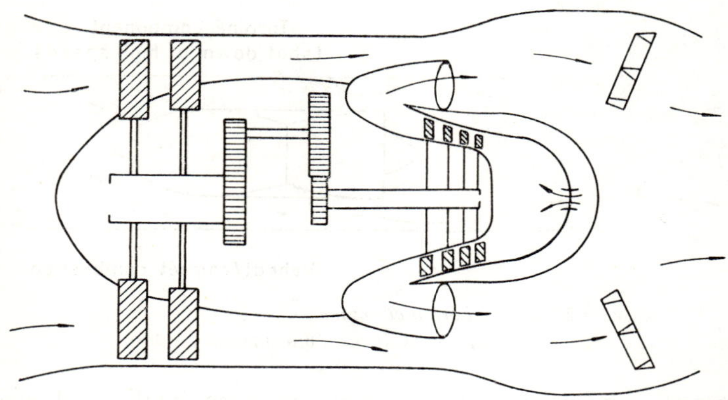

FIGURE 3.23 *The turborocket engine*
(Courtesy Rolls-Royce Ltd., Aero-Engine Division)

consists of a high mass flow compressor of relatively low pressure ratio driven through a reduction gear by a compact multi-stage turbine. The novel feature is that the gases which drive the turbine are generated in a combustion chamber which is really similar to the reaction chamber of a liquid fuel rocket engine. Fuel and oxidizer are injected in the liquid phase into this combustion chamber where they react to give hot, high pressure gases.

Because of turbine blade material limitations, the combustion temperature must be kept well below that which would result from stoichiometric combustion. This could be achieved by running with either excess fuel or excess oxidant. It is more useful to operate with excess fuel as the fuel which does not burn will pass through the turbine and it will then mix with the compressor delivery air.

If some means of flame stabilization is provided the excess fuel can be burnt downstream of the turbine. Figure 3.23 shows the ducting out of the turbine which ensures good mixing between turbine exhaust and compressor delivery air, and also shows the gutters used to hold the flame. Under some operating conditions it might be worth while to inject extra fuel in the exhaust pipe and operate the engine with reheat.

A description of the turborocket, in ref. 4, shows that its outstanding advantage is its low weight but against this must be set the fact that some of the oxidizer used, being supplied in liquid form, is carried by the aircraft. It should therefore be debited against the specific fuel consumption when discussing the efficiency. Because the space available for the turbine is low it must be of small diameter and must operate at a high rotational speed. This will not be compatible with the compressor requirements and hence a reduction gear is needed. Although none of the problems in design is insurmountable, the task of developing a practical turborocket would be a formidable one. It is not clear that the advantages it offers over its competitors such as the turboramjet, at least as the present state of knowledge, would justify the effort.

Besides the power units discussed above, proposals have been made at various times for various other thermodynamic cycles which offer particular advantages in meeting certain specialized needs. A number of these show promise in high speed flight applications especially in cases where liquid hydrogen is used as the fuel because this may act as a very low temperature sink for rejecting heat. For example the Inverted Turbojet described by Mordell, *et al.* (ref. 5) took in air which was first cooled by passing it through a heat exchanger containing the liquid hydrogen fuel before being compressed. Although this and other thermodynamically possible cycles are feasible on paper there is little prospect of them reaching the hardware stage in the near future.

4 Gas Turbine Compressors

4.1 INTRODUCTION

The first component of the turbine engine to be considered in detail is the compressor. From the work of the last chapter on propulsion cycles it has been seen how much the overall engine performance is influenced by the performance of the compressor. One which is low in efficiency or which does not achieve the required pressure ratio will cause serious losses in cycle efficiency. For aircraft engines there are the added requirements that a high performance must be obtained in a machine where bulk and cross-sectional area are to be kept to a minimum. This requirement of high air-flow per unit frontal area makes the design problem a very difficult one.

Two main classes of compressor, the axial flow and radial flow (or centrifugal) types, are in current use in aero-engines. Although most of the early British experience with turbojets was on centrifugal compressor types, the axial flow offered advantages of better efficiency, higher pressure ratio and greater flow per unit frontal area, and now most new engines are designed with axial flow compressors. The mechanics and operating characteristics of each type are described in this chapter with greater emphasis being placed on the axial flow machine.

In dealing with compressors it should always be borne in mind that with these machines the air is being made to flow from a low pressure to a high and that this is against its natural direction of flow—spontaneous fluid flow is always from a high to a low pressure. Because of this, great care must always be taken to treat the air as "gently" as possible in a compressor. Whereas in nozzle flow, with the gas accelerating, fairly drastic changes in flow area and direction are permissible, the diffusing flow in the passages of a compressor must be accomplished with relatively gradual changes in area. This fact was not appreciated when the first attempts were made to run turbines backwards and operate them as compressors and so early axial compressor designs were generally unsuccessful.

4.2 AXIAL FLOW COMPRESSORS

As has been said above, the history of the present-day high performance aero-engines is tied up very closely with the development of the axial flow compressor. The axial machine had a long history with virtually no success until the team at RAE under Dr. Griffiths started work in the late 1920's and set out to put the design on a sound aerodynamic footing. Results came slowly and in Britain the main stream of jet propulsion concentrated on the Whittle type engine, but by the 1940's successful axial flow compressors were being designed on the basis of the RAE (later NGTE) work. The first British axial engines to fly were the Metro-Vick designs which were later to be taken over by Armstrong-Siddeley, and within a few years all the country's aero-engine manufacturers (and there were quite a few in those days) were concentrating their efforts on the axial flow type.

Coming now to the layout of the axial flow machine, compression of the air takes place through a number of stages. Each stage (Figure 4.1) consists of a row of rotor blades attached to the rim of a rotating

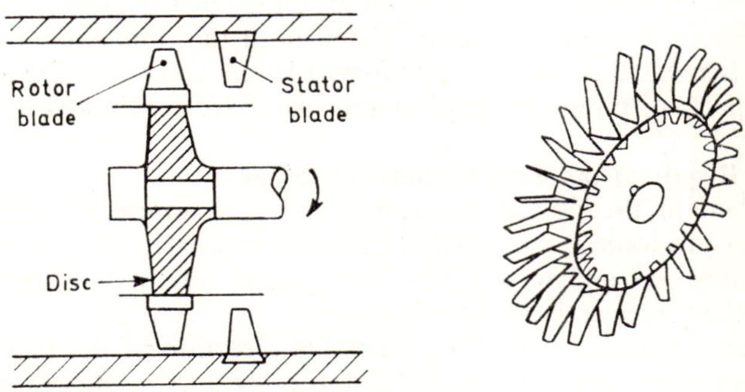

FIGURE 4.1 *Layout of an axial compressor stage*

disc, followed by a row of stationary stator blades attached to the compressor casing. In passing through the rotor, work is done on the air and it is accelerated to a high absolute velocity; the function of the stator row is to accept this high velocity air and decelerate it, so increasing its static pressure. As a general rule also, when passing through the rotor row the air velocity *relative to the rotor blades* will decrease so that through this row, too, there will be some increase in the static pressure of the air.

With the pressure rising through both rotor and stator blade rows, it will clearly be of interest to define what percentage of the total effect is contributed by the rotors. Following steam turbine practice the factor which describes this is the "degree of reaction" of the stage. To be precise, the degree of reaction is defined as the fraction of the stage static pressure rise which occurs in the rotor—a 50% reaction stage will have equal static pressure rise in rotor and stator rows. Now, losses in blade rows will rise as the amount of diffusion of the gas in the row is increased so the aim will be to limit the amount of diffusion in any one row. This means that most compressor designs for aircraft work will be designed for degrees of reaction of 50–70%; in some industrial applications higher reactions, even in excess of 100% may be used. In addition, as will be seen later, the degree of reaction may vary along the length of a blade from root to tip.

Inevitably the change in velocity of the air through a compressor stage results in losses of stagnation pressure and increases in entropy—the skill of the designer lies in keeping these losses to a minimum. It must be stressed here that the *efficiency* of a stage depends on the level of the losses and on the rate of work input per unit mass flow rate. Consequently comparatively high losses may be acceptable with a highly loaded stage, without an undue loss in stage efficiency. (This factor was not fully appreciated until the time came to build highly loaded stage of transonic form.) In the work that follows the two aspects of power absorption and losses are considered in turn.

Power Requirements for a Compressor Stage

To calculate the power absorption by a compressor stage, it is necessary to look in some detail at the air velocity changes through the blade rows and to use Newton's second law of motion to calculate the forces acting in the blades.

Figure 4.2 shows a view in the radial direction of the blading of a compressor stage—rotor plus stator—with the stators of the previous stage included for completeness. The velocity triangles have been drawn for the inlet and outlet of each blade row. The various velocities are found in the following manner:

V_0 at angle α_0 (all angles being measured from the axial direction) is the absolute air velocity entering the stage.

This is compounded with the blade speed u to give velocity V_1 at α_1 relative to the rotor blades at stage inlet.

In passing through the rotor, the air velocity relative to the blades falls to V_2 and the air is deflected to an angle of α_2 to the axial direction. α_2 is less than α_1.

The outlet velocity from the rotor V_3 is the vector sum of V_2

and *u*. V_3 is an absolute velocity; since it is larger than V_3 we see that the rotor has accelerated the air.

Finally, the stator row decelerates the air to V_4 and deflects it towards the axial direction ready for the next stage.

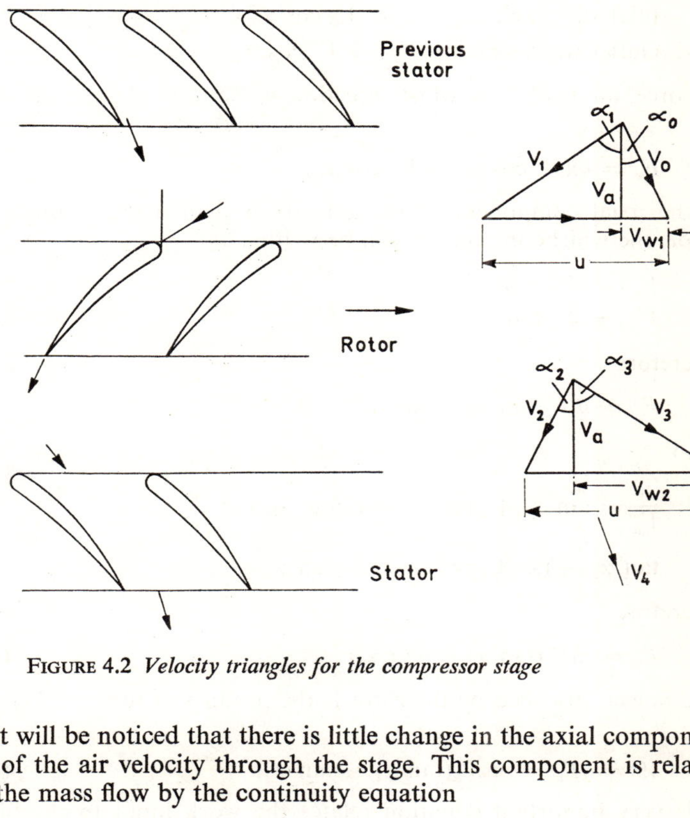

FIGURE 4.2 *Velocity triangles for the compressor stage*

It will be noticed that there is little change in the axial component V_a of the air velocity through the stage. This component is related to the mass flow by the continuity equation

$$\dot{m} = \rho A_N V_a \tag{4.1}$$

where A_N is the flow area perpendicular to the axis of the compressor. Although A_N will be decreasing through the compressor, at the same time there will be an increase in density and the two effects will counteract each other so keeping V_a sensibly constant.

To find the power required to drive the rotor we must calculate the force acting on the blades in the direction of their motion. (The axial components of forces are of no interest from the power absorption point of view.) Following turbine practice, velocities in the direction of motion of the blades are known as the *whirl* components of velocity and the useful force on the blades is the rate of change of whirl momentum.

Using the notation of Figure 4.2 we see

Inlet whirl velocity $V_{w1} = V_0 \cos \alpha_0$

Outlet whirl velocity $V_{w2} = V_3 \cos \alpha_3$

Force on blades in whirl direction = Rate of change of whirl velocity

$$F_w = \dot{m}(V_3 \cos \alpha_3 - V_0 \cos \alpha_0)$$

If the axial component of the velocity is unchanged through the rotor, the whirl components can be written as

$$V_{w1} = V_a \tan \alpha_0$$

$$V_{w2} = V_a \tan \alpha_3$$

Therefore

$$F_w = \dot{m}V_a(\tan \alpha_3 - \tan \alpha_0)$$

Also

$$\frac{u}{V_a} = \tan \alpha_0 + \tan \alpha_1 = \tan \alpha_2 + \tan \alpha_3$$

$$\tan \alpha_3 - \tan \alpha_0 = \tan \alpha_1 - \tan \alpha_2$$

Therefore

$$F_w = \dot{m}V_a(\tan \alpha_1 - \tan \alpha_2) \qquad (4.3)$$

The power absorbed by the rotor is the product of force and speed, so

$$\dot{W} = uF_w = \dot{m}uV_a(\tan \alpha_1 - \tan \alpha_2) \qquad (4.4)$$

This very important equation relates the work input to the blade speed, to the axial velocity and to the deflection of air by the rotor from α_1 to α_2. It assumes that the power can be calculated from mid-radius conditions so implying that conditions are relatively uniform along the blades. However, tip and root effects *are* important as they reduce the possible power absorption from that calculated from equation 4.4. A "work done factor" Ω is introduced to allow for these effects so that the complete power equation then becomes

$$\dot{W} = \Omega\dot{m}uV_a(\tan \alpha_1 - \tan \alpha_2) \qquad (4.5)$$

On the basis of experimental evidence, Howell and Bonham (ref. 1) showed that the work done factor varied with the number of stages in a compressor, being 0·98 for a single stage and falling to 0·87 for a six stage machine.

Very often it will be necessary to find the stage temperature rise for a compressor as this gives the pressure ratio if the efficiency is known. To find the temperature rise the steady flow energy equation 3.10 is combined with equation 4.5 to give

$$\dot{W} = \dot{m}c_p(T_{01} - T_{04}) = \Omega\dot{m}uV_a(\tan\alpha_1 - \tan\alpha_2)$$

Writing the stage temperature rise as ΔT_s gives

$$\Delta T_s = \frac{\Omega u V_a}{c_p}(\tan\alpha_1 - \tan\alpha_2) \qquad (4.6)$$

For consistency of units c_p must be measured in joules/kg-deg C. (Note that in equation 4.6 the negative sign for temperature rise which, by convention, is associated with the compression process has been ignored. The velocities and angles give the magnitude of the temperature change. Note also that as V_0 and V_4 will be approximately the same, equation 4.6 will give either static or stagnation temperature rise.)

If the compressor isentropic efficiency is known (at the preliminary design stage experience should allow its value to be estimated within a few per cent) the pressure ratio may be calculated from the following expression

$$\frac{p_{02}}{p_{01}} = \left(\frac{T_{02}'}{T_{01}}\right)^{\gamma/\gamma-1} = \left(1 + \eta_c\frac{\Delta T_s}{T_{01}}\right)^{\gamma/\gamma-1}$$

$$= \left(1 + \frac{\Omega u V_a \eta_C(\tan\alpha_1 - \tan\alpha_2)}{c_p T_{01}}\right)^{\gamma/\gamma-1} \qquad (4.7)$$

(T_{02}' is the temperature resulting from an isentropic compression.)

Example 4.1 The air entering the seventh stage of an axial flow compressor has a stagnation pressure of 300 kN/m² and a stagnation temperature of 410 K. At inlet to the stage the air velocity vector is inclined at an angle of +15° to the axial direction and the axial component of the velocity has a value, which remains constant through this stage, of 160 m/s. The rotor blade speed at mid-radius is 190 m/s and in passing through the rotor blade row the air is deflected through 30° towards the axial direction. In passing through the stator blades the air is turned so that it leaves at the original air angle of +15°.

Calculate the stage temperature rise for a work done factor of 0·885 and the rotor and stator blade inlet angles for zero incidence.

Solution. The problem resolves to that of solving the mid-radius velocity triangles. Using the nomenclature of Figure 4.2 we are given

$$\alpha_0 = \alpha_4 = +15° \qquad u = 190 \text{ m/s} \qquad V_a = 160 \text{ m/s}$$

and $\alpha_1 - \alpha_2 = 30°$

Therefore

$$\tan \alpha_1 = \frac{u - V_a \tan \alpha_0}{V_a} = \frac{u}{V_a} - \tan \alpha_0 = 1{\cdot}189 - 0{\cdot}268 = 0{\cdot}921$$

$$\alpha_1 = 42{\cdot}7°$$

This is the required rotor blade inlet angle for zero incidence.

$$\alpha_2 = \alpha_1 - 30° = 12{\cdot}7° \text{ and } \tan \alpha_2 = 0{\cdot}225$$

The stage temperature rise may now be calculated using equation 4·6, namely

$$\Delta T_s = \Omega u V_a (\tan \alpha_1 - \tan \alpha_2)/c_p$$

Putting in the appropriate values, with c_p from Table 1.1 as 1·011 kJ/kg-K, gives

$$\Delta T_s = 18{\cdot}6°\text{C}$$

To find α_3, the stator blade inlet angle for zero incidence we see that

$$\tan \alpha_3 = \frac{u - V_a \tan\alpha_2}{V_a} = \frac{u}{V_a} - \tan \alpha_2$$

$$= 0{\cdot}964$$

$$\alpha_3 = 44{\cdot}0°$$

Once the velocity triangles for a stage have been fixed the degree of reaction can be calculated if certain simplifying assumptions are made. These are that the flow is isentropic, that it is incompressible and that the stage outlet velocity V_4 is approximately equal to the inlet velocity V_0. Given these assumptions the work input can be equated to the gas volume flow rate times the static pressure rise, so we can write for unit mass flow

$$\dot{W} = v\Delta p_{\text{stage}} = \Delta p_{\text{stage}}/\rho = u\Delta V_w$$

$$\Delta p_{\text{stage}} = \rho u \Delta V_w$$

Looking now at the flow through the rotor blades, as it is assumed isentropic the stagnation pressure is constant through the row so

$$p_1 + \tfrac{1}{2}\rho V_1^2 = p_2 + \tfrac{1}{2}\rho V_2^2$$

Rotor static pressure rise $\Delta p_{\text{rotor}} = \tfrac{1}{2}\rho(V_1^2 - V_2^2)$

From the geometry of the velocity triangles (Figure 4.2) we see

$$V_1^2 = V_a^2 + (V_a \tan \alpha_1)^2$$

and

$$V_2^2 = V_a^2 + (V_a \tan \alpha_2)^2$$

Therefore

$$\Delta p_{\text{rotor}} = \tfrac{1}{2}V_a(\tan^2 \alpha_1 - \tan^2 \alpha_2)$$

$$\text{Degree of reaction R} = \frac{\Delta p_{\text{rotor}}}{\Delta p_{\text{stage}}} = \frac{V_a^2(\tan^2 \alpha_1 - \tan^2 \alpha_2)}{2u\Delta V_w}$$

(ρ cancels out as the flow is assumed incompressible.)

For ΔV_w we can write $V_a(\tan \alpha_1 - \tan \alpha_2)$ and then cancel $(\tan \alpha_1 - \tan \alpha_2)$ into the numerator to give

$$\text{R} = \frac{1}{2}\frac{V_a}{u}(\tan \alpha_1 + \tan \alpha_2) \qquad (4.8)$$

Also, as $u/V_a = \tan \alpha_0 + \tan \alpha_1$ we can write

$$\text{R} = \frac{1}{2}\frac{V_a}{u}\left(\frac{u}{V_a} - \tan \alpha_0 + \tan \alpha_2\right)$$

$$= \frac{1}{2} + \frac{V_a}{u}(\tan \alpha_2 - \tan \alpha_0) \qquad (4.9)$$

This final equation shows that if the velocity triangles are symmetrical with $\tan \alpha_0 = \tan \alpha_2$ and $\tan \alpha_1 = \tan \alpha_3$ the stage will be one of 50% reaction. For non-symmetrical velocity triangles the actual reaction will be found from equation 4.8.

Example 4.2 Calculate the degree of reaction at mid-radius conditions for the compressor stage of Example 4.1.

Solution. To find the degree of reaction a straightforward substitution of the appropriate figures into equation 4.8 or 4.9 is needed. Let us use 4.8, namely

$$\text{R} = \tfrac{1}{2}\frac{V_a}{u}(\tan \alpha_1 + \tan \alpha_2)$$

Referring back to Example 3.1 we see tan $\alpha_1 = 0{\cdot}914$ and tan $\alpha_2 = 0{\cdot}220$. Therefore

$$R = \tfrac{1}{2}\frac{160}{190}(0{\cdot}921 + 0{\cdot}225) = 0{\cdot}483$$

The triangles are clearly not far from being symmetrical in which case the degree of reaction would have been exactly 0·50.

Losses in the Axial Compressor Stage

Gradually it is becoming necessary to view the flow through an engine in greater and greater detail. When looking at the overall performance of the engine it was possible to ignore everything which took place between the inlet and exit planes. When looking at the power absorption of a compressor stage it was sufficient to consider the velocity changes through a blade row with little regard to how the

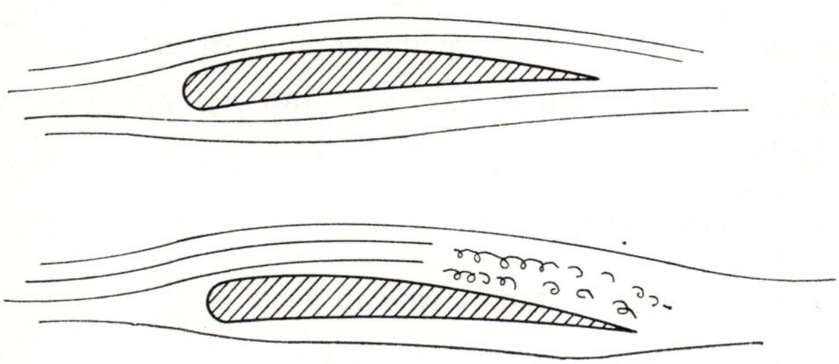

FIGURE 4.3 *Boundary layer separation*

changes came about. Now, in order to study the losses in a stage it is necessary to look more closely at the details of the actual flow. The conditions immediately adjacent to surfaces become important.

As stated in Chapter 1, p. 54, any flow of a real, viscous fluid in a duct will result in a loss in stagnation pressure. This loss is caused by the downgrading of kinetic energy to thermal energy due to the effects of viscous friction. It is in the boundary layer immediately adjacent to the solid surface that the viscosity effects are mainly concentrated as it is there that the velocity gradients mainly occur. For this reason the boundary layer thickness is of very great interest because it can be related directly to the loss in stagnation pressure of the fluid. When a fluid is decelerating and is trying to make its way against an increasing static pressure, the boundary layer may separate leaving a wide wake of disturbed air and a correspondingly large loss in stagnation pressure. Figure 4.3 shows the forms of boundary layer and wake for attached and separated flow. The losses associated with a separated boundary layer are so great that it may safely be assumed that a compressor would never operate steadily with any appreciable amount of separation.

Losses in the compressor stage are customarily classed under the two main headings of primary and secondary losses. Primary losses are those arising from the straightforward boundary layer growths over blades—they are like the profile drag of a wing and are due to two-dimensional flow phenomena. Secondary losses on the other hand are much more associated with three-dimensional effects, with the flow in corners and at tips of blades as in these regions vortices are set up which dissipate energy. Of the two types, the secondary losses are much more difficult to analyse.

In the work that follows the various factors which affect the primary losses in a stage are first considered before looking more briefly at the question of secondary losses. Use will be made of a pressure loss coefficient $\bar{\omega}$ defined as

$$\text{Pressure loss coefficient} = \frac{\text{Loss in stagnation pressure}}{\text{Inlet dynamic head}}$$

$$\bar{\omega} = \frac{p_{01} - p_{02}}{p_{01} - p_1} \tag{4.10}$$

A number of alternative ways of defining the loss, such as, for example, a drag coefficient C_D may be encountered in the compressor literature but $\bar{\omega}$ will be used here.

Primary Losses

As the primary losses in a blade will depend on its geometry and its operating conditions, much useful information can be obtained from the testing of "cascades" of blades—sets of fixed blades grouped at the appropriate distance from each other through which air can be blown. With cascades care is needed to isolate the effects of the end plates on the blades but successful test techniques have been developed over the years and the study of cascade flow has contributed greatly to the overall understanding of compressors. In particular, in the early days of axial compressor design, a great deal of work on cascades was carried out and the effects of a large number of variables were studied.

The most important piece of information for the designer is the loss at design conditions when the blades are operating at or near their optimum air flow angles. Clearly the loss will depend in some way on the loading of the stage—if a blade row is required to produce large deflections or extensive diffusion in a short distance then high losses can be expected. Many attempts have been made to define some limiting condition which should not be exceeded if excessive losses are to be avoided. One of the most useful loading parameters is the diffusion factor D derived at NACA by Lieblein, Schwenk and Broderick (ref. 2). This factor is defined by the expression

$$D = 1 - \frac{V_2}{V_1} + \frac{\Delta V_w}{2V_1} \cdot \frac{s}{l}$$

for a cascade of blades of chord length l and spacing between blades s, with the remaining nomenclature as in Figure 4.4. It is obtained by

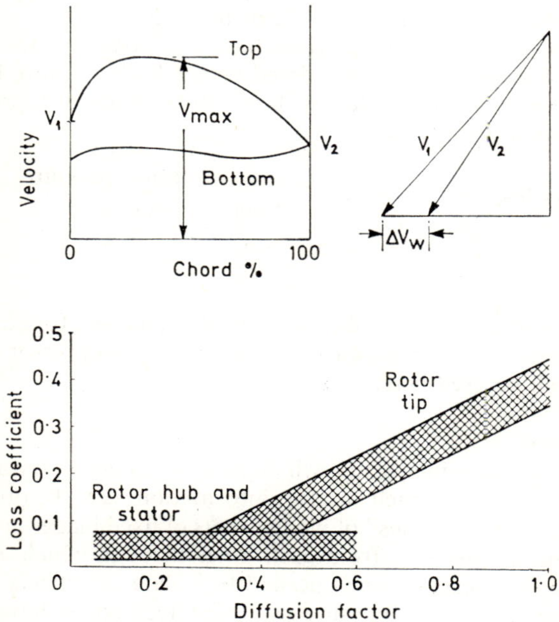

FIGURE 4.4 *The "diffusion factor"*
(Courtesy N.A.S.A. ref. 2)

determining when separation is liable to occur on the suction side of a blade. Experimental results, shown in Figure 4.4 (bottom) show that the loss coefficient for rotor blades increases rapidly if the diffusion factor exceeds a value of 0·45 at the tips of the blades. This information gives the designer a very valuable guide as to the permissible blade loading in a stage.

Later, American work by Lieblein and Johnsen described in ref. 3 led to the derivation of a further loading parameter, the equivalent diffusion ratio which allowed blade losses to be correlated even more closely.

The loss coefficient for a given set of blades will vary with the Mach number as, at least with conventional subsonic aerofoil section blading, there will be some critical Mach number at which part of the flow round the blades will become sonic. At the trailing edge of the blade the flow *must* be subsonic and so, somewhere over the suction surface (Figure 4.5(a)), the air flow must decelerate

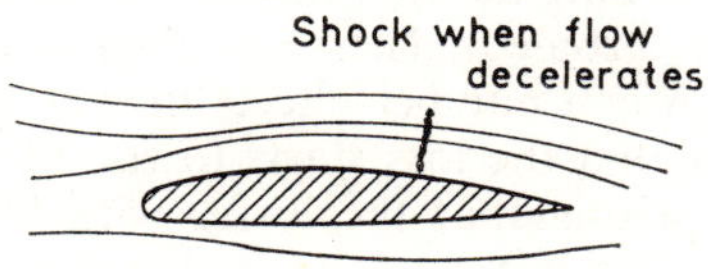

FIGURE 4.5(a) *Transonic flow around an aerofoil*

from supersonic to subsonic and so must pass through a normal shock. This shock will entail a loss in stagnation pressure and may induce separation of the boundary layer. For these reasons the variation of loss coefficient through a cascade as a function of Mach number will be of the form shown in Figure 4.5(b) with

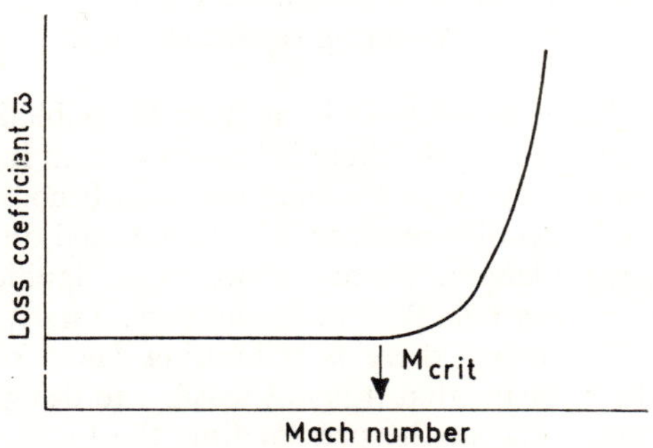

FIGURE 4.5(b)

virtually no variation until the critical Mach number is reached after which there is a steep increase in the loss.

For many years the Mach number limitation remained as a barrier to the progress of axial flow compressor and the absolute Mach number at a blade row was not allowed to exceed a value of about 0·75. With improved manufacturing techniques it became possible to design thinner blades—the thickness/chord ratio is the main factor determining the critical Mach number of an aerofoil—and so to push up the speed limits. Eventually compressors were designed specially to operate at transonic conditions. As was stressed earlier the efficiency of a stage is a function of both losses and specific power input; thus an acceptable transonic stage can be designed which accepts comparatively high losses taken in conjunction with a high specific power input. A complete issue of the American Transactions of the A.S.M.E. (ref. 3) is devoted to a review of the development and early history of the transonic and supersonic compressor.

Another important factor governing the loss through a blade row is whether the flow is laminar or turbulent. These so-called Reynolds number effects differ from the Mach number effects in that it is at *low* values of Reynolds number that the loss starts to rise in value. At first sight it may seem surprising that the losses with laminar flow are higher than with turbulent but the explanation lies in the separation of the boundary layer. After a boundary layer has separated it may re-attach itself to the surface and its ability to do so will depend on the momentum of the boundary air. A turbulent boundary layer, with its greater mixing, can re-attach itself more easily. (In recent years a number of aircraft, such as the Belfast and the Boeing 707, have been fitted with "vortex generators", small sheet protuberances projecting into the air stream and setting up turbulence in the boundary layer. These have the purpose of re-energizing the boundary layer, preventing separation, and so reducing drag.)

Experiments show that if the Reynolds number for a blade row falls below a value of about 3×10^5 there is liable to be an increase in loss and that if it falls below 1×10^5 the losses can become very serious. These values of Reynolds number (Vl/ν) are based on blade chord as the characteristic length. There are two main significances in the restriction on permissible Reynolds numbers. Firstly, with conventional values of velocity there is a limiting *lower* size on compressor blades which means that it is not feasible to design very small axial flow compressors—at the present time the limitation is about 2 kg/s mass flow. Secondly, as altitude increases, the value of the kinematic velocity ν rises so, for a given size of compressor and mass flow, the Reynolds number will fall. This means that at some altitude the performance of the compressor will start to deteriorate and that small engines will run into "Reynolds number troubles" at lower altitudes than will large engines. With some engines the change in performance may occur with almost dramatic suddenness.

All the effects discussed above are essentially concerned with the "design point" performance of the compressor but the designer will also be interested in losses away from the design point. The most significant parameter here will be the angle of incidence, the angle between the air entry velocity and the blade angle. Figure 4.6 shows

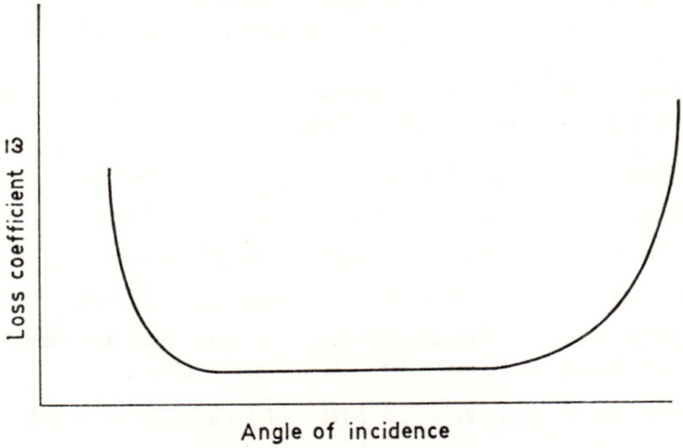

FIGURE 4.6 *Variation of loss coefficient in a cascade with incidence*

the form of the variation of loss coefficient in this case. Over a wide range the loss is virtually constant but at high positive and negative angles (greater than about 12°) the blades stall with extensive breakaway. At these conditions there is a massive increase in the loss coefficient. The deflection of air through the blade row will also vary with angle of incidence, the deflection increasing linearly over a range of about $-3°$ and $+12°$ incidence. It is at the point where breakaway occurs, as shown by the increase in loss coefficient, that the deflection versus incidence curve will start to fall off, again indicating that the blades are stalled.

A number of other factors such as the surface finish of the blading and the flow conditions previous to the blade row may have significant effects on the loss coefficient but as a rule such matters are not amenable to a general analysis.

The example which follows shows how information on the loss coefficient can be used to find the efficiency of a stage. For convenience the same stage velocities have been used as in the previous two examples.

Example 4.3 For the compressor stage of Example 4.1 there is a loss in stagnation pressure through each row of rotors and stator blades equal to 6% of the dynamic head at inlet to the blade row. Calculate the stage efficiency implied in this loss coefficient.

Solution. This is a most important example as it lays particular stress on considering how the blades "see" the air behaving. It emphasizes that the stagnation pressure of a given particle of air varies with the reference velocity and that the only true absolute pressures (which do not depend on the reference plane) are the static ones.

For convenience the most important velocities through the stage, as calculated from the velocity triangles, are listed below.

Absolute velocity at inlet to the stage $V_0 = 166$ m/s

Velocity relative to rotor at rotor inlet $V_1 = 218$ m/s

Velocity relative to rotor at rotor outlet $V_2 = 164$ m/s

Velocity relative to stator at stator inlet $V_3 = 223$ m/s

It was given that the stagnation pressure and temperature at inlet (i.e. relative to V_0) are 300 kN/m^2 and 410 K.

Assume mean values of $c_p = 1{\cdot}011$ kJ/kg-K and $\gamma = 1{\cdot}4$ for the stage.

The first step is to calculate the *static* temperature and pressure of the air at inlet to the rotor

$$T_0 - T = \frac{166^2}{2 \times 1\,000 \times 1{\cdot}011} = 13{\cdot}8$$

Thus static temperature = 396·2 K

$$T_0/T = 1{\cdot}0349$$

Thus $p_0/p = 1{\cdot}128$ and therefore

Static pressure = 266 kN/m^2

This air is approaching the rotor blades at 218 m/s so that relative to the blades

$$T_{01} = 396{\cdot}2 + \frac{218^2}{2 \times 1\,000 \times 1{\cdot}011} = 420{\cdot}6 \text{ K}$$

$$(T_0/T)_1 = 1{\cdot}060$$

Therefore $(p_0/p)_1 = 1{\cdot}226$, $p_{01} = 326$ kN/m^2 and $p_{01} - p_1 = 60$ kN/m^2

As far as the air is concerned, its flow through the blades passage is purely a duct flow with no external work and no heat transfer so the stagnation temperature is constant at 420·6 K. There will be a loss of stagnation pressure equal to 6% of the inlet dynamic head so

$$p_{02} = 326 - 0{\cdot}06 \times 36 = 324{\cdot}8 \text{ kN/m}^2$$

From the rotor outlet velocity of $V_2 = 164$ m/s it is possible to calculate the air static conditions at rotor outlet.

$$(T_0 - T)_2 = \frac{164^2}{2 \times 1\,000c_p} = 13{\cdot}3°\text{C}$$

Therefore $T_2 = 406{\cdot}4$ K

$$(T_0/T)_2 = 1{\cdot}0326$$

Therefore $(p_0/p)_2 = 1{\cdot}119$ and $p_2 = 290$ kN/m²

Air with this static temperature and pressure is approaching the stator blades at 223 m/s so the stagnation conditions relative to this vector may be found.

$$(T_0 - T)_3 = \frac{223^2}{2 \times 1\,000c_p} = 24{\cdot}6°\text{C}$$

Therefore $T_{03} = 431$ K

$$(T_0/T)_3 = 1{\cdot}0605$$

Therefore $(p_0/p)_3 = 1{\cdot}228$,

$$p_{03} = 355 \text{ kN/m}^2 \text{ and } p_{03} - p_3 = 65 \text{ kN/m}^2$$

Through the stator blade row, stagnation temperature will be conserved at 431 K while the stagnation pressure will fall by 6% of the inlet dynamic head to 350 kN/m². These outlet conditions are at the same velocity as the stage inlet conditions because we are given $V_4 = V_0$.

The stage isentropic efficiency can now be calculated as the stagnation pressure has been increased from 300 kN/m² to 350 kN/m² while the stagnation temperature has risen by 21°C from 410 K to 431 K.

For this stagnation pressure ratio, the isentropic stagnation temperature ratio is $(350/300)^{0{\cdot}286} = 1{\cdot}046$. Therefore

Isentropic temperature rise = 18·85°C

Stage isentropic efficiency = 18·85/21·0 = 0·895

Two small points are worth noting. Firstly, the stage temperature rise does not tie in with the previously calculated value as no "work done" factor is included. Secondly, as $V_4 = V_0$ the same answer would have been obtained if the efficiency had been calculated in terms of static values rather than stagnation figures.

Secondary Losses

Mention must be made of a review paper by Lakshminarayana and Horlock (ref. 4) on secondary losses in cascades and axial flow turbomachinery. This paper deals very fully with the whole topic and gives a wealth of references to the research literature. Very briefly the losses are considered as being due to cascade effects—flow in corners and spanwise flow for example—and those due to relative movement between blades and adjacent walls. A number of empirical relationships are put down which allow losses to be calculated for given conditions even though it may not be possible to separate out individual effects. For example, a typical equation for a cascade would express a loss coefficient (or a drag coefficient) as a function of cascade loading and such geometric factors as the aspect ratio and pitch/chord ratio.

When considering effects due to the relative motion of blades and casing it is possible to distinguish a "scraping" effect which becomes significant at high speeds. This loss will depend on the tip clearance and would be expected to increase as the tip clearance was reduced. The other tip clearance effect, the straightforward leakage from pressure to suction side across the end of the blade will *decrease* as the tip clearance is reduced. As a general rule the latter is more important and tip clearances should be kept as low as mechanical considerations will permit. At one time attempts were made to reduce tip clearance to a minimum by coating the inside of the compressor casing with a soft material and actually allowing the blades to rub and cut themselves out a clearance. Difficulties arose because the rotor did not stay exactly concentric with the casing at all times.

Radial Equilibrium (R.E.)

When the equation for the power absorbed by a stage was developed the flow through the blades was assumed implicitly to be two-dimensional and lying in the plane at right angles to the axis of the blades. Only the whirl and axial components of the velocity were considered as only these are needed for the calculations of power or axial thrust. However, the whirl component will have associated with it an intense centrifugal acceleration field—the value of the centrifugal acceleration V_w^2/r may reach many thousands of times that of the gravitational acceleration. In these circumstances, there will be a strong tendency for a radial component of flow velocity to be set up. This would be undesirable as it would lead to a maldistribution of the flow with radius, and an unnecessary increase in the Mach number of the air flowing over the blades. The radial flow can be reduced if a suitable static pressure variation along the blade length is provided and conditions of radial equilibrium (abbreviated to R.E.) are then said to exist.

Though it is easy enough to state the problem of R.E. and suggest how equilibrium might be achieved, the actual design process is very difficult and is fraught with uncertainties. In this section the equations governing R.E. are first developed and then some of the design choices that are available to the engineer are shown. To achieve equilibrium it is necessary to make a compressor with blades which are to some degree twisted and which may have varying degrees of reaction along the blade length. When it is realized that the largely incalculable growths in boundary layers through a compressor may quite well nullify the R.E. design it will be appreciated why the

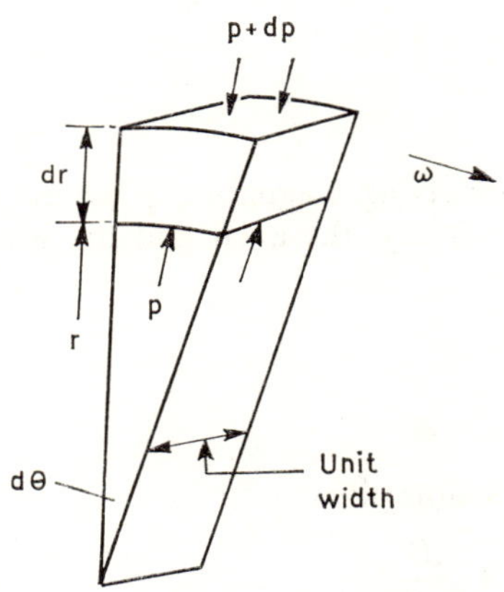

FIGURE 4.7 *Forces on a gas element in a centrifugal field*

requirements are often ignored in the later stages of compression—without apparently affecting compressor performance very badly.

As a start, consider the forces acting on an element of air of unit width as shown in Figure 4.7. If the air has no radial component of velocity then the difference in static pressure must be balancing exactly the centrifugal force. We can write the pressure force F_p as

$$F_p = (p + dp)(r + dr)d\theta - prd\theta - 2(p + \tfrac{1}{2}dp)dr \sin \tfrac{1}{2}d\theta$$

If $d\theta$ is small, $\sin \frac{1}{2}d\theta$ is approximately equal to $\frac{1}{2}d\theta$ and the equation can be expanded to give

$$F_p = prd\theta + pdr \,.\, d\theta + rdp \,.\, d\theta + dpdr \,.\, d\theta - prd\theta - pdr \,.\, d\theta - dp \,.\, dr\tfrac{1}{2}d\theta$$

This will simplify to

$$F_p = dp \,.\, d\theta(r + \tfrac{1}{2}dr)$$

so as $dr \to 0$ we get

$$F_p = dp \,.\, rd\theta \tag{4.11}$$

The centrifugal force is given by

$$\begin{aligned} F_c &= \rho(rd\theta \,.\, dr)V_w^2/r \\ &= \rho V_w^2 dr \,.\, d\theta \end{aligned} \tag{4.12}$$

where ρ is the fluid density.

For R.E. therefore, $F_p = F_c$ and so

$$dp \,.\, rd\theta = \rho V_w^2 dr \,.\, d\theta$$

$$\frac{1}{\rho}\frac{dp}{dr} = \frac{V_w^2}{r} \tag{4.13}$$

Consider now the equations involving stagnation pressure p_0. As there are two components of velocity, the axial and the whirl, we can write

$$p_0 = p + \tfrac{1}{2}\rho(V_a^2 + V_w^2)$$

$$\frac{p_0}{\rho} = \frac{p}{\rho} + \tfrac{1}{2}(V_a^2 + V_w^2)$$

Differentiating with respect to radius gives

$$\frac{1}{\rho}\frac{dp_0}{dr} = \frac{1}{\rho}\frac{dp}{dr} + V_a\frac{dV_a}{dr} + V_w\frac{dV_w}{dr} \tag{4.14}$$

$$\left(\text{Remember } \frac{d}{dr}(V^2) = 2V\frac{dV}{dr}.\right)$$

Writing the expression for $\frac{1}{\rho}\frac{dp}{dr}$ from equation 4.13 gives

$$\begin{aligned} \frac{1}{\rho}\frac{dp_0}{dr} &= \frac{V_w^2}{r} + V_a\frac{dV_a}{dr} + V_w\frac{dV_w}{dr} \\ &= V_a\frac{dV_a}{dr} + \frac{V_w}{r}\frac{d}{dr}(rV_w) \end{aligned} \tag{4.15}$$

Now it will be reasonable to assume that p_0 does not vary with radius so equation 4.15 becomes

$$V_a\frac{dV_a}{dr} + \frac{V_w}{r}\frac{d}{dr}(rV_w) = 0 \tag{4.16}$$

A similar expression can be derived from the stagnation enthalpy equation assuming that the flow is isentropic and that stagnation enthalpy does not vary with radius.

Equation 4.16 above is the basic equation which must be satisfied if R.E. is to be achieved in the blade rows of a compressor—with the assumptions as stated in the development of the expression. The fact that this equation is to be satisfied does not restrict unduly the choices for the compressor design as a large variety of forms of whirl velocity variation with radius are available. For any functional variation of *whirl* velocity with radius it is possible to calculate the corresponding *axial* velocity variation and when this is done the velocity triangles can be drawn for each blade section. Looking at the velocity triangles it will be then possible to see what drawbacks arise from the particular R.E. design chosen and it will be possible to check such points as the degree of reaction variation along the blade, the gas velocities relative to the blade (to see whether Mach number limitations are being exceeded) and the degree of twist of the blades. This last may in fact be the ultimate controlling parameter as the blade required aerodynamically may be so twisted as to be too complex to manufacture.

For a full summary of the problems of designing for R.E. and of the main methods available, the reader is referred to Chapter 5 of Professor Horlock's standard textbook on axial flow compressors (ref. 5). At this stage it is sufficient to look at three design methods and see what each implies.

(*a*) *Free Vortex Flow*

Free vortex flow is the flow which occurs under natural conditions in such phenomena as whirlpools and cyclones. With this type of flow the whirl component of the velocity varies inversely with radius so that near the core very high velocities will result—obviously the conditions of free vortex flow will break down as the radius becomes very small. If the requirement that rV_w = constant is fed into equation 4.16 it will be seen that $d(rV_w)/dr = 0$ and so the axial velocity V_a will not vary with radius.

A stage designed to satisfy free vortex conditions will have the advantage that the work distribution with radius is constant but against this must be set the disadvantage that the degree of reaction will vary along the blade and the required air deflections in the rotor may be very high at the root.

(*b*) *Constant Reaction Blading*

A start for a design may be made by stipulating that the degree of reaction must be constant at all radii. When this is done it is found that the distribution of whirl velocity with radius is of the form

$$V_w = ar + \frac{b}{r} \text{ behind the rotor}$$

and

$$V_w = ar - \frac{b}{r} \text{ behind the stators}$$

(a and b are constants which are calculated from the mid-radius velocity triangles). Using these conditions the variation of V_a with radius can be derived from equation 4.16. With constant reaction blading the work distribution is the same at all radii but highly twisted blades must be used.

(*c*) "*Exponential*" *Design*
The whirl velocity variation with radius may be defined by

$$V_w = a \pm \frac{b}{r}$$

to give what is known as the "exponential" design method. When compared with the other forms described, it is found that the exponential design gives less twisted blades and lower Mach numbers at entry to stator and rotor and so is a preferable design method.

Although three different methods of design have been mentioned it must be emphasized again that a number of other possibilities exist, any of which may give a "best" design under certain conditions. It is in this field of compressor work as much as anywhere else that experience comes to the aid of the designer. The whole question of flow patterns in the highly complicated machinery of an axial compressor is so complex that it is not feasible to attempt to analyse the flow for all conditions and so solutions which are to a certain extent empirical must be the order of the day.

The example which follows shows how the root and tip velocity triangles can be found for a typical compressor stage designed for free vortex flow conditions.

Example 4.4 The axial compressor stage of Example 4.1 is to be designed with free vortex flow at entry to an exit from the rotor. For this stage the mean radius is 250 mm, and the blade height is 100 mm. Calculate the rotor blade inlet and outlet angles at root and tip assuming constant axial velocity across the annulus.

Solution. By the definition of free vortex flow we know that $V_w r$ is constant. At mid-radius the inlet whirl V_{w1} is $V_a \tan 15° = 42{\cdot}8$ m/s while the outlet whirl V_{w2} is $V_a \tan 44° = 154$ m/s.

At the tip therefore

$$V_{w1} = 42{\cdot}8 r_m/r_t = 34{\cdot}2 \text{ m/s}$$
$$V_{w2} = 154 r_m/r_t = 123 \text{ m/s}$$

V_a is constant across the annulus so at the tip

$$\alpha_0 = \tan^{-1} 34{\cdot}2/160 = 12{\cdot}1°$$
$$\alpha_1 = \tan^{-1} (190 - 34{\cdot}2)/160 = 44{\cdot}2°$$
$$\alpha_2 = \tan^{-1} (190 - 123)/160 = 22{\cdot}7°$$
$$\alpha_3 = \tan^{-1} 123/160 = 37{\cdot}6°$$

Similarly at the root we have

$$V_{w1} = 42{\cdot}8 r_m/r_t = 51{\cdot}4 \text{ m/s}$$
$$V_{w2} = 154 r_m/r_t = 185 \text{ m/s}$$

The angles are therefore found as

$$\alpha_0 = \tan^{-1} 51{\cdot}4/160 = 17{\cdot}8°$$
$$\alpha_1 = \tan^{-1} (190 - 51{\cdot}4)/160 = 40{\cdot}9°$$
$$\alpha_2 = \tan^{-1} (190 - 185)/160 = 1{\cdot}8°$$
$$\alpha_3 = \tan^{-1} 154/160 = 43{\cdot}9°$$

Clearly this implies a highly twisted blade especially at the trailing edge where the angle between tip and root varies by 23°. The very high deflection at the root section of over 39° will also be noted. If the degree of reaction at the root is calculated from equation 4.8 it will be found that it has fallen to 0·36 from the mid-radius value calculated previously of 0·48.

The Design of the Compressor

Having looked at three fundamental aspects underlying the performance of axial flow compressors—the questions of power absorption, losses and radial equilibrium—it is now possible to go on to see how a compressor is designed. This resolves eventually to the problem of fixing sizes and shapes for the hardware so, besides needing to look at the aerodynamic side, it is also necessary to satisfy strength, weight and economic criteria. As with all design processes, compromise will be needed at each stage as requirements are so often contradictory. (To take one example, blades should be made as thin as possible from the aerodynamic point of view but stressing calls for a thick blade.)

Each team of engineers working on axial compressors will have its own backlog of experience and its own preferred approach to the task of designing a compressor. What follows cannot, therefore, be thought of as being in any way a design manual but rather it is an attempt to show some of the conflicts that exist and how they may be recognized.

As a general rule the first specification for a new engine will give a required mass flow and a target cycle pressure ratio. Depending on the eventual application of the engine, a decision will be made on whether to go for a highly loaded compressor which may be light and compact or whether to be more conservative and aim rather for a high efficiency with some weight penalty. At this early stage, decisions are still at a policy level rather than being based on purely technical considerations but they can be backed up by extensive optimization studies into the various parameters. Decisions made now will have a profound effect on the whole future of the engine, so great care is needed not to rush things unduly. From a knowledge of other components such as the combustion and turbine systems, the decision will be made as to whether the compressor will be constant outer diameter, constant inner diameter or constant mid-diameter. The constant outer diameter layout is most common nowadays.

With a given target pressure ratio and with an idea of the required efficiency, the overall temperature rise is calculated and so, for the selected stage loading (tied in with the efficiency), will come an estimate of the number of stages required to achieve the compression. This is a very important figure as it decides, to a large extent, the length of the compressor. If it is found that too many stages are needed and that the compressor will be inordinately long, it may be worthwhile to raise the loading of the early stages and, for example, change from subsonic to transonic design. In addition, if too many stages are used in one shaft, handling difficulties can be expected and some means must be provided to improve the surge characteristics. The most fundamental step would be to split the compression into two sections and make the engine a two-spool design, but if this complication is unjustified the designer must be prepared to incorporate some surge palliatives at a later stage in the design. Once having chosen a general stage temperature a more detailed consideration of the variation of stage rises through the compressor will be decided upon.

The next big step will be to choose the cross-sectional area for the compressor. By this time, the required loading will have given some idea of what axial velocity components will be permissible and in addition the density variation through the compressor will be known.

To achieve the necessary flow areas, the designer will have some freedom in balancing up the two main linear dimensions of the compressor blading, the mean radius and the blade height. Regarding the mean radius, it must be realized that the rotational speed of the compressor is the same as that of the turbine and that the turbine designer wishes to run as fast as possible, while the compressor man wants to keep the speed down. In the compressor design this

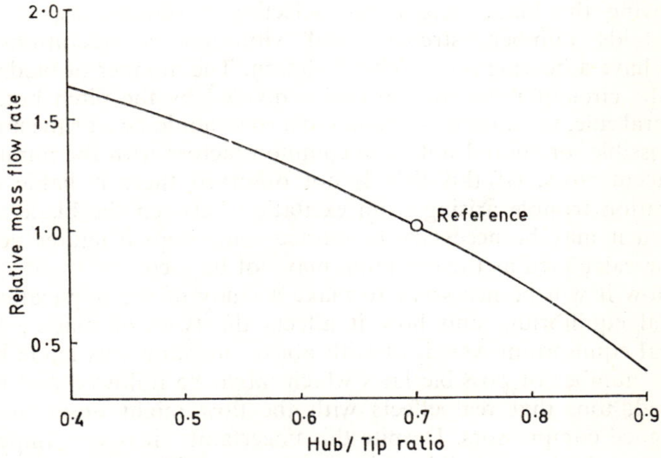

FIGURE 4.8 *Effect of hub/tip ratio on compressor flow*

means there will be an incentive to use a low value for the mean radius. The required flow per unit area can then be achieved by using long blades, that is by using a low value of hub diameter to tip diameter, the H/T ratio. There is a limit to what is feasible in this direction as a low H/T ratio will mean a wide variation in loading between root and tip, and a highly twisted blade to allow for radial equilibrium effects. Over the years there has been a progressive reduction in the H/T ratio possible and whereas a few years ago 0·6 was a low value, modern engines use values of 0·45 and lower. Figure 4.8 shows what increase in flow per unit frontal area is possible for given velocity and density if the H/T ratio is reduced.

The process of determining the main dimensions of the compressor is known as "fixing the scantlings" and once it has been done the design of the stage in more detail can be started. In doing this, velocity triangles will be drawn for mid-radius in each stage and a check will be made that limiting conditions of Mach number or

Reynolds number are not being exceeded. It is at this stage too that the preferred blade row loading parameter (such as the diffusion factor) will be calculated. Some adjustment of conditions may be needed to see that this factor is within the permissible limits.

At about this point in the proceedings the calculations will be done which will say how many blades will be wanted in each row. To find this, the optimum pitch/chord ratio will be calculated from diffusion factor considerations and the chord will be found from knowing the blade height and selecting a suitable aspect ratio. Reynolds number, strength and vibration considerations may also have a bearing on the chord chosen. The number of blades will be the circumference at mid-radius divided by the pitch but, as a general rule, the number of blades in a row should be a prime number if possible, or should not have common factors with the number in adjacent rows. (If this rule is not observed there is liable to be vibration trouble arising from excitation between the blade rows.) Again it may be necessary to retrace some steps if figures near to those calculated as the optimum may not be used.

Now it will be necessary to make a study of the implications of radial equilibrium and how it affects the twist of blades. When radial equilibrium was dealt with above, mention was made of the large number of possible laws which might be followed and it was hinted, too, that real effects with the flow might upset the best designed compressors. Despite this uncertainty, it is very important to make the best possible estimate of these effects and to try to see how the compressor design would be influenced. A close look will, therefore, be taken at the conditions and velocity triangles away from the mid-radius to see if, for example, loadings became excessive and to take any necessary correcting action. The degree of reaction variation along the blades will also be checked.

With each step that is taken with a design and with each dimension that is fixed, it becomes more and more difficult to make subsequent alterations. Design has been likened to a converging process wherein freedom of movement is progressively curtailed. In the classic example, the only thing that can be changed at the end without widespread repercussions is the colour of the paint! This is, of course, an exaggeration, but the fact remains that the work involved in re-doing a design reaches prohibitive proportions and it may in the end be worthwhile to accept some erosion of limiting conditions rather than to alter things from the beginning. The good designer is one who can keep a design from solidifying too early in the process.

So far no mention has been made of the blade profile which will be used. A number of aerofoil sections have proved particularly popular for compressor work, such as the C7 in Britain and the NACA.65 in the United States. Details of these sections and the way in which they can be superimposed on base lines—the camber lines—are available in standard textbooks such as Horlock (ref. 5). In a paper devoted to blade profiles (ref. 6) Carter made a case for taking great care in the choice of profiles and advocated the use of

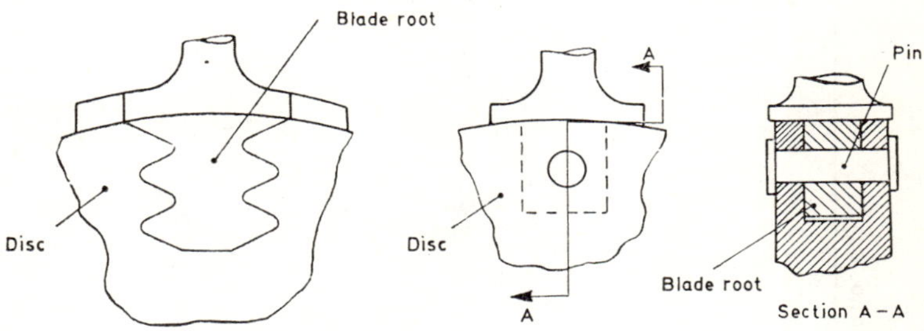

FIGURE 4.9 *Methods of compressor blade attachment to discs*

varying profiles at different stages through the compressor. His work suggests that quite appreciable changes in the shape of the performance characteristics could come from seemingly negligible changes in blade shape.

Finally, in the aerodynamic design of a compressor, the variation of the various velocities and angles will be smoothed so that there are no sudden variations from stage to stage. An attempt may also be made to group stages together so that the same basic blade shape is used for three or four rows. If this can be done, there may be worthwhile savings in production costs for the compressor.

Once the desired gas flow pattern at design has been determined, the task passes to the specialist designer who deals with the stressing of the blades and the details of the mechanical construction. These men have the job of transforming a table of vectors into metal and of designing a component which will operate without trouble for thousands of hours. Their job is a complicated one of stacking the blade sections in the best way to keep stresses to a minimum and to avoid such factors as the flutter of blades. Decisions must also be made on the method attaching the blades to the rotor discs (Figure 4.9 shows alternative fir-tree and pin methods), and on whether any special provision is needed for improving the handling of the compressor. If the surge line is lower than planned (see next section) some degree of variable geometry or inter-stage air-bleed may be necessary.

4.3 THE AXIAL COMPRESSOR OPERATING CHARACTERISTICS

The overall operating characteristics of an axial flow compressor can best be presented by a map of the performance known as the compressor characteristic. This shows the variation of compressor pressure ratio with mass flow at constant values of compressor speed.

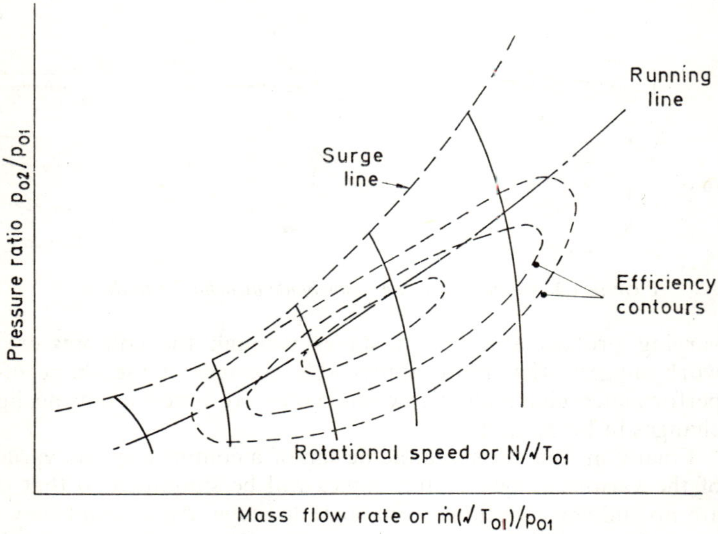

FIGURE 4.10 *The axial compressor characteristic*

The other main parameter of interest, the isentropic efficiency, can be shown by drawing contours of constant efficiency on the characteristic. Figure 4.10 shows a typical characteristic for an axial compressor, as would be obtained from rig testing. It will be seen that if the compressor delivery is restricted with the compressor running at constant speed, the mass flow will decrease while the pressure ratio builds up—this is only what would be expected. At high rotational speed the *range* of mass flow at any speed falls very much and the constant speed lines become nearly vertical.

For any constant speed line there are definite limitations on the range of pressure ratio possible for stable operation. At the low pressure ratio end, the position is reached where the pressure ratio produced is insufficient to pass the mass flow through the outlet casing—the compressor is said to be choked. The limit of stable operation at the upper end is set by surge of the compressor, a violent breakdown in flow which may wreck the compressor in a very short time. Surging is described in greater detail below (p. 150) so at this stage it is sufficient to note that the surge line shows an operating limitation on the compressor.

As the compressor is an aerodynamic machine its performance is affected greatly by variations in the air conditions at inlet. It will be fairly obvious by now that the velocity triangles into a stage determine the stage performance, from the points of view of both power absorption and efficiency. For a given geometry of velocity triangles the mass flow will vary directly with the intake air density. Consequently the compressor characteristic as drawn in Figure 4.10 will only be usable for specified intake conditions as on the day of the test. Fortunately, by applying the methods of dimensional analysis, scaling factors can be deduced which allow the compressor performance under any set of intake conditions to be found.

Early work in Britain on the application of dimensional analysis to centrifugal compressors was carried out at Farnborough in connection with the supercharging of piston aero engines. A classic paper by Capon and Brooke (ref. 7) describes their work. In this paper it is shown that a characteristic which is plotted in the form of pressure ratio p_{02}/p_{01} against a mass flow function $\dot{m}\sqrt{T_{01}}/Ap_{01}$ for lines of constant rotational speed function, $ND/\sqrt{T_{01}}$ will be valid for a wide range of inlet stagnation pressure and temperature p_{01} and T_{01}. In these expressions A is a characteristic area and D is a characteristic length both of which will be constant for a given compressor and so can be omitted. Figure 4.10 shows the usual form of characteristic as plotted in non-dimensional form.

A fuller application of dimensional analysis shows that Reynolds number should also be brought in as a variable but, as mentioned above on p. 134, compressor performance is relatively insensitive to Reynolds number changes provided that the value does not fall below some critical value.

(In Chapter 1, various non-dimensional flow parameters were developed in terms of the stagnation to static pressure ratios in a gas stream, see p. 34. The mass flow parameter $\dot{m}\sqrt{T_0}/Ap_0$ has appeared before and the compressor expression of $ND/\sqrt{T}$ is in fact equivalent to the non-dimensional velocity parameter $V/\sqrt{T}$.)

An alternative way of plotting characteristics in a non-dimensional form is to correct them always to a standard set of intake conditions, generally ISA sea level conditions of 288 K temperature and 101·3 kN/m^2 pressure. The characteristic then appears as pressure ratio (non-dimensional in any case) against $(\dot{m}\sqrt{T_{01}}/p_{01})(101{\cdot}3/\sqrt{288})$

for lines of constant $(N/\sqrt{T_{01}})\sqrt{288}$ (efficiency is already non-dimensional). A characteristic plotted in this way represents the true picture on a standard day and gives a much clearer picture on days not far from standard. An engineer will appreciate the physical meaning of 10 000 rev/min much more than $N/\sqrt{T_{01}}$ of 370.

When a compressor is matched to a combustion system and turbine to make a gas generator an equilibrium running line for the unit can be drawn on the compressor characteristic. Details of how the running line is calculated are given in Chapter 8 where it is shown that the position and shape of the line are determined mainly by the gas swallowing capacity and power characteristics of the turbine. A typical running line is shown on Figure 4.10 and it will be noted that for the greater part of the operating range the line runs roughly parallel to the surge line. The engine designer will aim to run as close to the surge line as feasible as this means that the pressure ratio is high so the engine cycle is efficient. However, as is also shown in Chapter 8, the engine running line *rises* during acceleration so great care must be taken not to sail too close to surge.

The effects of cold weather operation or of flying at altitude, where the intake temperature is low, can be seen by looking at the running lines on the characteristics. If an engine is operated at a constant rev/min then as T_{01} falls $N/\sqrt{T_{01}}$ will rise, so that the operating point will move up the running line towards the right. This will result in a higher pressure ratio and increased non-dimensional mass flow so the engine cycle efficiency will improve. There is a tendency with some compressors however for the surge line to curve over at high N/T_{01} and this may combine with an upward rise of the running line to give a dangerous "high-$N/\sqrt{T}$ surge". If an engine *were* liable to meet this condition, it would become apparent in flight testing and special protective controls would be fitted.

Surge

It must be said from the start that compressor surge is a highly complex phenomenon which is even now far from being fully understood. The flow instability of a single stage can be understood without too much difficulty but when a number of stages make up a compressor, a variety of interactions are possible. The aim here is to discuss the problem in general terms and show how the danger of surge may be reduced.

Surging of a compressor manifests itself as a breakdown of the flow through the compressor so that the air flow falls to a very low value or may even flow in reverse. As the power absorption of the compressor will fluctuate violently, severe shocks will be imposed on the driving shaft and as the air flow is reduced at a fairly high power input (with, of course, very low efficiency) the air temperature will increase rapidly. Compressors' blades have been known to melt when a compressor has been held in surge inadvertently. If the compressor is running as part of an engine, the fuel flow may not be able to follow the rapid changes in air flow and there is a real danger that the engine might be burnt out.

Surging may result if the air mass flow through the compressor is reduced so causing the blades to stall. The reason for the stall at

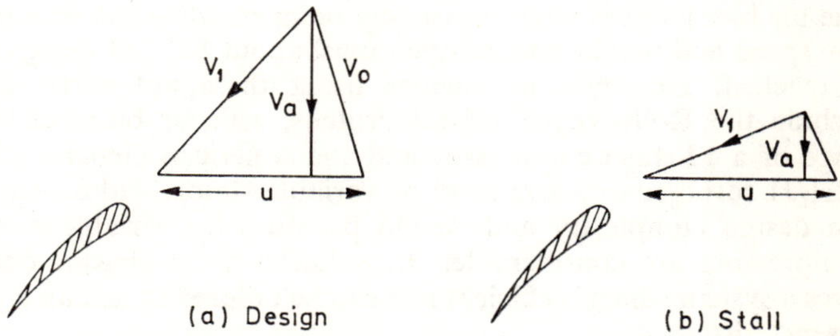

FIGURE 4.11 *Velocity triangles at design and stall*

reduced air flow can be seen by looking at the velocity triangles for a single stage. Figure 4.11(a) shows the triangles for a stage at the design point conditions and from previous work it is known that the air will strike the blades at a small positive angle. Suppose now that with the blade speed unchanged the mass flow is reduced. A diagram as in Figure 4.11(b) will result, the blade speed u being unchanged but the axial velocity V_a being smaller. It is clear from the diagram that if, to a first order, the outlet angle from the previous stage is the same, the net effect will be to increase considerably the angle of the air on to the rotor blades. The angle might exceed the stalling angle for the blade section in which case the flow would fall still further, the stage would operate very inefficiently and a flow breakdown might occur.

It would be a gross over-simplification to say that the above analysis explains the surging of an axial compressor. Even if one stage should stall, there is no certainty that the flow through the rest of the compressor will be so disturbed as to cause a complete breakdown in flow. (If an early stage should stall the later stages can provide a very useful steadying effect.) The compressor surge line may also be affected by other factors such as by a bad velocity distribution into the intake due to yawing motion of an aircraft. Any factor indeed which is liable to alter the angles of the air meeting the blades must be watched carefully.

As an engine will generally run fairly close to the surge line during acceleration, it may be necessary or desirable to incorporate some form of surge-mitigation device in the design. Two common means employed are to physically alter the angles of one or more stator blade rows (with obvious effects on the velocity triangles) or to bleed off some air after it has passed through the early stages of the compressor. By increasing the air flow through the early stages, the axial velocity is increased and the stages move away from stall. The air bleed valves would normally be open when the engine is at low speed and would remain open until about 80% of design speed is reached. Examples of engines using these anti-surge devices include the Rolls-Royce Bristol Proteus, with air bleed at the 8th stage of a 12-stage compressor and the American General Electric GE1/J1 turbojet with five rows of variable stator blades. Although the design complexity and weight penalties for variable geometry compressors are considerable, the solution is an elegant one and gives a system which is efficient and can be tailored to suit an engine's characteristics.

The anti-surge measures of bleed or variable geometry were originally introduced as palliatives to improve the handling of high pressure ratio compressors—though variable geometry has since been developed to be a powerful device in its own right. A more fundamental approach to the problem of achieving high pressure ratios is to design a multi-spool engine in which compression takes place in two or more axial compressors. These compressors are arranged in series but are mechanically independent. Each compressor finds its own equilibrium speed; it is found that as overall pressure ratio falls the low pressure spool runs down more rapidly than do the others. The Olympus was the first two spool turbojet and the Rolls-Royce RB 211 is an example of an engine using three spools.

One other compressor phenomenon concerned with non-steady flow deserves special mention and that is "rotating stall". Rotating stall is a problem not so much because it limits the possible working ranges of the compressor, but because it is liable to lead to blade vibration and perhaps breakages. Briefly, when rotating stall is occurring in a compressor, a number of "cells" of low energy air will be present in the gas stream and the cells will rotate in the same direction as the compressor but at about half compressor rev/min.

As compressor blades catch up with and pass through the cells the air angles and air loads on the blades will vary and may excite vibrations. The cells may be detected while the engine is on test by placing special instruments known as hot wire anemometers in the air stream and recording their output on high speed recorders.

An explanation for the mechanism of propagation of rotating stall was given by Stenning and Kriebel (ref. 8). Looking at a cascade of blades as in Figure 4.12 it is suggested that some small disturbance

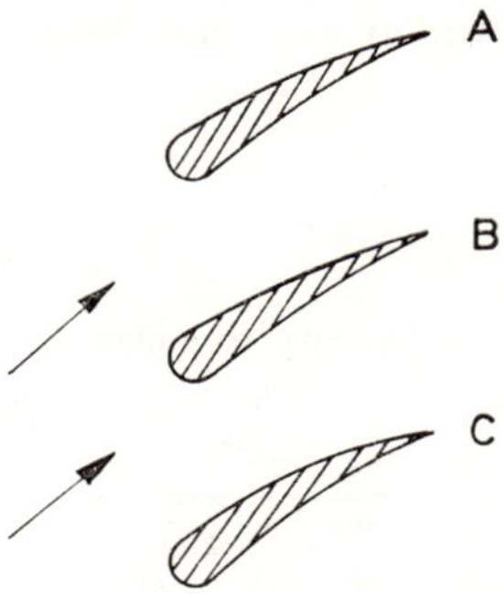

FIGURE 4.12 *A mechanism of rotating stall propagation*

may upset the flow past a blade (e.g. blade B) causing it to stall. The breakaway from the suction surface of B will reduce the flow capacity of the passage between A and B so that the air which should pass through the passage will spill past the adjacent blades. Looking at the diagram, it will be seen that this will cause the air angle on A to increase while that on blade C will decrease. This will tend to cause blade A to stall in its turn but the incidence on C will decrease and move it away from stall.

The sequence of events will be followed up through the row of blades so that each blade in turn will stall but the very fact of its stalling will improve matters for the preceding blade. In this way an isolated cell will be set up. Obviously there will be a limit to how far this self-regulating process can work and finally a state will be reached where a blade row gets in the stalled condition and stays there.

4.4 CENTRIFUGAL COMPRESSORS

There can be no real doubt that Sir Frank Whittle was correct in his decision to use the centrifugal compressor in his original jet engines, despite the fact that research effort was showing that efficient axial flow compressors could be made. Essentially, the centrifugal machine was much more robust and it could be designed fairly safely to give a pressure ratio which would allow a turbojet cycle to operate. In the early days, too, jet engines were not of such a size that the large outside diameter of the centrifugal was a problem—the required engine air mass flow could be handled with an acceptable frontal area. Eventually though the axial was bound to win on the grounds of lower diameter as well as by being capable of giving higher pressure ratio and efficiency, but even now the centrifugal type is by no means dead. Not only are very many engines of older design still in use (the Dart is one of the most widely used turboprops in the

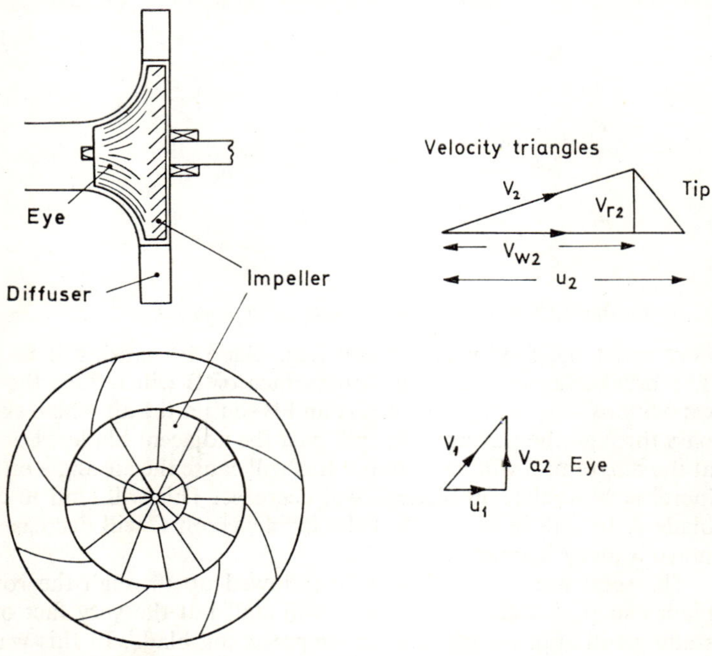

FIGURE 4.13 *Layout of a centrifugal compressor*

world) but in small sizes the axial flow cannot be made efficient. There is, therefore, ample scope for further development and refinement of the centrifugal machine.

Figure 4.13 shows the layout of a single-stage centrifugal compressor. Air enters at the eye of the rotating element, the *impeller*, and flows out to the impeller tip by virtue of the centrifugal field. In passing through the impeller the air is accelerated to a high velocity and the high velocity air is then fed to a *diffuser* where it decelerates with a resulting increase in static density. As with the axial, it is only in the rotor that there is a transfer of external work to the working medium. Through the diffuser, the stagnation temperature and enthalpy remain constant although there is, of course, a loss in stagnation pressure. The impeller shown in Figure 4.13 is of the single-sided type but the classic Whittle and Rolls-Royce designs used a double-sided impeller, which looked like two single-sided ones joined back-to-back.

To find the power required to drive the impeller, use is made again of Newton's laws of motion and the velocity diagrams are drawn for the eye (inlet section) and tip of the impeller. Looking at the diagrams in Figure 4.13 it is assumed in the first instance that the air enters in the axial direction. It will be seen that the air leaves the impeller tip with a whirl component of velocity V_{w2} which is less than the tip speed u_2. The force applied to the air in the whirl direction at the impeller tip for a mass flow of $\dot{m}$ kg/s with V_w being measured in m/s is

$$F_w = \dot{m}V_{w2}$$

and the power will be

$$\dot{W} = \dot{m}u_2V_{w2} \tag{4.17}$$

The "slip factor", denoted by σ, is the ratio of V_{w2} to the tip speed* so equation 4.17 can be rewritten in simpler form as

$$\dot{W} = \dot{m}\sigma u_2{}^2 \tag{4.17a}$$

As before the equation for power required can be used to derive expressions for compressor temperature rise and pressure ratio if the intake temperature and efficiency are known. These equations are

$$\Delta T_0 = \frac{\sigma u^2}{c_p} \tag{4.18}$$

and

$$p_{02}/p_{01} = \left(1 + \frac{\eta\sigma_c u_2}{c_p T_{01}}\right)^{\gamma/\gamma-1} \tag{4.19}$$

What will determine the magnitude of the slip factor? Detailed study of the flow in the impeller passages suggests that a general eddy motion takes place between the vanes (rather as shown in Figure 4.14). Because of the direction of rotation of this eddy there

* Strictly speaking, this is only true for radial-vaned impellers but in practice all impellers for aero-engine use are of this type.

will be a component of the air flow backwards at the tip so that V_{w2} is less than u_2 as is observed. The "strength" of the eddy will depend on the geometry of the space between the vanes and on the blade speed.

For a typical compressor as used in an aircraft turbine engine, σ would have a value of about 0·9.

The power equation 4.17 ignores any friction and windage losses for the impeller but these may be significant under some conditions. Testing will be needed to find the extent of these losses as they will clearly depend to a large extent on the configuration of the complete

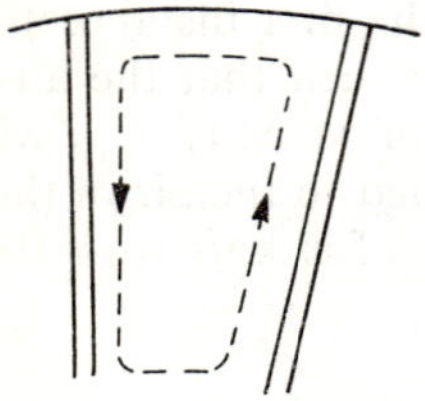

FIGURE 4.14 *Relative eddy motion between vanes of an impeller*

compressor. For example, a double-sided impeller should be better than a single-sided as there is no large flat back to be subjected to frictional forces. A power factor of the order 1·05 may need to be applied as a multiplier for equation 4.17 to take account of the friction and show the true power required to drive the compressor.

In designing the eye of the impeller, care must be taken to choose the angles of the inlet vanes so that high angles of incidence on the blades are avoided. To find the angles needed, velocity triangles must be drawn at a number of radii. These diagrams will then show that even though the air velocity may be axial at all radii (and this would be a first assumption in the design) the *relative* velocity into the eye will vary with radius. It will be necessary to twist the blades to allow for this and to avoid high angles of incidence or high Mach numbers.

Turning now to the question of non-axial entry flow, pre-swirl may be imparted to the air by means of aerofoil shaped vanes in the intake duct. Now the purpose of the impeller is to increase the angular momentum of the air as it passes through it and the power required is the product of the rate of increase in angular momentum and the angular velocity. If the entry is axial then the angular momentum of the air at inlet is zero and the power is actually derived in equation 4.17 above. If, on the other hand, pre-swirl has been imparted to the air, then the power absorption by the stage will be altered. Pre-swirl in the direction of rotation of the impeller—taken by convention as positive—will *reduce* the power absorption while negative pre-swirl will *increase* the work capacity. Provided there are no drastic changes in compressor efficiency these power changes will give corresponding changes in the compressor temperature rise and pressure ratio as indicated by equations 4.18 and 4.19. The actual *amount* of power reduction will depend on the form of whirl velocity profile at the impeller eye but if at some representative mid-radius the blade speed is u_1 and the whirl component of the velocity is V_{w1} we can write the equation

$$\dot{W} = \dot{m}(u_2 V_{w2} - u_1 V_{w1}) \tag{4.20}$$

(The use of variable pre-swirl vanes has proved a potent method of improving the acceleration characteristics of small gas turbines as it becomes possible to vary the pressure ratio of the engine without having to accelerate the rotating assembly. Objections to the system are that the vanes must introduce some loss in pressure all the time and that the inducer blade angles can only be correct at one setting of the pre-swirl vanes.)

Looking in rather more detail at the velocity triangle at impeller tip it is seen that besides the two vectors of u_2 and V_{w2} which have been dealt with before, two other vectors are of interest to us. They are the radial component V_{r2} and the absolute velocity at outlet V_2. The *radial* component corresponds somewhat to the axial velocity component in an axial flow stage because it is this component which determines the cross-sectional area required for the flow. In this instance the flow area will be the product of the impeller tip width and the tip circumference. The *absolute* velocity at exit gives a measure of the Mach number into the diffuser and of the kinetic energy which is available for conversion to static pressure in the diffuser passages. The angle between V_2 and the tangential direction is the angle at which the diffuser vanes must be set if zero incidence onto the blades is required.

The actual design of the diffuser will be determined very largely by the design of the remainder of the engine as the delivery air must be turned through two right angles until it is flowing in an approximately axial direction. If space is limited it will be necessary to use a diffuser with a large number of passages as this allows the equivalent diffusion angle to be controlled more closely. A multi-passage diffuser has the disadvantage that it is more liable to suffer from instabilities and break-away of the flow as the operating conditions vary.

Before doing an example to illustrate the performance calculations for a centrifugal compressor a comment is needed on the efficiency level and limiting pressure ratio of this type of compressor. As a general guide the peak isentropic efficiency for a well designed centrifugal is of the order of 75–80% and this is five points or more below that of an axial machine which is designed for the same pressure ratio and mass flow. Fundamentally this difference arises because the centrifugal is a much more "violent" machine—it tries to do everything at once, and the velocities and losses are bound to be much greater. Many years ago, Dr. S. G. Hooker drew an analogy between compressing air and sweeping water up a slope. He likened the axial compressor to a man sweeping the water with a succession of small rapid movements of the brush, whereas the centrifugal was a man trying to sweep up the lot in one mighty swoosh. This pictorial analogy gives an idea of why the losses are so much higher with the centrifugal. Given very careful design of flow passages, losses *can* be reduced but there are economic and technical limits on what can be achieved.

The pressure ratio limitation resolves in the end to being a materials problem as the stresses in an impeller are ultimately dependent on the tip speed. By using higher strength materials it is possible to achieve pressure ratios of the order of 6:1 but, beyond that, increases in pressure ratio can only be achieved by using two-stage compression. Although this has been done with outstanding success on a number of occasions, there are formidable design problems in ducting air efficiency from first stage delivery to second stage inlet.

Example 4.5 A centrifugal compressor has a single-sided impeller of 0·6 m diameter, the impeller eye root diameter being 0·15 m and the tip diameter 0·30 m. At design conditions when the rotational speed is 14 000 rev/min air enters the eye of the impeller axially at 125 m/s with a stagnation pressure of 101·35 kN/m² and a stagnation temperature of 288 K. If the slip factor is 0·9 and the power input factor is 1·04 calculate the power required to drive the compressor. Find also the overall pressure ratio if the isentropic efficiency is 0·76. Estimate the Mach number of the air leaving the impeller.

Solution. It is first necessary to calculate the air mass flow rate and this can most conveniently be done by using the isentropic flow curves of Figure 1.11 (p. 36). It is given that $V_1 = 125$ m/s and $T_{01} = 288$ K. Therefore

$$\frac{V}{\sqrt{T_{01}}} = 7{\cdot}38 \text{ and } \frac{\dot{m}\sqrt{T_{01}}}{Ap_{01}} = 25{\cdot}6$$

$$A = \tfrac{1}{4}\pi(0{\cdot}3^2 - 0{\cdot}15^2) = 0{\cdot}053 \text{ m}^2$$

Thus $\dot{m} = 8{\cdot}11$ kg/s

Now to find the tip speed u:

$$u = \frac{\pi DN}{60} = \frac{\pi \times 0{\cdot}6 \times 14\,000}{60} = 440 \text{ m/s}$$

The power is calculated from equation 4.17, making use of a power input factor of 1·04:

$$\dot{W} = 1{\cdot}04 m\sigma u^2 = (1{\cdot}04 \times 8{\cdot}11 \times 0{\cdot}9 \times 440^2)/1\,000$$
$$= 1\,470 \text{ kW}$$

Using equation 4.18 the stage temperature rise can be calculated Taking a mean value of c_p of 1·01 kJ/kg–K we get

$$\Delta T = \frac{\dot{W}}{\dot{m}c_p} = \frac{1\,470}{8{\cdot}11 \times 1{\cdot}01} = 179°\text{C}$$

Therefore

$$p_{02}/p_{01} = \left(1 + \eta_C \frac{\Delta T}{T_{01}}\right)^{\gamma/\gamma-1} = \left(1 + \frac{0{\cdot}76 \times 179}{288}\right)^{1{\cdot}4/0{\cdot}4}$$
$$= 3{\cdot}88$$

The final part of the question asks for an "estimation" of the Mach number at outlet from the impeller. To find this involves knowing the stagnation temperature at outlet and the absolute velocity V_2 in Figure 4.16(b). The stagnation temperature can be found quite easily because the entire stagnation temperature in the compressor takes place in the impeller. Finding V_2 is more difficult as it involves finding V_{r2} the radial component of the impeller outlet velocity. This can be obtained from the mass flow and the flow area if the tip width of the impeller is known (the flow area perpendicular to V_{r2} is $\pi D \times$ tip width) and also the air density. But, to find the density, information is needed on the characteristics of the *diffuser*, as without this information the static pressure cannot be found. However, let us be realistic. Looking at Figure 4.16(b) it is clear that V_2 will be only slightly greater than V_{r2}. As an approximation let us assume that V_2 equals $V_{w2}/0{\cdot}95$ corresponding to an angle 18°. V_2 then has a value of $440 \times 0{\cdot}9/0{\cdot}95 = 417$. Therefore

$$\frac{V_2}{\sqrt{(T_{02})}} = \frac{417}{\sqrt{(288 + 178)}} = 19{\cdot}3$$

Using the isentropic flow curves of Figure 1.11 gives the Mach number as 1·08 for this value of $V/\sqrt{T_0}$. (It is not unusual for high performance centrifugal compressors to operate with supersonic conditions at outlet from the impeller.)

The Centrifugal Compressor Characteristic

The general form of the centrifugal compressor characteristic is very similar to that of the axial flow type described previously (p. 148). As in that case, it is usual to plot the performance in non-dimensional or corrected form as pressure ratio against mass flow for lines of constant rotational speed. At each speed the limit of stable operation as flow is reduced is reached at the surge line where there is a complete break-down in flow. Comparing the axial characteristic of Figure 4.10 with the centrifugal of Figure 4.15 it

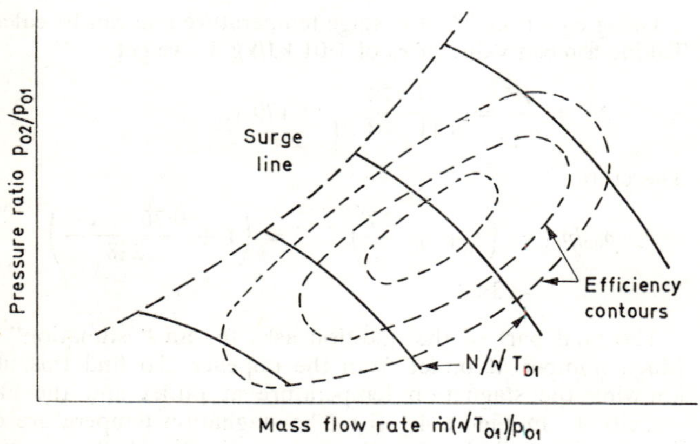

FIGURE 4.15 *The centrifugal compressor characteristic*

will be noticed that for the centrifugal the speed lines are much flatter. Whereas with the axial, particularly at the high-speed end, rotational speed determines the mass flow, with the centrifugal speed determines primarily the pressure ratio.

5 The Expansion System of Turbine Engines

5.1 INTRODUCTION

The high temperature, high pressure gases entering the expansion system of a turbine engine have a large amount of available energy with respect to exhaust conditions and this must be extracted as efficiently as possible. A high percentage of this energy needs to be fed back into the cycle to drive the compressor while the remainder is available for useful output, either by driving a power turbine or by forming a propelling jet. The need for good efficiency will be obvious when it is realized that 1% loss in turbine efficiency will mean about 2% loss in useful output of the engine. Additionally in the case of a turbojet a 2% loss in *gross* thrust at high forward speed could mean something of the order of 5% loss in net thrust.

When the work on compressors was begun it was emphasized that the compression process was essentially one that forced an airstream to flow against its natural direction; consequently great care was needed to handle the air gently as it flowed through the compressor. Conversely now, in the expansion system, the air is flowing in the direction of decreasing pressure—its natural flow—so the tendency to incur losses is very much less. The engineer takes advantage of this fact to design for very much higher specific work transfers in a turbine than in a compressor. It is not unusual to find cases where a compressor of eight stages is driven by a single stage turbine. At one time it was said that it was not possible to design a bad turbine but this facile view is no longer acceptable. The aim must always be to push designs to the limit, so that, if a component is happy doing a job, it must be made to work harder. In the end this means that everyone has a difficult job in making an efficient turbine.

The procedure in this chapter will be to look first at the turbine from the points of view of design, performance and cooling. Much of the work on the aerodynamic side of the turbines is closely related to that on axial compressors as discussed in Chapter 4. Then, after touching briefly on the jet pipe, the conditions in the propelling nozzle will be considered. With early engines, the expansion ratio across the final nozzle was comparatively low and simple configurations worked well. Now, with high flight speeds giving big nozzle expansion ratios and with the great need to avoid any losses in gross thrust, propelling nozzle design is one of the most important items in the whole engine design. Finally mention is made of the problems of silencing and thrust reversal.

5.2 TURBINE DESIGN AND PERFORMANCE

The great similarity in layout between axial flow compressors and axial flow turbines will be apparent from the diagram of a single stage turbine in Figure 5.1. As before, there is a set of fixed blades,

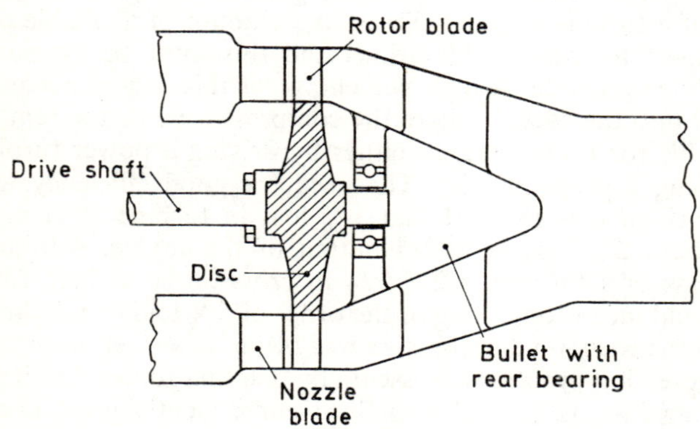

FIGURE 5.1 *Layout of an axial turbine stage*

known in this case as nozzle or turbine stator blades; again there are the rotor blades attached to the periphery of the turbine disc (the rotor blades are known as "buckets" in American parlance). If the gas flow passages are looked at in detail, the rates of change in area and direction will be seen to be much more drastic than for the compressor stage but this will not be immediately apparent at first glance. What *may* show the difference between a turbine and a compressor stage would be the fitting of shrouds on turbine blades. This is common practice for turbines but almost unknown for compressor blades. The reasons for shrouding are discussed in greater detail below (p. 172).

As with compressors the degree of reaction of a stage is an important design factor. The degree of reaction is defined as the ratio of design enthalpy drop through the rotor to the design enthalpy drop across the complete stage. Referring to Figure 5.2, which illustrates the flow through a stage on the enthalpy–entropy plane, it will be seen that the constant pressure lines are more or less parallel on the *h–s* diagram. To a close approximation therefore the degree of reaction may be expressed as the ratio of the rotor pressure drop to the stage pressure drop.

The extreme case of a zero reaction stage wherein all the pressure drop occurs in the nozzle blades is known as an "impulse turbine"

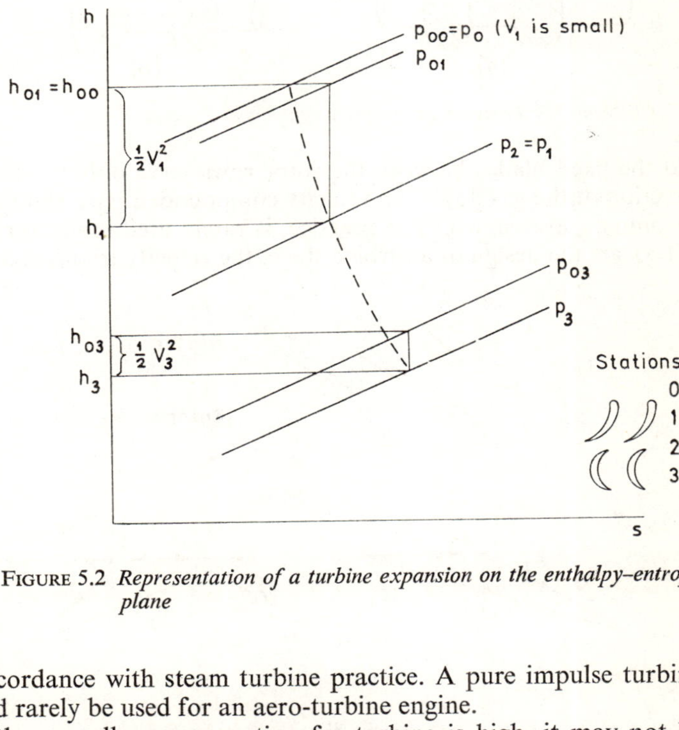

FIGURE 5.2 *Representation of a turbine expansion on the enthalpy–entropy plane*

in accordance with steam turbine practice. A pure impulse turbine would rarely be used for an aero-turbine engine.

If the overall pressure ratio of a turbine is high, it may not be feasible to operate efficiently and get the required output in a single stage. A multi-stage turbine would then be used and the choice would lie between a pressure compounded or a velocity compounded arrangement. These two possible types are shown in Figure 5.3. This figure shows that basically a pressure compounded arrangement consists of a number of individual single stages of the type described above, set up in series with all the discs on a common shaft. In the case of a velocity compounded, however, the entire pressure drop takes place through the first set of fixed nozzle blades

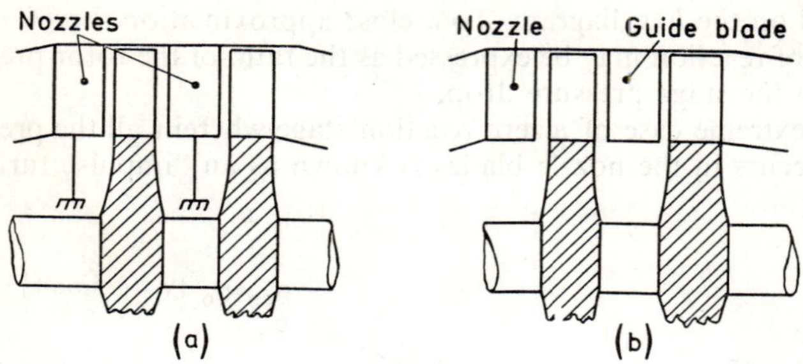

FIGURE 5.3 *Pressure and velocity compounded stages*

and the fixed blades between the rotor rows serve only to alter the direction of the gas flow. The velocity compounded type, though not uncommon in steam turbine practice, is rarely used in gas turbines.

To start the design of a turbine stage, the velocity triangles for the

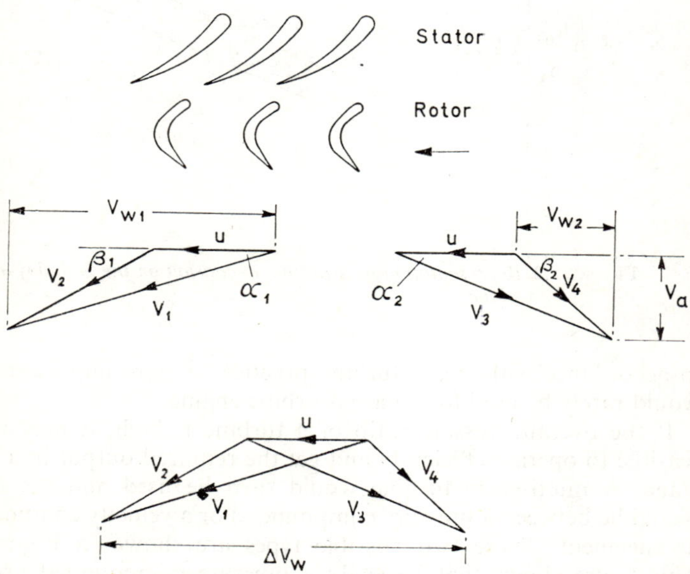

FIGURE 5.4 *Velocity triangles for a turbine stage*

mid-radius of the blades will be drawn. Figure 5.4 shows diagrammatically the blading of a stage viewed in the radial direction and also the velocity triangles at design conditions. The various velocity vectors are found as follows.

(a) Nozzle Outlet Absolute Velocity

Application of the steady flow energy equation to the flow through the nozzle blades allows V_1, the velocity at outlet from the blades, to be calculated from the known inlet conditions and enthalpy drop. The geometry of the blading will determine the angle of the gas at outlet from the nozzles and this gas angle will, in general, be close to the physical angle of the blades. There will probably, however, be small variations of gas angle with pressure ratio across the nozzle blade row.

In choosing the nozzle blade outlet angle, α_1, two conflicting requirements must be reconciled. For the highest specific work output, the gas should leave the nozzles at as small an angle as possible to the plane of the turbine wheel—α_1 should be as small as possible. On the other hand, the annulus area required for the flow varies inversely with the axial component of the velocity (for a given gas density) so in order to keep size down an appreciable axial velocity must be provided. The compromise value is generally of the order of α_1 equal to 20°.

(b) Blade Speed

The blade speed *u* comes simple from the mean diameter of the blading *D* and the rotational speed *N*. The rotational speed used by the turbine must be acceptable as a rotational speed for the compressor (the two are on one shaft) and a conflict may arise here as the turbine designer wants to operate at high speed while the compressor man wants to keep the rev/min down. With continual improvements in the state of the art for both compressors and turbines it becomes more easy to reconcile these conflicting wishes.

(c) Gas Inlet Relative Velocity

By closing the triangle started by the vectors of gas absolute velocity and blade speed the velocity of the gas relative to the blade V_2 is found. The magnitude of this velocity is, of course, less than the magnitude of the absolute velocity as the blades are "running away" from the gas leaving the nozzles. From the completed vector triangle, the angle of the gas relative to the rotor blades will be found, angle β_1. This angle is of very great importance as it is the angle at which the leading edge of the blades must be set if the angle of incidence is to be zero. (In steam turbine work it is sometimes spoken of as "the blade angle for shock-less entry of the steam"—and expression which is descriptive without being too exact.)

Passing now to the velocity triangle for the gas leaving the blade, the blade speed vector u is as before. The general case of a turbine which is not pure impulse will be considered and the other vectors are derived as follows.

(d) Gas Outlet Velocity Relative to the Blades

To find the velocity at outlet from the blades use can again be made of the steady flow energy equation but great care is needed to decide what is happening *relative to the blades*. As far as the gas is concerned, once it is moving through the space between two blades, the flow is exactly equivalent to that through a nozzle. A force will be produced on the blade by the changes in gas velocity (magnitude and direction) through the row but when considered relative to the blades, no external work is done. Considered from the blades' point of view the steady flow energy equation may be written in the same form as for a nozzle, i.e.

$$h_2 + \tfrac{1}{2}V_2^2 = h_3 + \tfrac{1}{2}V_3^2$$

(h_2 and h_3 are the static specific enthalpies at entry and exit to the rotor blade row as marked on the *h–s* plot of Figure 5.2).

Hence the gas outlet velocity is obtainable from

$$V_3 = \sqrt{[2(h_2 - h_3) + V_2^2]} \tag{5.1}$$

An efficiency for the flow through the blade row is implicit in the enthalpy drop as illustrated on the plot of Figure 5.2. This efficiency may be found from rig tests on cascades of turbine blades in which the loss in stagnation pressure across the blade is measured.

The angle of the gas outlet vector will be determined by the geometry of the blading and by the operating conditions. Nominally, if the blade passages were long enough and the change in direction sufficiently gentle it would be reasonable to expect that the gas would leave the blades at the outlet angle of the blades. In practice there will almost always be some deviation of the gas from this ideal case, and, as in the case of the nozzle blade row, the deviation will vary with the expansion ratio across the blades. In particular, if the blades are convergent and operating above the critical pressure ratio so that sonic conditions prevail at the section of lowest cross-sectional area, the gas expanding further in the free space may go supersonic and diverge (Figure 5.5). When this occurs, appreciable deviation of the gas outlet direction from the outlet blade angle may occur: rig tests of turbine blade cascades are again useful in determining these deviations.

(e) Gas Outlet Absolute Velocity

The absolute velocity of the gas at outlet from the stage V_4 is found by completing the velocity triangle of blade speed u and relative

FIGURE 5.5 *Supersonic flow from nozzle*

outlet velocity V_3. The kinetic energy corresponding to this vector represents a loss of possible output from the stage so V_4 should be as small as possible. For the cases of greatest interest, where the turbine is used in a turbojet, the kinetic energy corresponding to V_4 is not a complete loss to the cycle, as the gases have still to pass to a propelling nozzle. If the gas is diffused carefully, the "waste" kinetic energy from the turbine can be recovered partly.

It is desirable, also, that there should not be too much swirl of the gases leaving a final turbine stage, that is, the absolute outlet direction should be as nearly axial as possible. A high degree of swirl would mean that high absolute velocities were needed to provide the axial component of velocity which moves the gas down the jet pipe. In addition, streamlined fairings are generally fitted downstream of the turbine to support the exhaust cone, and if the gases are swirling when they leave the turbine they will meet these fairings at a high angle of incidence. Break-away might occur from the fairings with resultant high losses.

Once the velocity triangle for a stage has been drawn, an estimate of the power output can be made. The method is the same as that used in Chapter 4 for calculating the power input to a compressor stage, that is, the useful force acting on the blades is found by considering the rate of change of momentum of the gas across the rotor blade row. Using the nomenclature as in Figure 5.4 we have

Force on blades = Mass flow rate × Change in whirl component

$$F = \dot{m}(V_{w1} - V_{w2}) \tag{5.2}$$

And the power is

$$\dot{W} = Fu = \dot{m}u(V_{w1} - V_{w2}) \tag{5.3}$$

In this equation, due attention must be paid to the direction of the outlet whirl component. With triangles as drawn, the gas is being ejected in the opposite direction to the direction of motion of the blades so that its velocity has a *negative* whirl component compared with that of the inlet whirl. The two velocity terms in the brackets of equation 5.3 are, therefore, added together numerically.

Again, following compressor procedures, the temperature drop across the stage can be found by equating the power output as given by equation 5.3 with that coming from the steady flow energy equation. Writing these two equations together we get

$$\dot{W} = \dot{m}u(V_{w1} - V_{w2}) = \dot{m}c_p(T_{01} - T_{02})$$

Therefore

$$\text{Stage temperature drop } \Delta T = \frac{u(V_{w1} - V_{w2})}{c_p} \tag{5.4}$$

The axial components of the velocity through the stage are of use in calculating the required annulus area for flow. It will be appreciated that the continuity equation $\dot{m} = \rho AV$ can be applied to the flow through the stage taking components perpendicular to the plane of the turbine. In the interests of reducing the frontal area of the turbine it is desirable to increase the axial velocity component but this has the effect of reducing specific work output—α_1 would have to be increased. As mentioned previously, some compromise is, therefore, necessary. In addition the only "useful" forces on a turbine blade are those resulting from changes in the whirl components of velocity; changes in the axial component of velocity will cause axial thrusts which are undesirable and which must be resisted by the thrust bearings holding the shaft. Care is therefore needed in choosing the values of axial velocity.

Turning now to the question of the efficiency of a turbine, the losses that occur in a stage can be considered under two main headings. These are firstly the losses due to the thermodynamic irreversibilities of the expansion process—factors such as the friction and boundary growth effects which cause losses in stagnation pressure and show up as increases in entropy. Secondly there are the losses arising from leakage across the turbine and across the blade tips. These leakage losses can be highly significant especially if there is a large pressure drop across a stage.

Taking firstly the thermodynamic losses, a detailed study of the changes in state through a stage on the enthalpy–entropy plot can give a useful picture of what is happening. Looking at Figure 5.2 it is seen that in any given set of blades the stagnation enthalpy referred to the blades remains unchanged through the row but that there is a loss in stagnation pressure. This loss arises from viscosity effects which cause a thickening of the boundary layer on the blades. The state point through the row moves to show an increase in entropy. Great care must be taken in choosing the frame of reference for calculating the stagnation conditions in a row. If, for example, the stagnation enthalpy is calculated using *absolute* velocities through a row of *moving* blades, an incorrect answer is obtained. It is most important to put yourself in the position of the blades themselves in thinking about the relative velocities through a row. Remember that the gas itself is primarily interested in the *static* properties. Density, velocity of sound, etc., must always be based on static values. (A similar approach to that of Example 4.3 may be used with the turbine to illustrate these points.)

As a general rule, the pressure losses in a gas stream are a function of the square of the gas velocity so it would be expected that turbine stages with high velocities between gas and blades would be less efficient than those using lower velocities. This rough-and-ready rule *does* apply and it is found that an impulse turbine will generally have higher losses than a reaction type. Losses will also vary with the magnitude of the gas deflection through a blade row so that a stage designed for high specific work output will be inherently less efficient. This factor was brought out clearly in a paper by S. F. Smith (ref. 1). In the paper, results were plotted for a range of different turbines of efficiency in terms of a non-dimensional stage loading $\Delta H/u^2$ and a non-dimensional flow coefficient V_a/u. The plot was of the form shown in Figure 5.6 and contours of constant efficiency could be drawn on the graph as shown. With advances in turbine technology the actual figures on a plot such as this will improve but the general picture—that high stage loading means a lower efficiency—will remain unchanged.

Besides the pressure losses due to friction, further losses will occur if the flow at any point passes through a shock wave in decelerating from supersonic to subsonic conditions; flow through a shock invariably gives a loss in stagnation pressure. Once again, therefore, the stage loading comes in as a variable, as shock losses are more liable to occur in highly loaded stage.

The losses discussed so far have been those arising at the design operating conditions of the turbine. At design conditions, the blade angles will be such that the gas will meet the blades at approximately zero angle in order to keep the losses as low as possible. As soon as the turbine moves away from the design conditions, the gas angles will change so appreciable angles of attack—positive or negative—

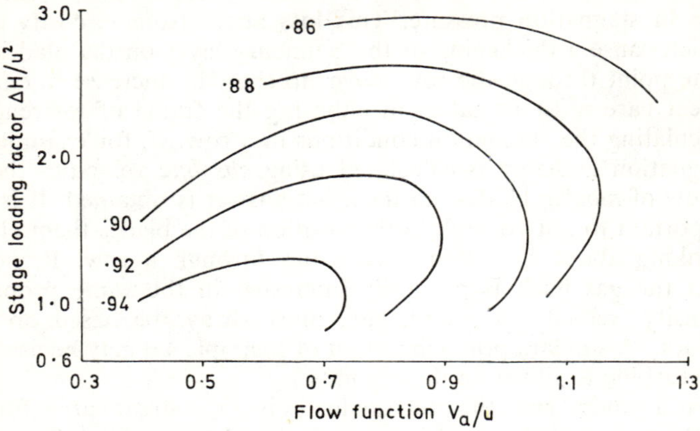

FIGURE 5.6 *Efficiency variation for a turbine stage*
Efficiency contours are plotted through many test points.
(Courtesy Royal Aeronautical Society, ref. 1)

may result. With these larger angles losses would increase and, as would be expected, the turbine efficiency will fall.

From these considerations, an overall turbine characteristic can be plotted and this may take the form shown in Figure 5.7. Here non-dimensional curves have been drawn of a mass flow function $\dot{m}\sqrt{T_{01}}/p_{01}$ a power output function $\dot{W}/p_{01}\sqrt{T_{01}}$ and the efficiency versus turbine expansion ratio for lines of constant non-dimensional speed $N/\sqrt{T_{01}}$. (The efficiency lines are shown for convenience but are actually implicit in the other two curves.) These curves show the general form of characteristics encountered in turbine work and in particular they show that the mass flow function levels off when the choking pressure ratio is reached. It will also be seen that the power output is relatively insensitive to changes in turbine speed at given pressure ratio.

In most modern engines multistage turbines are used and very often the stages are mechanically independent so that the running speeds of various stages are different. It is important in these circumstances to know how the work split between stages will alter as the overall pressure ratio changes. The matching of the two stages can be checked by tying in the non-dimensional *exit* mass flow rate for a

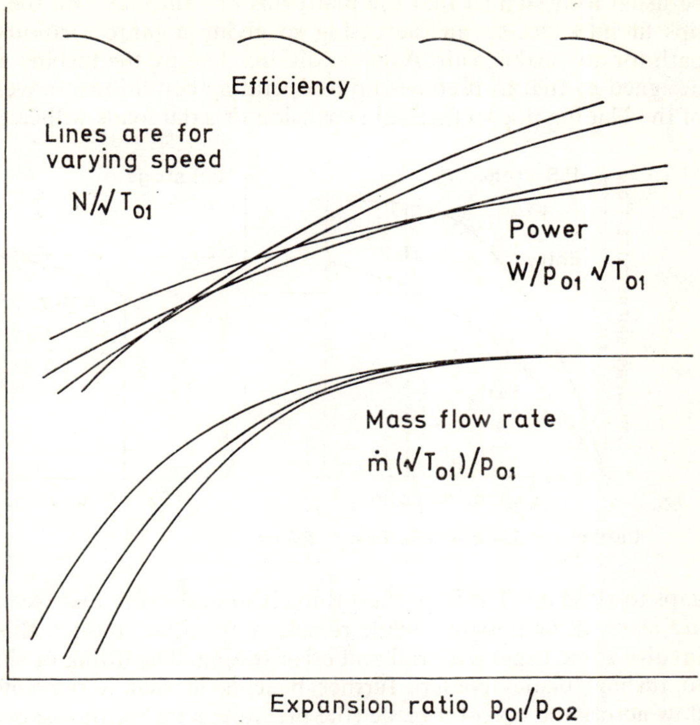

FIGURE 5.7 *The overall turbine characteristic*

high pressure stage with the non-dimensional *entry* mass flow rate for the following low pressure stage. Figure 5.8 shows the picture which is obtained; at design conditions both HP and LP are operating at about the same pressure ratio but at part load the HP pressure ratio is much higher than the LP. This means that the first stage would supply an increasing percentage of the total power output as the pressure ratio was reduced—the low pressure turbine unloads itself more rapidly than the high pressure one. If the turbines are each driving a compressor (as happens with a two spool turbojet) this will mean that the LP speed will fall much more rapidly than the HP speed. There may be important repercussions on the acceleration characteristics of an engine.

Turning now to the other source of loss in turbine efficiency, that due to leakages, the only way to keep this to acceptable proportions is to design the whole turbine and casing with very great care. It will be usual to design so that the platforms or "shrouds" on the blade tips fit into recesses in the casing so giving a more tortuous flow path for any leakage air. As an additional feature the turbine will be designed so that as it comes up to operating conditions, movements of the blading due to thermal expansion or axial loads will cause the

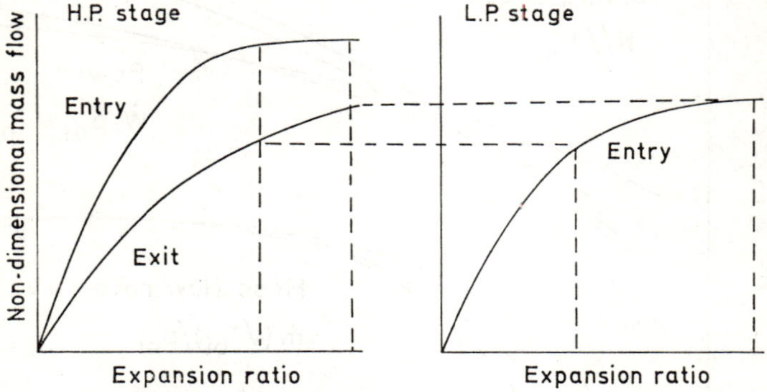

FIGURE 5.8 *Multistage turbine matching*

gaps to close up. The important thing is to make sure that clearances are as small as possible, while remaining positive, though this may involve some expensive trial and error testing. The fitting of shrouds to turbine blades confers further benefits in that it prevents any flow across the tips of a blade (the pressure is higher on the concave side than on the convex) and, if the shrouds are butting against each other, more rigid blading results.

The preliminary design of a turbine stage will concentrate mainly on the calculations of conditions at mid-radius and of checking that velocity diagrams and loading are satisfactory. Fairly soon in the design however it will be necessary to start taking account of conditions at other radii along the blade. As with the axial compressor stage design, the two factors which make this necessary are the need to allow for blade speed variation with radius and the need to provide some measure of radial equilibrium in the flow. One of Sir Frank Whittle's most important contributions to turbine design was his appreciation of the problem of radial equilibrium in a turbine. In his autobiography he described the impact of his vortex flow concepts on the steam turbine practice of the 1930's.

It is not necessary here to repeat the theoretical work on radial equilibrium covered for the compressor stage in Chapter 4 as, in this respect, the turbine is very much a reversed compressor stage.

As the whirl components of velocity in a turbine are higher than in a corresponding compressor stage the radial equilibrium problem will be more acute so, if equilibrium is to be achieved, there must be an appreciable variation in conditions from root to tip. This may necessitate a highly twisted blade which introduces problems of stressing and of manufacture. In addition, with a wide variation in conditions the degree of reaction will change along a blade and the condition may even be reached where the gas flow through the rotor blades at the root section is decelerating. This would be an undesirable state of affairs as decelerating flow is always more difficult to handle than accelerating flow. All in all, there is much to be said for compromising with a design and accepting something less than complete radial equilibrium, particularly as the gases delivered to the turbine from the combustion chamber will not have a completely uniform profile.

On first thinking about turbines, the tendency will generally be to concentrate attention on the blades themselves and to consider them as aerodynamic shapes in a gas flow. An alternative approach is possible, though, which concentrates attention on the passages between the blades and this second view came into prominence following the publication of work by Stannitz in the United States, in the early 1950's. Stannitz was trying to use numerical methods of mathematics to design flow channels to give certain Prescribed Velocity Distributions (PVD) along the walls of the channel. His work (ref. 2) gave theoretical results for certain standard cases which were in good agreement with flow conditions as measured experimentally. The work is clearly applicable to turbines if attention is directed to the space between the blades. These can be treated as passages which are required to alter gas velocities in some required manner.

Although Stannitz's original work was strictly applicable to inviscid incompressible fluids the techniques could be applied without excessive errors to the flows of real gases through turbines. The PVD design methods are especially suited to solution by computers which are able to handle the numerical methods involved. When experimental turbines were designed by this means (and the design involved some compromise of shapes to meet stressing as well as aerodynamic requirements) it was found that on test they gave about the same peak efficiency as conventionally designed turbines but that the efficiency away from the design point was higher. The gains achieved were well worth while.

So far, the turbine design problems discussed have been basically similar to those of axial flow compressors with, perhaps, a slightly different emphasis due to the higher stage loadings. However, besides being satisfactory from the gas flow point of view, turbine blading must also be designed to withstand very much higher temperatures. Essentially the aero gas turbine is a high temperature machine and there is always a great incentive to work at a high turbine entry temperature. Any increase in temperature gives a higher specific output from a turbojet and so reduces bulk and weight for a given thrust. There is, of course, a performance penalty due to the increased specific fuel consumption but, for very many applications, the gain in specific output outweighs this disadvantage.

At the temperature corresponding to turbine entry conditions all materials suitable for blading are affected by creep. By "creep" is meant the continuous extension of a component under loads even though the load may be less than that which would cause fracture. Quite obviously it would be unacceptable to allow appreciable creep in the life of a turbine blade as this would alter all clearances and eventually allow rubbing of blades on the casing. A balance must, therefore, be struck between operating temperature and the loads on the blades to give an acceptable life for the turbine.

The attack on the problem of meeting high cycle temperatures takes two main lines. Improvements in performance can be obtained by using new and better materials or by using existing materials and providing some method of cooling for the blade. The question of materials for turbine blades is considered more fully in Chapter 12 (p. 341) so at this stage only the principles of blade cooling are considered.

On the question of cooling turbine blades, it is worth noting that cooled turbines have been designed from the very earliest days of jet propulsion and that a large variety of systems have been tried with varying degrees of success. A review of the historical background is included in the paper by Ainley (ref. 3) and schemes using water, air and liquid metal are described. Even though water and liquid metals have advantages under certain circumstances, air is so readily available in turbine engines, that it is the general choice nowadays as a cooling medium. Air tapped from compressor delivery is at suitable temperature and pressure for cooling purposes and the quantities required do not impose unacceptable performance penalties.

Having decided to use air cooled turbine blades, the designer now has the task of seeing whether adequate cooling can be provided. The problem is to keep the blade material at some temperature appreciably below the gas temperature and this will imply a heat flow from gas to blade. When steady conditions appertain, the rate of heat flow *into* the blade from the gas stream must equal the rate of heat transfer to the cooling air plus any heat flow through the blade roots to the turbine disc. Now, in heat transmission work the rate of heat flow from a gas to a solid surface depends primarily on the *film coefficient h* which defines the heat flow rate per unit area for a given temperature drop. This film coefficient will be a complex function of the gas properties and the local flow conditions around a blade but the general form of the variation of h around a blade at operating conditions will be as shown in Figure 5.9(a) with high

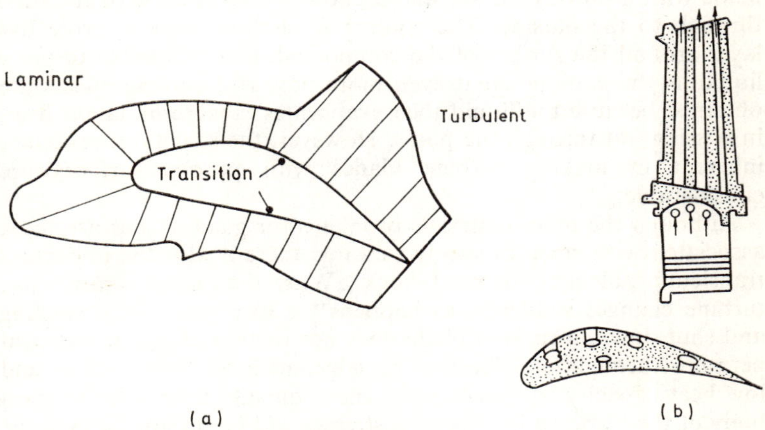

FIGURE 5.9 *Air cooling of turbine blades*

(*a*) Variation of heat transfer coefficient around a blade.
(*b*) A typical air cooled blade. (ref. 4).
(Courtesy Rolls-Royce Ltd., Aero-Engine Division)

values at the leading edge, near the stagnation point, and near the trailing edge where boundary layer separation may have occurred. It is unfortunate that such high film coefficients occur at leading and trailing edges as the material is thinnest there, so that the provision of cooling air passages is particularly difficult. However, by using sophisticated manufacturing techniques such as spark erosion and electro-chemical machining it is possible to make suitable blade passages. Figure 5.9(b) shows the section of a typical air cooled blade, in this case a forged blade for the Rolls-Royce Spey.

The practical difficulties of applying cooling to turbine blades for use in civil aircraft have been largely overcome and by this technique peak turbine entry temperatures have now been raised by something of the order of 200°C. Rolls-Royce have built up an extensive backlog of experience by now and a description of some of their achievements on the Conway, Avon and Spey engines is given by Hall in ref. 4. One of the most intriguing facts quoted in this paper is to do with the Spey blade which without cooling would have a life of 12 min whilst with cooling the design life is 10 000 hr.

Besides the simple system of providing cooling air passages through the blades, a number of other ideas for air cooling have been suggested at various times. One of the most interesting proposals has been to use what is known as "transpiration cooling." In this, a component is made with a porous surface and a cooling fluid is allowed to effuse through to the outside. The cooling fluid thus forms a protective layer around the surface of the component. If it is feasible to use a liquid as the cooling agent even more powerful cooling effects are obtained because the liquid then evaporates, absorbing latent heat, in passing out through the pores. However, the structural problems inherent in making turbine blades with porous surfaces are considerable.

Although the main materials problems for gas turbines are those associated with creep in steady running, there is also the important transient problem of "thermal shock". When the gas temperature in a turbine changes suddenly, as happens for example during starting and shut down, a turbine blade does not remain at a uniform temperature throughout. The trailing edge, with its light section and low heat capacity will react much more quickly than will the main body of the blade and so thermal stresses will be set up. To take the shut-down case, the trailing edge will tend to contract and so will be in tension until the main bulk of the blade has reached the same temperature. Repeated stressing like this is liable to result in hairline cracks in the trailing edge of the blades.

Throughout the field of aircraft propulsion, there are always gains to be made through taking care with details of design. Nowhere is this more true than in the case of the turbine, and big improvements in engine life can sometimes be made at no sacrifice to performance or weight by making relatively small changes in components. Bearing in mind also that there is always a heat flow through the blade roots into the turbine disc it is clear that any reduction in disc temperature will give a corresponding drop in blade metal temperature. The attachment of blade to disc is another potential source of benefit and extended roots have at times been used to reduce blade temperatures.

5.3 THE JET PIPE

All too often in the past the jet pipe has been dismissed lightly in engine design as being very much a minor component of the engine.

In fact, the jet pipe has a vital part to play in the engine's operation and it may have appreciable effects on the performance and weight. The design is dictated to a large extent though by the design of the aircraft and the actual location of the engine. For a pod-mounted engine following the classic American pattern, there are few troubles in keeping the jet pipe compact. In other aircraft, such as the Vulcan with its buried Olympus engines or the Vanguard with its Tynes, the pipe may be long and may have to wind past parts of the aircraft.

Two main problems arise in the design of jet pipes. There is a requirement that losses in stagnation pressure should be kept to a minimum as they will lead inevitably to a loss in thrust. And there is the problem of reducing the heat loss from the pipe. Not only would a heat loss be bad as representing a loss in stagnation temperature and so in thrust, but there is also the question of protecting the aircraft structure from overtemperature. Usually the jet pipe will pass close to aircraft structural members and it is desirable that these should be kept as cool as possible. Thorough lagging of the jet pipe is therefore necessary with, of course, the lowest possible weight penalty.

The pressure loss due to friction in the pipe may be estimated by assuming a friction factor f and using either the Darcy formula or, more accurately, the Fanno line approach (p. 46). The example on Fanno flow of Chapter 1 (see p. 47) showed the importance of keeping the Mach number low if high stagnation pressure losses are to be avoided.

The jet pipe is an important instrumentation station for the turbojet and provision for this must be made at the design stage. The gas temperature downstream of the turbine gives a good measure of the turbine entry temperature especially if the final nozzle is choked so it is convenient to use the jet pipe temperature for "driving" and rating purposes—jet pipe temperature is displayed on the pilot's panel and he is told what limits are to be observed. Great care must be taken to get a true average temperature of the gas stream and it is usual to take the average reading from a number of thermocouples. On some engines jet pipe pressure is also measured and is used as a general indication of the engine's thrust.

5.4 THE FINAL PROPELLING NOZZLE

This section will deal specifically with the problems of the propelling nozzles for turbojet engines. The design conditions are more severe in this case than for either turboprops or for turbofans as the pressure ratio across the nozzle must inevitably be highest for the turbojet. In addition, the penalties for inefficiencies are more severe for the turbojet than for the others so there is a greater incentive to design with care.

The general aspects of the propelling nozzle were discussed in Chapter 3 where it was shown that the ideal flow through the nozzle was isentropic and that due to the expansion a high velocity gas stream was produced. Nozzle efficiency was defined as the ratio of the actual enthalpy drop to the isentropic drop and a thrust coefficient was introduced (p. 88) to take account of the fact that the gas

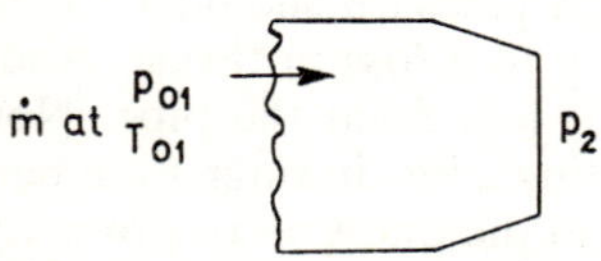

FIGURE 5.10 *Thrust of a convergent nozzle*

leaving the nozzle was not entirely uniform and flowing in the axial direction. Mention was also made of the fact that the effective nozzle exit area will always be slightly less than the physical area.

When dealing with isentropic flow in Chapter 1 it was pointed out that if the nozzle expansion ratio p_{01}/p_2 were greater than the critical value then a convergent–divergent shape of duct is required. The nozzle outlet velocity and thrust can then be calculated using the methods of equations 3.21 and 3.22. For reasons which will be given below many turbojets are fitted with final nozzles which are *convergent only* despite the fact that they operate normally above the choking pressure ratio. The thrust from an overchoked convergent nozzle is made up of two components, the momentum change component and a pressure component. To see how these are derived consider a convergent nozzle passing $\dot{m}$ kg/s of gas with entry stagnation pressure p_{01}, entry stagnation temperature T_{01}, mean specific heat c_p and mean specific heat ratio γ (Figure 5.10).

In the nozzle the gases will expand to sonic conditions and the static pressure and temperature at nozzle outlet p_2 and T_2 will be found from the critical relationships (equations 1.31 and 1.32) as

$$p_{01}/p_2 = (T_{01}/T_2)^{\gamma/\gamma-1}$$

The exit velocity is sonic so $V_2 = \sqrt{(\gamma R T_2)}$ and hence the momentum component of the thrust can be calculated in the normal way.

As the nozzle is overchoked the nozzle exit static pressure p_2 will be higher than the ambient pressure p_3 so there will be a pressure component of thrust equal to the static pressure difference times the nozzle exit area, i.e.

$$\text{Pressure component of thrust} = A_2(p_2 - p_3)$$

By using the continuity equation the required area for flow can be calculated as

$$\dot{m} = \rho_2 A_2 V_2 = \frac{p_2 A_2 V_2}{RT_2}$$

Therefore

$$A_2 = \frac{\dot{m} R T_2}{p_2 V_2}$$

If the appropriate calculations are carried out it is found that the gross thrust produced by the nozzle which is convergent *only* is always less than the gross thrust which can be obtained by isentropic expansion in the ideal convergent/divergent nozzle. Table 5.1 shows how the two thrusts compare at different expansion ratios.

TABLE 5.1 *Gross Thrust of Convergent and Convergent–Divergent Nozzles*

Design nozzle expansion ratio	4	8	12	16	20
X_G (conv.–div.)/X_G(conv.)	1·007	1·062	1·09	1·12	1·13

Looking at the results in the table we see firstly that the thrust penalty through using a nozzle which is convergent only is small at low pressure ratios. The normal subsonic jet operates with a nozzle pressure ratio of the order of 4·0 and the performance gains of going to a convergent–divergent are not worthwhile. Conversely however, at high pressure ratios quite appreciable increases in thrust are possible by adding the divergent portion. In addition, the gain are in *gross* thrust and there will be much bigger percentage improvements in net thrust. This is doubly important as high nozzle pressure ratios are usually associated with high forward speeds where the net thrust is the difference between two large quantities, a large gross thrust and a large intake momentum drag. With a supersonic aircraft operating at about 2·5 Mach number the pressure ratio across the propelling nozzle will be of the order of 20 and a 1% loss in gross thrust will be worth about $2\frac{1}{2}$% in net thrust.

For a high performance engine it is highly desirable that a convergent–divergent propelling nozzle should be provided, but a number of design problems arise straight away. The most immediate problem is that the ratio of exit to the throat area for the divergent portion can only be at its correct value for one nozzle pressure ratio. (There is actually a low pressure ratio where the flow is complete subsonic which will satisfy the same area ratio but this will be well away from the design condition.)

If the overall pressure ratio is less than the design value a shock system will form inside the divergent portion of the nozzle and will lead to appreciable losses. There is also a risk that the shocks will be unstable and they may cause vibration of the jet pipe. As a rule the efficiency of a convergent–divergent nozzle is much more sensitive to variations from design conditions than a simple convergent one would be.

The next main problem with the design of a convergent–divergent nozzle arises if reheat is fitted to an engine. When reheat is brought into operation there is a large increase in the volume flow through the nozzle and so its area must be increased if the jet pipe pressure is to be maintained at its original value. Looking at it from the point of view of the non-dimensional flow parameters, $\dot{m}\sqrt{T_0}/Ap_0$ is constant for a choked nozzle so that if T_0 goes up A must be increased. This means that it is necessary to design a nozzle with a variable area throat as well as having convergent and divergent portions. Quite obviously the mechanical design problems would be formidable to say the least.

Luckily it has been found that a number of aerodynamic techniques show promise of giving acceptable solutions to the design problems. One of the neatest of these solutions is the ejector nozzle shown in Figure 5.11(a). With the ejector nozzle the gases leave a mechanical nozzle which is convergent only and which may be made of variable area to allow reheat. Surrounding this nozzle is a stream of secondary air which is controlled so that its static pressure permits the gases to expand as if they were in a divergent portion. Control of the secondary air can give flow patterns equivalent to divergence through varying effective area ratios if required. There is the penalty that the secondary air for the ejector must be accelerated up to flight speed and so gives a momentum drag. In many cases though, it is possible to use the air which is required for jet pipe cooling or intake boundary layer control as the source of secondary air. A number of variations on the theme of aerodynamic control of the efflux are described by Migdal and Horgan (ref. 5). This paper described various arrangements which had been investigated by the Pratt & Whitney Company and showed in particular that one with mechanical doors to control the secondary air (shown in Figure 5.11(b)) gave a good performance over a wide range of flight Mach numbers. Yet another design, evolved at the National Gas Turbine Establishment and using a centre-body with a cylindrical outer shroud was described by Herbert (ref. 6). Only extensive flight experience over the next few years will show which of these types is most promising.

Before finally leaving the propelling nozzle it should be mentioned that the gases leaving an engine will have a profound effect on the

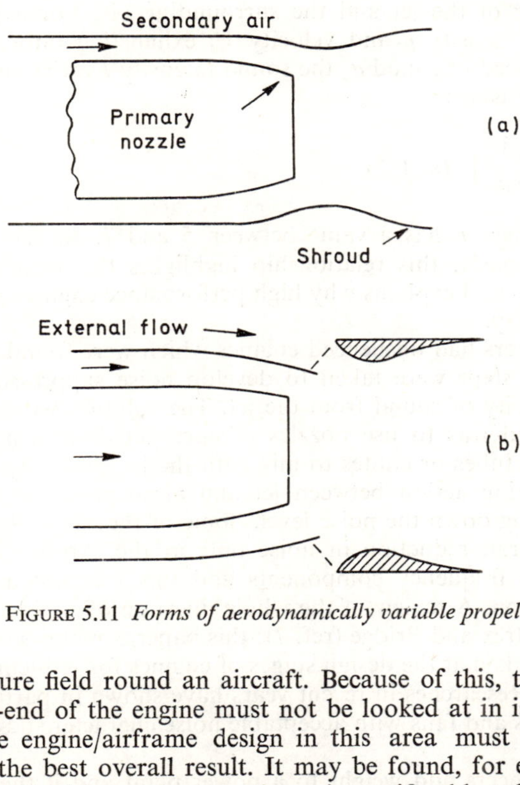

FIGURE 5.11 *Forms of aerodynamically variable propelling nozzle*

pressure field round an aircraft. Because of this, the design of the back-end of the engine must not be looked at in isolation, but the whole engine/airframe design in this area must be integrated to give the best overall result. It may be found, for example, that by cutting back a nozzle some thrust is sacrificed but the corresponding drag change more than counterbalances this effect. Great judgement is needed in these matters as so often the engine and airframe makers are friendly rivals!

5.5 SILENCERS AND THRUST REVERSERS

In Chapter 2, when dealing with the social acceptability of engines, mention was made of the noise problem. With advanced technology turbofan engines it was shown that the predominant noise problem is that arising from the compressor but with low bypass ratio turbofans and turbojets exhaust noise is still the main source of trouble. The noise nuisance is so serious that, at least for civil aircraft, some attempts *must* be made to silence the engine. This entails the addition of some device to the final nozzle.

The actual mechanism of the sound generation of a jet is the turbulent mixing of the jet and the surrounding air. For a jet of mass flow $\dot{m}$ of density ρ and velocity V_J exhausting into air of density ρ_0 and speed of sound a_0 the sound intensity I varies according to the relationship:

$$I \propto \left(\frac{\rho_0}{\rho}\right) \left(\frac{V_J}{a_0}\right)^n (\dot{m}_J V_J^2)$$

where the exponent n has a value between 5 and 7. As has been mentioned previously, this relationship highlights the significance of the jet velocity and explains why high performance engines are so exceptionally noisy.

Early jet airliners had unsilenced engines which were found to be unacceptable so steps were taken to develop noise suppressors to reduce the intensity of sound from the jet. The solution which was generally adopted was to use nozzles of corrugated form and to take air through tubes or chutes to mix with the jet efflux. By these means the shearing action between jet and atmosphere could be reduced so cutting down the noise level. Some of the early silencers achieved an overall reduction in noise only at the expense of an increase in high frequency components and this was not a fully acceptable solution. A review of the whole question of engine noise is given by Greatrex and Bridge (ref. 7); this paper explains some of the steps being taken at the design stages of engines for reducing the noise nuisance. Researches in recent years have shown in particular how compressors and fans with acceptable noise characteristics may be designed.

Inevitably silencers add weight to a power plant and if they are of the chute variety, there is bound to be some performance loss. For the installation of the Olympus 593 in the Concorde airliner the silencer has been designed with retractable obstructions so that under altitude cruise conditions there is no thrust penalty. Clearly there will be weight and economic prices to pay for this increased complexity but on balance it must in the end be worthwhile.

When jet airliners were first flown it was realized that with their clean configuration and relatively high landing speed the landing run was liable to be considerably longer than with piston engined aircraft. Wheel brakes had to work very much harder to give acceptable landing distances and there were correspondingly higher rate of wear and tear on brakes and tyres. The problem becomes particularly acute if the runway is wet or icy when hard use of the brakes is not feasible. Before long it was realized that some form of thrust reverser was needed and the obvious step is to deflect the jet efflux in a forward direction. A simple clamshell-type arrangement with pivoted doors has proved suitable. On touch down the engine rotational speed is reduced to idling, the thrust reversers are operated, and the engine is then opened up to high power. In this way progressive braking, which is independent of the runway condition, is obtained. Care is needed to ensure that the exhaust products from one engine do not enter the intake of an adjacent engine and so cause damage. This would be particularly liable to occur in an aircraft like the Boeing 707 where the outboard engines are further back than the inboard. There is the additional risk of debris being picked up from the runway and finding its way into the intakes.

6 Combustion Systems for Turbine Engines

6.1 INTRODUCTION

Comparing the combustion system of a gas turbine with the other main components of the engine it is found that the processes involved are an order of magnitude more complicated. Any analysis which is to be carried out must consider the aerodynamic problems as before but, in addition, it must also take into account the fact that a chemical reaction is taking place in the chamber. This means that the actual composition of the working fluid will vary from place to place in the combustion zone and that under transient conditions there will be a time variation in composition as well as a spatial variation. The rate at which a reaction takes place to form certain products will depend on both the rate at which the reactants are mixed and on the chemical reaction rates for the process. Mixing rates will be a function of geometry while reaction rates are governed by the initial conditions and on the concentration of the various reactants. Given all this complexity, it is not really surprising that even although successful combustion chambers have been designed for very many years now, some of the basic mechanisms are not yet fully understood and much design is still carried out on an empirical basis.

This lack of full comprehension of all the processes would be serious if it were not for the fact that, all in all, turbine combustion systems work very well. Even though there is room for improvement to cut down the size and weight of systems they are not unacceptably bulky and the combustion chamber is not often the component with the largest diameter in the engine. On the performance side, the losses that *do* arise in the combustion process do not penalize the cycle efficiency excessively. The position is very different from that with axial compressors, for example, where a sound understanding of the basic mechanism was essential before an acceptable component could be made. Empirical rules and a gradual evolution and development of design have ensured that satisfactory combustion chambers have always been available. This is meant in no way to decry the fundamental studies which are being carried out and which are leading to better and better understanding of why chambers work as they do.

The aim in this chapter is to discuss the basic needs of the combustion system and to show how successful systems have evolved, and how the design can be approached. The various performance criteria by which real systems are assessed are then dealt with before passing to the particular problems of fuel injection and ignition. Consideration is then given to the reheat system which is used to boost the thrust of high performance engines. Finally some of the fundamental studies being carried out into the combustion process are mentioned briefly.

Naturally with a topic like combustion a single chapter can do little more than introduce the subject. For further reading mention should be made of Professor Spalding's work. His textbook (ref. 1) gives a full background to the work while his paper (ref. 2) sets out to give a practical design method for the combustion engineer.

6.2 BASIC REQUIREMENTS FOR THE COMBUSTION SYSTEM

The combustion system of an aero-engine is designed to release the chemical energy of the fuel completely in the smallest possible space and with the minimum of pressure loss. At outlet, the gases should have a reasonably uniform velocity and temperature distribution and the peak values of temperature must lie within the limits imposed by the turbine blade material. An aero-engine combustion system should give efficient, stable operation over a wide range of altitudes and flight speeds and it must be possible to relight in flight after shutting down an engine. From the mechanical point of view the system must be reliable and it should have a life equal to the overhaul life of the complete engine.

Almost without exception current aero-engines are designed to operate on hydrocarbon fuels such as kerosine and petrol. Certain characteristics of these fuels exert a profound influence on the layout of practical combustion systems. The most important of these characteristics are as follows:

1. *Limiting flame velocity*. It is not generally appreciated that the velocity of propagation of a flame through a mixture of hydrocarbon fuel and air is comparatively low. The flame speed will vary with mixture strength and with temperature and pressure but for most cases of interest it will be about 2–4 m/s. Now the average air

velocity through a combustion chamber of an acceptable size will be very many times this value—of the order of 50 m/s. A sheltered region with recirculation of air and low average velocities must therefore be provided if the flame is to be held in one place.

2. *Stability limits with variation in fuel/air ratio*. Stable burning of hydrocarbon fuels will only take place within fairly close range on either side of the stoichiometric value of fuel/air ratio. It is therefore necessary to ensure that the mixture strength in the sheltered region is fairly close to this stoichiometric value. The range of stable operation will vary with the absolute pressure in the combustion chamber so it will decrease as altitude increases until some limiting altitude is reached. In addition, any contamination of the air in the sheltered region by combustion products or water vapour (such as might result from flying through a rain storm) will narrow the stability limits.

3. *Temperature resulting from stoichiometric combustion*. The temperature obtained from the combustion of a stoichiometric mixture varies with the initial temperature but is of the order of 2 000°C. This temperature would be achieved in the sheltered region which is necessary for stable combustion but it is far higher than the maximum gas temperature which can be withstood by turbine blade materials. A large measure of excess air to dilute the products of combustion is needed to reduce the gas temperature to a value acceptable to the turbine.

4. *Fuel evaporation characteristics*. Before it can be burned, a liquid fuel must first be evaporated. The evaporation process will be helped if the fuel presents a high surface area to the air and if the fuel has an appreciable velocity with respect to the air. Both these desirable attributes can be obtained if the fuel is injected at high velocity into the air as the aerodynamic forces will then cause the fuel to break up into a large number of droplets. Alternatively, evaporation can be obtained by a straightforward addition of heat to the fuel rather as is done with a blow-lamp.

Any practical turbine engine combustion system must be designed with these fuel characteristics in mind. In addition, if an engine is to be successful it should be capable of accepting a wide variety of different fuels of the hydrocarbon type. An engine which will only burn fuels of high volatility, for example, would present problems to the operator of ensuring that such fuels were always available.

6.3 EVOLUTION OF A PRACTICAL COMBUSTION SYSTEM

The evolution of a classical pattern of combustion system to meet the requirements listed above was described by J. S. Clarke (ref. 3). The steps involved in building up the system are shown in Figure 6.1(a)–(d) which are based on illustrations in Dr. Clarke's paper. It

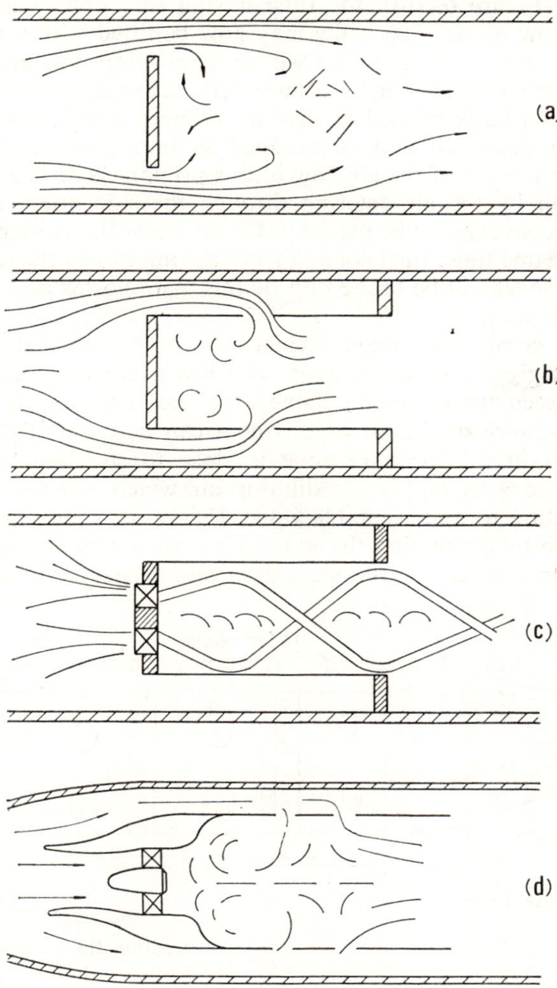

FIGURE 6.1 *Evolution of primary combustion zone*
(From ref. 3, Courtesy J. Lucas Ltd.)

will be seen that the first step (Figure 6.1(a)) is to set up a circular baffle in a duct. A highly turbulent wake will be generated by the baffle with a recirculation pattern around the edge and a dead region immediately behind the baffle. Clearly a baffle like this will introduce a pressure loss in the air flow.

An improved pattern results if a flame tube is built on to the baffle plate (Figure 6.1(b)), the tube having side holes which set up a cross-flow of air. The baffle may now be fitted with a swirler —an arrangement of vanes which will set up a spiral flow along the axis of the tube as shown in Figure 6.1(c). Finally, swirler and flame tube can be combined to give a "primary zone", the whole set-up being tidied up and streamlined as in Figure 6.1(d). The predominant feature of the air flow in this primary zone is a strong vortex set up by the air entering through the side holes. A fuel injector can conveniently be placed at the centre of the swirler in the nose of the flame tube. Fuel coming from the injector in the form of of a conical sheet will be broken up by this main vortex and so will be finely atomized.

The main combustion requirements have now been met as this arrangement gives a sheltered zone with low average velocity and a degree of recirculation of the flame which helps to keep it alight. By suitable choice of the air hole sizes it can be ensured that the fuel/air ratio in this primary zone is close to stoichiometric. It is now necessary to supply the dilution air which will reduce the gas temperature to an acceptable value. Figure 6.2 shows how this may be done by continuing the inner flame tube arrangement and providing large holes for the secondary and tertiary dilution air.

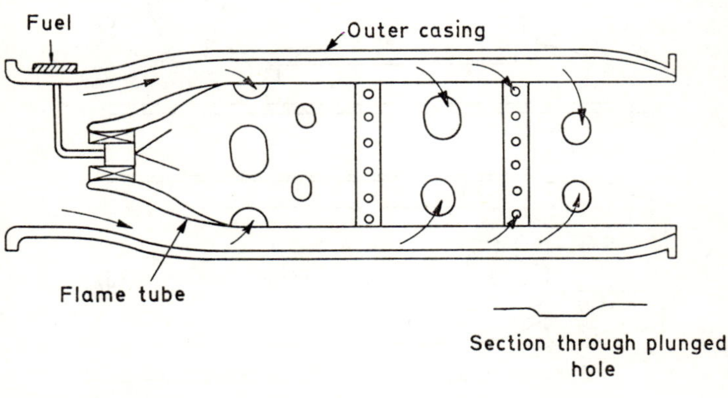

FIGURE 6.2 *The "conventional pattern" combustion chamber*
(Courtesy J. Lucas Ltd.)

An outer casing encloses the whole and forms the duct for the dilution air. Examination of the inner flame tube will show that the main air holes will normally be of the "plunged" type, shown in the detail sketch in Figure 6.2, rather than being plain cut-outs in the sheet metal. A plunged hole will pass a higher flow for a given pressure drop and gives a deeper penetration of the dilution air into the bulk of the hot gases. In addition to the main airholes, a large number of smaller plain holes will usually be fitted which will provide a layer of cooling air to keep down the flame-tube metal temperature.

The arrangement whereby the complete combustion system of an engine is split up into a number of individual chambers or cans was a popular one and was widely used for many years. A picture of an engine of the older generation will show the distinctive tubular cans disposed around the engine. By using this arrangement, development of the combustion system was facilitated, as rig testing could be carried out using only a small fraction of the total engine air and fuel flow. Production of a number of small cans is easier than making a single large system and replacement of damaged components is much cheaper. Against this must be set the fact that the individual cans are very wasteful of space and the available cross-sectional area for flow is used inefficiently. (This is particularly the case for axial-flow compressor engines; with a centrifugal compressor type the diameter was large and there was ample space to fit in the cans.) Further, there are flow area variation problems at entry and exit of this type of combustion system. Both compressor outlet and turbine inlet are essentially annular in shape and it is troublesome to convert to a number of distinct pipes and then revert to the annular shape.

The ideal configuration of combustion chamber is a fully annular system as shown in Figure 6.3; this makes the fullest use of the cross-sectional flow area available and blends in well with both compressor and turbine. A number of early turbojets of the axial compressor type such as the German Jumo 004 and the Metropolitan-Vickers F.2 had annular combustion chambers but they lay outside the mainstream of British combustion work which was concentrated on the individual can type. After a lapse of years the annular is again becoming popular and is used in many new engines such as the Rolls-Royce Spey and R.B. 211.

From Figure 6.3 it can be seen that the annular chamber shown there is again of two-part construction with an outer casing and a single large "flame tube". As before a primary zone of relatively sheltered air is situated at the forward end of the flame tube and the high temperature gases leaving this primary zone are diluted by cool air to a temperature acceptable for the turbine. Because of the flow pattern around the outside of the flame tube, the outside casing remains comparatively cool.

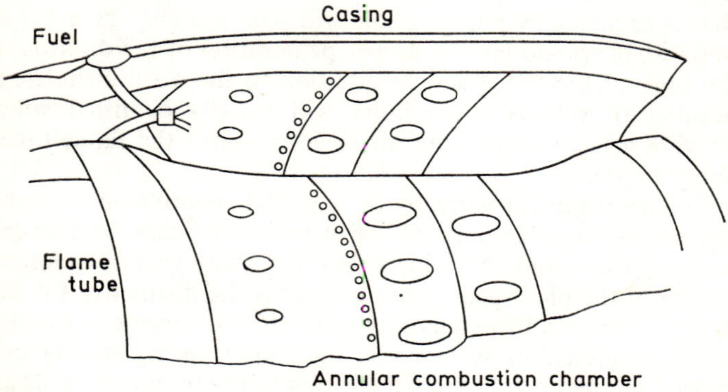

FIGURE 6.3 *The annular combustion chamber*

One other compromise arrangement of combustion system has been used quite extensively and that is the one known as the tubo-annular or cannular type. As the name suggests, this type (Figure

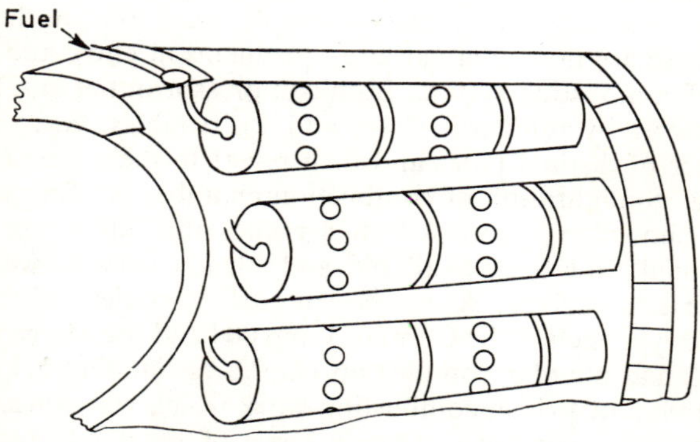

FIGURE 6.4 *The cannular combustion chamber*

6.4) consists of a number of individual cylindrical flame tubes enclosed within a single completely annular outer casing. Compared with the individual chamber type there is appreciably better utilization of cross-sectional area, while the fact that a number of small flame tubes are used eases the development and production problems as compared with the fully annular chamber. However, there will be a disturbing lack of symmetry in the flow areas around the flame tubes and this may result in flows which are difficult to analyse or which include stagnant regions.

6.4 THE PERFORMANCE AND DESIGN OF COMBUSTION CHAMBERS

Before considering in greater detail how a combustion chamber is designed, it is necessary to look more closely at the performance requirements. Although the ideal combustion chamber would have 100% combustion efficiency and negligible pressure loss, it is clear that no real system can do more than approach these standards. At the preliminary design stage of an engine it will be necessary to set certain performance targets for the combustion system (and, of course, the other components), these targets representing compromises between performance and bulk. The object of the development programme for the engine is to see that these targets are achieved and that they are then improved upon.

The main performance aspects of the combustion system are

(i) Efficiency of combustion
(ii) System pressure loss
(iii) Outlet temperature distribution
(iv) Stability and light-up limits.

Each of these will now be considered in turn.

(i) Combustion Efficiency

The combustion efficiency is defined as the ratio of the enthalpy rise of the gas stream to the enthalpy of the reaction of fuel and air. Any drop in combustion efficiency will give a pro-rata rise in the engine's specific fuel consumption, so clearly a high efficiency is essential.

To achieve high combustion efficiency it is necessary to ensure that the fuel and air have both adequate *time* and adequate *space* to mix and react. It is clear that the aero-engine trends of reducing bulk and increasing average velocities will make the problem of burning efficiently more and more difficult. However, it is found that provided the intensity of combustion is kept within certain limiting values (which are discussed more fully on p. 197) and provided the fuel is atomized adequately, it is possible to obtain combustion efficiencies of 98–100% at design conditions with conventionally designed combustion chambers.

As soon as conditions are reduced from the design values there is a danger that the combustion efficiency will start to fall. In part this will be due to changes in the fuel spray pattern and possibly poorer atomization at reduced fuel flows. The changes in air conditions at combustion chamber entry will be more important however. With lower temperature and pressure of the air, mixing will be less intense and the reaction times for the combustion process will increase.

When a chamber is operated at reduced pressure it is found that there is a tendency for the flame to grow longer so that it may extend downstream of the normal burning zone. If this happens, there is a risk that secondary air may "quench" the combustion before the reaction has been completed. This is particularly liable to happen at high altitude conditions where the secondary air temperature is much lower so quenching can occur more easily.

An important side effect of poor combustion efficiency is that free carbon may be formed which can be deposited on the flame tubes or the turbine nozzles. Carbon deposits cause trouble because they give rise to hot spots in the engine and because they upset the correct air flow pattern. In bad cases it has sometimes been necessary to change the combustion chambers for an engine half-way through the normal overhaul life. Sooting up of flame tubes can only be overcome by a slow development process wherein air and fuel flows are adjusted until there is no quenching and no risk of unburnt fuel coming near cools surfaces.

(ii) System Pressure Loss

Consideration of the pressure loss in a combustion system shows one of the clearest instances of the conflict between size and performance. Any attempt to reduce the size of a chamber will give an increase in velocities and so a rise in pressure losses. When a gas turbine is being designed for industrial use it is feasible to fit a large combustion chamber and so to achieve a low loss. For an aero-engine on the other hand some compromise must be reached. The duty of an engine will influence the decision and a pure lift engine, for example, might accept a relatively high loss in the interests of low bulk and weight.

Pressure losses in a combustion system may be defined in a number of ways. From the cycle calculation point of view the most convenient statement is of the "percentage loss in combustion inlet stagnation pressure" as this tells at once the extent by which the overall cycle pressure ratio is penalized by combustion losses. However, it is more usual in aerodynamic work to express pressure losses in terms of the dynamic head at some station—generally the inlet to a component. In line with this convention, a "pressure loss factor" λ is often used in measuring the combustion system losses. This factor is defined by the equation

$$\lambda = \frac{\Delta p_0}{\frac{1}{2}\rho_1 V_m^2} \tag{6.1}$$

where Δp_0 is the overall loss in stagnation pressure, ρ_1 is the air density at *inlet* to the combustor and V_m is the air velocity based on *inlet* density and *maximum* cross-sectional area of the chamber. When the basic pressure loss factor is found from a cold flow test of a combustion system the air density will be nearly constant through the chamber but, of course, when burning is taking place, there will be appreciable density changes.

Of the various factors which contribute to the overall pressure loss during combustion, the most important is that due to the baffling—the obstructions which are placed in the gas stream in order to stabilize the flame. Any form of baffling must set up a wake and this automatically means a loss in stagnation pressure. For a combustion chamber of the type described as being of conventional pattern, it will be seen that *all* the air passing through the chamber must go through one or other of the holes in the flame tube. The flow through each hole may be considered as a straightforward orifice flow with a pressure loss which will be a function of mass flow, density, hole area and discharge coefficient. At the design point, once a target pressure distribution through the flame tube has been chosen, the hole sizes to pass the flow can then be calculated.

Stabilization of the flame and dilution of the products of the primary zone combustion both involve mixing processes and gas flows which are non-axial in direction. In each case the turbulence set up will result in a loss of stagnation pressure between inlet and exit stations. As before there must be a compromise involving the overall bulk of the system because if the mixing is to be carried out in a very confined space it must be more intense. The extent of this mixing loss can again be estimated from a cold flow test and it will be included in the overall friction loss for the chamber.

When dealing with the case of Rayleigh flow in Chapter 1 (p. 42) it was shown that heat addition to a constant area, frictionless gas stream led to a drop in stagnation pressure. This loss is often referred to as the "fundamental loss due to heat addition" and it will be a function of inlet Mach number to the chamber and the temperature ratio between inlet and outlet.

Generally the fundamental loss due to heat addition in an aero-engine will be low compared with the losses due to friction and mixing. Considering the particular case of a pure lift engine however, it would be expected that a lift engine would have a low pressure ratio compressor and a high turbine inlet temperature, as well as having the highest possible velocities in order to keep down bulk. The low pressure ratio will imply a low compressor delivery temperature which will mean a higher Mach number for a given velocity. Also the high turbine entry temperature will combine with the low compressor delivery temperature to give a high temperature *ratio*. All these effects combine to suggest that the fundamental loss for a lift engine may be significant.

If these various effects are combined, the pressure loss characteristics may be presented in the form shown in Figure 6.5. In this graph pressure loss factor is plotted against stagnation temperature

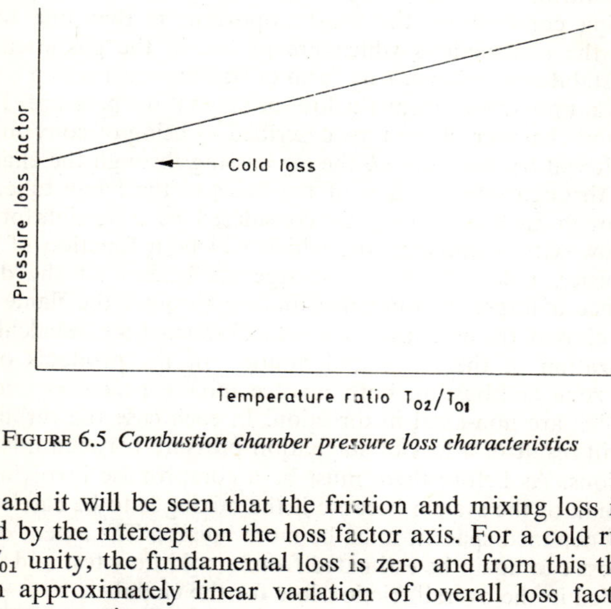

FIGURE 6.5 *Combustion chamber pressure loss characteristics*

ratio and it will be seen that the friction and mixing loss is represented by the intercept on the loss factor axis. For a cold run, with T_{02}/T_{01} unity, the fundamental loss is zero and from this there will be an approximately linear variation of overall loss factor with temperature ratio.

(iii) Outlet Temperature Distribution

The two combustion chamber performance aspects considered so far have been important because of their effects on the overall engine *performance*. On the other hand the outlet temperature distribution is a combustion parameter whose effect is mainly on engine *life*. A poor temperature distribution with marked hot spots can have serious effects on the mechanical condition of the turbine stator and rotor blades. It is the hot spots which are important—whereas the overall engine performance is decided by the mean gas temperature, the blades are mainly affected by the highest temperatures to which they are subjected.

In general the aim will be to achieve a uniform temperature at combustion chamber outlet although there would be some advantages in trying to bias the temperatures so that they were higher at the turbine blade tip sections where stresses are lower. A measure of the quality of the distribution is made by the Temperature Distribution Factor (TDF) which is the ratio of maximum-minus-mean temperatures to mean temperature. The TDF will normally be found by carrying out a detailed temperature survey of a combustion chamber outlet on a rig test. If the distribution is bad, the combustion engineers will tailor it by cutting and changing the hole pattern in the flame tube.

Unfortunately it is only possible to reproduce certain standard conditions with rig testing and measurements of the temperature distributions on engine test is difficult. In addition, variations in flight conditions may have appreciable effects on the gas flow pattern in the combustion chamber. Much of the evidence on TDF in service must therefore come from examination of the engine after long periods of running and the development programme of the combustion chamber must continue until all hot spots have been eliminated. The ultimate criterion is whether the turbine has an acceptable life.

(iv) Stability and Light-up Limits

For any given combustion chamber it will be possible to plot a stability loop which will be of the form shown in Figure 6.6. The region of stable burning is bounded by rich and weak extinction limits and as the mass flow rate through the chamber is increased at given inlet conditions, the two limits will approach each other finally meeting at a limiting mass flow. Even before this limiting condition is reached, burning is liable to be very rough, and at mass flows above the limiting value sustained combustion is impossible. A reduction in chamber pressure will result in a narrowing of the stability limits and a decrease in the limiting mass flow. In every case though the limiting condition will occur at a fuel/air ratio near to the stoichiometric in the primary zone.

The extinction limits are most liable to be encountered in practice during transient operation of an engine. When, for example, the pilot suddenly closes the throttle, the fuel system will react much more rapidly than will the engine. With the sudden fall in fuel flow at an unchanged air flow there will be a danger of reaching weak extinction. During an acceleration on the other hand, the fuel/air ratio will already be richer than for normal steady running. If the compressor should stall or surge with a sudden fall in air flow

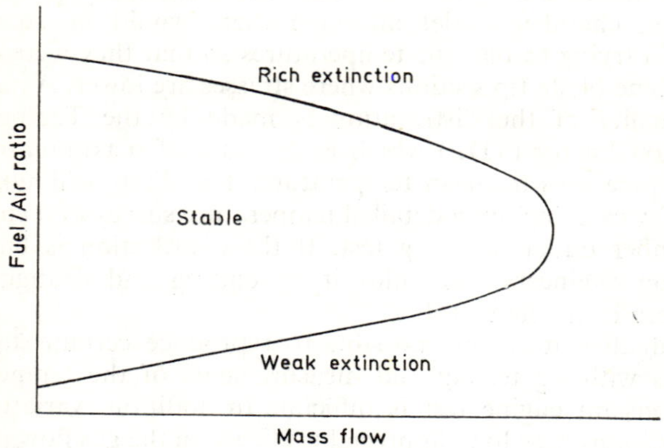

FIGURE 6.6 *The stability loop*

extinction is likely. These cases are normally guarded against by building special safety devices into the fuel system; an acceleration control will almost always be fitted although devices to limit the minimum fuel flow are less common.

Brief mention should be made of the fact that stability limits are impaired if the engine air is contaminated by water vapour, such as might occur when an aircraft is flying through heavy rain. This problem first attracted attention when various engines started to malfunction when flying through heavy storms in the Tropics. In the course of the investigations, combustion testing revealed a deterioration in the stability loop when a combustion chamber is fed with a water vapour/air mixture. Under these same flight-conditions, there was a chance also of compressor instability causing the combustion system to move towards the rich extinction limit. The problem of contaminated air has now been placed on a sounder theoretical basis—see, for example, Clark and Odgers (ref. 4).

Associated with the problem of stability limits is that of relighting an engine in flight after it has been shut down or has gone out due to some malfunction. After a shut-down, the engine will windmill at some rotational speed which will depend on the flight conditions and there will be a high mass flow of cold air through the combustion system. It should be appreciated that, as there is no heating of the air, its density at the turbine stators will be high—much higher than usual—and so the swallowing capacity of the turbine in terms of air mass flow will be large. Consequently the whole level of velocities through the combustion chamber will be higher than in normal running. This, combined with low pressure and temperature at combustion entry, make the flight relight problem a difficult one. Although ideally it should be possible to relight an engine at any flight condition this is not often feasible in practice. A limiting altitude will generally be stated above which relighting cannot be carried out with certainty.

Having looked at the performance characteristics in some detail, we will now consider how an actual design is carried out. Let us think about the design at the stage where the engine performance at design conditions has been specified—such cycle figures as the compression ratio and compressor efficiency, the turbine entry temperature and the mass flow—but when the combustion designer has a blank sheet of paper in front of him. We will try to see what factors influence the actual process of choosing the size and the shape of the chamber and, as usual, we will be interested in the conflicts and compromises which arise.

Certain decisions on the layout of the combustion system will be affected by "political" as well as by technical considerations. For example, a company may be anxious to gain experience with a fully annular chamber and might bias the choice in that direction. Or a firm may have particular interests in one particular form of fuel injection so will choose this in preference to others if there is little to be lost by the choice. If a number of alternatives are open, preliminary designs may be made for the different cases with a view to making more logical decisions. For our purposes it is sufficient to appreciate that these non-technical considerations exist and that they must be given due weight.

As a first step in choosing the dimensions of the chamber, the volume required for combustion will be calculated. This is decided by the "combustion intensity", a parameter connected with both efficiency and stability limits. Combustion intensity is the rate of heat release per unit volume per atmospheric pressure in the combustion space, and is normally quoted in British gas turbine work in the units of Chu/ft^3-h-atmosphere. The SI units will be MW/m^3-atmosphere. Because of the stability limit effects, if an engine is to operate over a wide range of altitudes and flight conditions a relatively conservative combustion intensity of the order of $2{\cdot}5 \times 10^6$ Chu/ft^3-h-atmosphere would be chosen—equal to 47 MW/m^3-atmosphere.

On the other hand an engine with a more limited range of flight conditions (such as a lift engine or a take-off booster) would be designed with an intensity of perhaps four times this value. As flight conditions change so the intensity will vary. Calculations for a typical turbojet show that the intensity at a representative altitude cruise would be three or four times the value at sea level take-off conditions.

For the next step, that of calculating the combustion chamber length, the important criteria are that the flame should be contained within the flame tube at all conditions and that the residence time for reactants and products should be sufficient to allow complete reaction always. Provided these conditions are satisfied the aim must then be to design the shortest possible combustion chamber as this will help to keep down the overall length of engine. In carrying out the actual calculations, a start can be made by choosing an average velocity in the chamber which experience has shown to give acceptable combustion. Conventional engines work with average velocities of the order of 50 m/s while lift engines can go up to 80 m/s or more. By dividing this velocity into the gas volume flow calculated at mean combustion conditions an area for flow can be found. This in turn can be divided into the total volume as calculated previously to give a length of combustion space.

Finding the length required for combustion is only part of the story as the combustion engineer is responsible for the flow from the compressor delivery station onwards. The velocities at outlet from the compressor will generally be 150 m/s or more, which is far too high for combustion inlet, and a diffusing section must be provided to reduce the velocity. In addition some compressors have a poor outlet velocity distribution under certain conditions and this can play havoc with the combustion. It is the responsibility of the combustion engineer to accommodate poor distributions and correct them as necessary.

The saying "No man is an island, entire of itself" is never more true than in the case of design of an aero-engine. At this early stage the volume, cross-sectional area and length wanted for combustion will have been calculated by the combustion engineer and passed to the project designer. He in turn will be collecting information from all the other specialists and trying to reconcile their needs. A mechanical designer for example will be working on the coupling shaft between compressor and turbine and his aims will be to have a short hollow shaft of the largest possible diameter in order to avoid vibration troubles. These needs may conflict with those for combustion so that the project designer has the interesting, if slightly unenviable, task of trying to give each person as nearly as possible what he wants.

Let us assume that the initial bargaining has been completed and that everyone is content with how things have been agreed. Let us assume too that a decision has been made to use a cannular layout of chamber with a cross-section as in Figure 6.4. To illustrate the type of question that will now be asked consider what arguments may be put forward for deciding what number of flame tubes to use.

A first consideration must be to design for low weight. Suppose that there are N tubes of diameter d and length l and that they may be treated as being plain cylinders. The volume V of the tubes will be

$$V = \tfrac{1}{4}\pi d^2 N l$$

If the weight of the end-caps is neglected the total weight of all tubes will be given by

$$\text{Weight} = K_1 \pi N d l$$

where K_1 is a constant depending on the wall thickness and the specific weight of the flame tube material.

Hence the weight per unit volume is

$$\frac{W}{V} = \frac{4K_1}{d} \tag{6.2}$$

This equation shows that for a given length and volume the lowest overall weight is obtained by using a *small* number of large diameter flame tubes.

Consider now a second important requirement, that the outside diameter of the combustion chamber should be low. For a cannular layout of chamber there will be a certain inside diameter D_i for the annular space which will be determined by the need to accommodate a shaft and provide a minimum stiffness in the carcase of the engine. In a limiting case when there is no clearance between flame tubes and air casing the outside diameter D_o will be

$$D_o = D_i + 2d$$

The total flame tube cross-sectional area A will be

$$A = N\tfrac{1}{4}\pi d^2$$

and this has of course been determined previously. Therefore

$$d = \sqrt{\frac{4A}{\pi N}} = \frac{K_2}{\sqrt{N}}$$

where K_2 is a constant based on the required flow area. Combining these equations shows the overall diameter to be

$$D_o = D_i + \frac{K_2}{\sqrt{N}} \tag{6.3}$$

From this equation we see that to get the least outside diameter we should use the *largest* possible number of flame tubes.

These two simple analyses, resulting in equations 6.2 and 6.3, illustrate the clash that results from specifying different initial criteria. In actually selecting the number of flame-tubes to be used, a number of other factors must be taken into account. For example, with a conventional pattern chamber, with a fuel sprayer at the nose of each flame tube an optimization into the number of sprayers will also be involved. Again, as pointed out earlier, there are advantages from the development and production points of view of using a number of smaller flame-tubes. The important point to appreciate is that the designer must take into account as many as possible of these considerations before finalizing the design.

This work has attempted to show how the design of a combustion chamber is started and how the main dimensions are chosen. From this stage onwards the actual detailed calculation of hole sizes and positions will be based very largely on the designer's own experience. It is easy to say that the flow through each hole in a flame tube may be treated as flow through a simple orifice but the real expertise, which comes from a long background of research and development work, lies in knowing just where to place the holes if they are to be most effective. Each engine will pose its own problems and there is little in the published literature which will help in laying down general rules.

6.5 THE FUEL INJECTION PROCESS

As both space and time for combustion are limited in an aero-engine, it is of great importance to ensure that the fuel is supplied in a suitable form for burning. The requirement is that evaporation should be rapid and complete. In fact, with the vapourizing system described later the fuel actually comes into the combustion space in the vapour phase. It is more usual though to spray the fuel at high pressure into the combustion chamber. Initially the high velocity between fuel and air causes high drag forces which break up the liquid into fine droplets. Evaporation from the surface of these droplets takes place partly by convective mass transfer and partly by heat pickup by convection and radiation. The high surface area resulting from fine atomization will help these processes.

With a combustion chamber of the conventional pattern (Figure 6.2) fuel is projected across the main vortex flow of the air in the primary zone. To achieve the high velocities the fuel pressure at the sprayer must be high and it is desirable that the fuel should emerge in the form of a conical sheet of liquid. The cross-flow between air and fuel will cause a rapid break-up of the sheet and the general high degree of turbulence will ensure rapid and thorough mixing.

A type of sprayer which gives the conical sheet form of spray is the swirl atomizer shown in Figure 6.7. Fuel enters tangentially through a number of holes at the rear of the cylindrical portion—

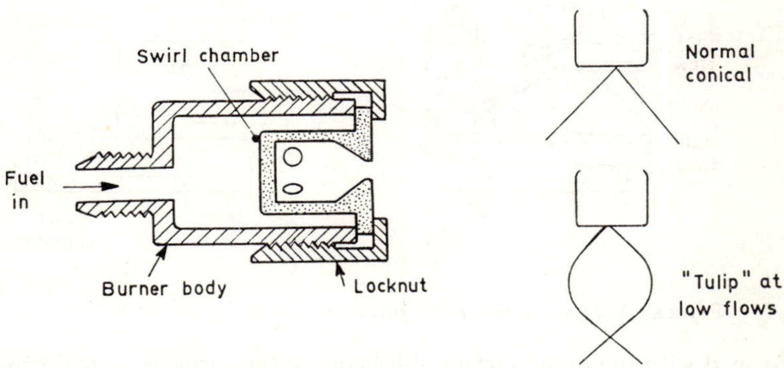

FIGURE 6.7 *Swirl atomizers*

the swirl chamber—and a very vigorous vortex motion is set up. The tangential component of this flow persists as the fuel leaves the outlet orifice so that the resultant spray is in the form of a conical sheet of fuel at design conditions.

The object of the atomizer is to break up the fuel into small droplets and a measure of how effectively it does this is given by the Sauter Mean Diameter (SMD) of the spray. Sauter mean diameter is defined as the drop size which has the same surface to volume ratio as has the complete spray. Normally the SMD is quoted in microns (1 micron = 10^{-6} metre) and the SMD of a spray will be a function of the fuel pressure and the surface tension of the liquid. In addition the SMD will be reduced if fuel is injected into a turbulent air space rather than into still air.

Although the SMD is the most significant *average* figure describing a spray, it is important too that there should be a good *range* of droplet sizes as this affects the distribution of fuel in the primary vortex. Small droplets have a high drag to mass ratio and so are soon deflected from their paths while large ones have sufficient impetus to penetrate right across the vortex. Figures quoted by J. S. Clarke (ref. 3) show that droplets of 300 microns in diameter were able to pass almost undeflected across a typical vortex. If this should happen, there is a danger of losing combustion efficiency and of forming carbon deposits on the flame tube.

Reduction in sprayer pressure drop will lead to a progressive increase in SMD until a point is reached, generally with about 60 kN/m^2 pressure drop, when the fuel will no longer come out as a cone. Instead a "tulip" shape as sketched in Figure 6.7 will be

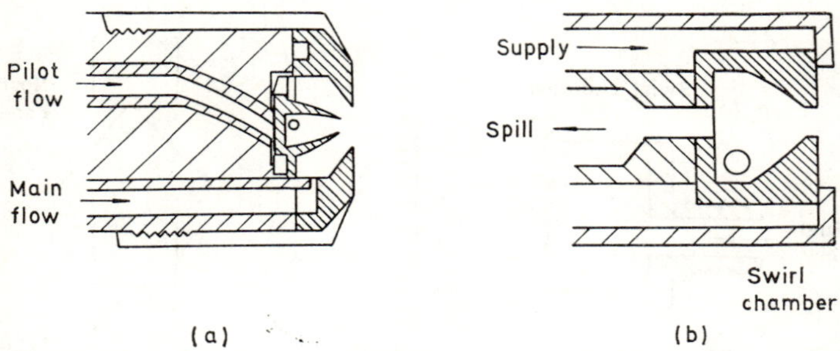

FIGURE 6.8 *Types of wide range fuel atomizers*

formed with the outer surface still being in the form of large drops. Combustion efficiency will start to fall-off for spray Sauter Mean Diameters of 150 microns and above and when the tulip type of spray is formed it will be difficult to maintain stable burning.

The fall-off in quality of spray from the swirl atomizer at low flows is an embarrassment, as these conditions are liable to be encountered during normal operation. The flow characteristics of a simple swirler are the same as those of a plain orifice, with pressure drop varying as flow squared; normally the ratio of maximum to minimum fuel flow for an aero-engine is about 30:1 so this would imply a pressure drop ratio of 900:1 which is beyond the capacity of existing pumps. Two main ways of getting over this problem have been used in Britain and they are illustrated in Figure 6.8(a) and (b). That shown in Figure 6.8(a) is the "Duplex" type injector made by Lucas Ltd. which is a sprayer which uses two separate swirl chambers. At starting and at low flows fuel is only admitted to the small primary swirler but once the primary line pressure reaches a value of about 2 000 kN/m^2 the main swirl chamber is brought progressively into use. The system shown in Figure 6.8(b) is a spill type injector as developed by Dowty Fuel Systems Ltd. With this type, the swirl chamber pressure is kept at a high value at all operating conditions so ensuring a good spray, but excess fuel to that required by the engine is bled back to pump suction from the rear of the swirl chamber. Both Duplex and Spill systems with their double piping are more complicated to install than a simple swirl-type injector with only one swirl-chamber and if an engine does not operate over too wide a range of fuel flows the single system will be preferred.

Before leaving the topic of fuel sprayers mention should be made of systems using upstream injection. If fuel is sprayed *against* the direction of air flow rather than with it, higher velocities between fuel droplets and air will result, so giving better break-up of the fuel sheet and more rapid evaporation. In addition, at the point where the droplets change direction prior to blowing downstream with the gas stream, their absolute velocity will be zero and this should help flame stabilization. For these reasons upstream injection has been tried at various times during the development of the aero-gas turbine. Although the advantages suggested do appear, there are a number of drawbacks. An obvious one is that the sprayer and its fuel supply pipe will now be downstream of the flame front, so care will be needed to ensure that heating of the fuel in the pipe and sprayer does not cause it to decompose and block the passages. In addition, the whole arrangement is less neat than that using downstream spraying.

As suggested at the beginning of this section, an alternative means of evaporating the fuel would be to evaporate it by supplying heat. Work on this type of vapourizing system was started by Armstrong-Siddeley Motors Ltd. and has been continued by its successors. Among the well-known engines using vapourizing burners are the Sapphire (built in the United States as the J.65), the Viper and the Mamba.

A typical vapourizing system is illustrated in Figure 6.9. The essential feature is the use of a "walking-stick" type fuel injector. Fuel and air enter at the forward end of this tube which, being in the combustion zone, is very hot, and in passing down the tube the fuel evaporates completely. The mixture of fuel vapour and air leaves the walking-stick "handle" in an upstream direction and is in such a condition that it burns immediately on coming out, so helping to maintain the injector tube at high temperatures.

The obvious question is, which gives the better combustion system, the spray atomizer or the vapourizing system? Unfortunately it is very difficult to get a clear-cut answer to this as so many other factors are liable to complicate the picture. It would be expected that the vapourizing system would give at least as good efficiency and comparable, if not better stability limits, while having the advantage of managing with lower fuel line pressures. On the other hand, special provision is needed for starting a vapourizing system before the tubes get heated up (a small spray atomizer is used) and the spray system looks neater and lighter. At present Rolls-Royce,

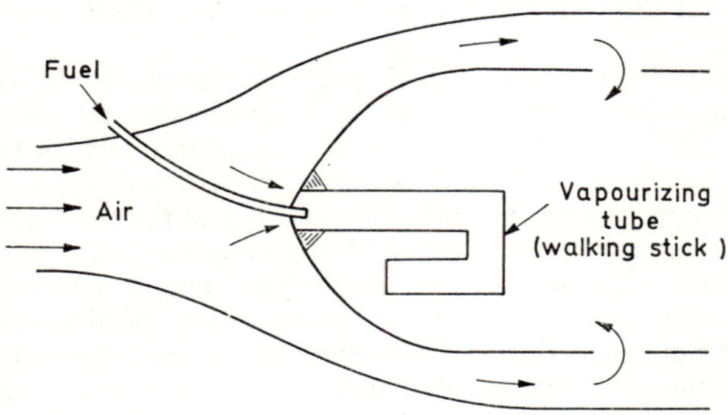

FIGURE 6.9 *The vapourizing fuel injector*
(Courtesy Rolls-Royce Ltd., Bristol Engine Division)

Bristol Engines who have wide experience of both systems are still continuing with design and development of the two distinct types.

6.6 THE IGNITION PROCESS

One of the real advantages of the gas turbine compared with the piston engine is that the ignition problem is greatly reduced. The spark plug of the high power aero-engine was always a comparatively weak link in the engine as it was required to perform an arduous continuous duty under extremely difficult conditions. For a gas turbine on the other hand, with its continuous combustion, ignition is only needed when the engine is first started, while the igniter itself is exposed to lower temperatures and pressures and is less liable to suffer fouling-up from additives in the fuel.

The igniters used for the early jet engines were derived directly from existing sparking plugs but were fitted with enlarged electrodes. They were fed with a continuous stream of high tension pulses from a coil and trembler system which was switched on for the duration of the starting cycle. On some engines where ignition was rather more difficult than in others a torch igniter was used. With this arrangement the sparks from the plug first fired fuel from a small starting atomizer and this pilot flame would then light up the fuel from the main spray nozzles.

As knowledge of the ignition process was accumulated, it was realized that more reliable ignition could be obtained if the ignition energy was supplied in the form of distinct pulses of energy with appreciable time intervals between the pulses. The studies led to the development of the High Energy Igniter, a device in which electrical energy is stored in a condenser and discharged in pulses across the surface of a special plug. Each pulse has a duration of the order of 150 μsec and the electrical energy at the plug is a few joules in magnitude and pulsing takes place at the rate of about one or two discharges per second. Losses in the system are surprisingly high and, even with careful design, more than half the energy stored in the condenser is dissipated before reaching the plug. Of the energy which *does* arrive at the plug, only the fraction which can be turned into thermal energy contributes to the ignition. An appreciable proportion is lost in setting up shock waves in the gas and as acoustic energy.

Supplying thermal energy by an electrical discharge is only the start of the ignition process because, once a reaction has been initiated between fuel and air, it must be propagated throughout the combustion space. There will at first be a delicate energy balance in the initial burning region as it will be gaining energy by spreading into the unburnt mixture but losing it by radiation and by encountering cold gas and cold metal. Some years ago a remarkable technical film was made at the Lucas Combustion Research Laboratories in connection with ignition studies. In the test set-up, a Perspex combustion chamber model was used, the fuel being cut off immediately ignition was obtained so that the Perspex had not time to melt. The film was shot at very high speed and when run through at normal speed it was possible to study the light-up in detail. A fascinating picture could be seen of the first flame region waxing, then waning until it almost went out before finally growing once more and spreading through the flame tube.

If an engine is fitted with a combustion system made up of individual tubular chambers, or if it has a cannular type, it will be usual to fit igniters in two or three of the flame tubes only. Small interconnecting tubes then join up the other flame tubes. As soon as one flame tube lights up the pressure in that section will build up (due to the lower gas density at the turbine) so the flame will pass through the interconnectors and ignite fuel in the other cans. More reliable ignition can be achieved in this way as a flame can travel completely round the combustion chamber and any individual flame tubes may be lighted from either side. With a fully annular system there *is* only one space so ignition in one part will spread through the chamber.

6.7 REHEAT SYSTEMS

In Chapter 3 it was shown that the specific thrust of a turbojet or turbofan could be boosted by burning fuel downstream of the turbine—a process known as reheat or afterburning. As there are then no turbine materials problems to worry about, it is theoretically feasible to raise the final nozzle inlet total temperature to a very high value, of the order of 2 000 K. Parts of the engine such as the tail-pipe and final nozzle which are exposed to these temperatures are not as highly stressed as rotating components and they can be made heavier to withstand the heat flows. Design of a reheat system poses a number of interesting questions both on the combustion and on the mechanical sides. As a general rule the operating life required of an afterburner is very much less than that of the remainder of the engine as thrust boost is only used occasionally for short periods at any one time.

A number of design requirements for a combustion system, such as the need to provide a sheltered zone with near stoichiometric burning, were stated in Section 6.2, and these still hold for the burning of fuel in the tail-pipe. In some ways the problem is eased because of the high temperature level in the system but against this must be set the fact that the pressure level is very much lower. The actual value will depend on the particular engine but it will almost always be appreciably less than half the main chamber pressure. This low pressure level is a dominant factor in determining the performance of the reheat system; as stability depends on pressure level it may be difficult to operate successfully over the required range of conditions.

When starting the design of a reheat system, the first problems are those of space. The velocities generally used in the tailpipes of engines are unacceptably high for entry to a combustion system, especially one with a temperature ratio of 2:1 or more. If these velocities and Mach numbers were accepted the pressure loss due to heat addition would be so high that excessive performance losses would result. A diffusing section for the flow must be used which reduces the Mach number to an acceptable level. Design of the diffusing section must strike a balance between excessive bulk and weight (especially as space in the airframe may be limited) and too much pressure loss.

Baffles will be required to hold the flame in the reheat section and the pressure loss resulting from the baffling will reduce the engine thrust for any given final nozzle entry temperature. As the baffles will be a permanent fixture in the engine they will give some thrust loss even when afterburning is not in use. This is known as the "cold loss" and it should not exceed 3–4% of the normal engine thrust. Once more the diameter of the reheat pipe will come in as a design variable since a large diameter pipe with lower velocities past the baffles will give a lower cold loss.

In order to improve the stability ranges without using excessive baffling it is common practice to make provision for a pilot flame. With this arrangement part of the reheat fuel supply goes to a specially protected region where a high degree of baffling is used and the flame from this part acts as a powerful ignition source for fuel being sprayed in at the main baffles. A suitable position for this pilot flame might be at the end of the tail-cone downstream of the turbine.

Figure 6.10 shows the layout of a typical reheat system incorporating the main features discussed above. The diffusing section is shown

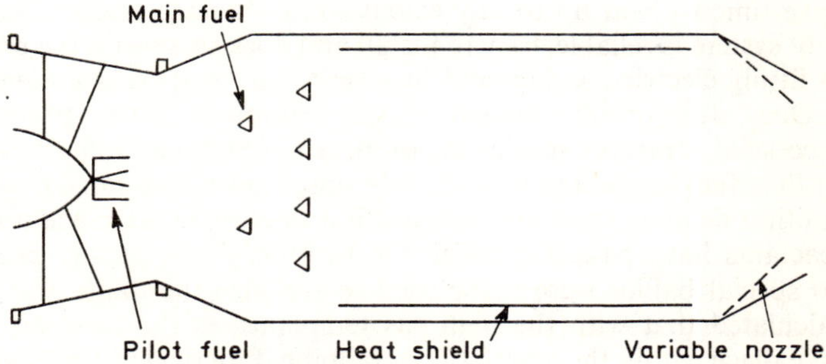

FIGURE 6.10 *Layout of a reheat system*

with the main fuel jets and baffles and the special baffle fitted at the tail-cone to hold the pilot flame. A variable area final nozzle of the type discussed in the last chapter must also be fitted.

When it comes to the detailed design of a reheat system, a great deal will depend on the past experience of a firm in the field. A paper by Edwards (ref. 5) described the work carried out by the de Havilland engine company in developing reheat for the Ghost engine. Among other things, the paper gave details of the problems that were overcome in devising a satisfactory fuel distribution system and showed in particular how fuel jets penetrated into a gas stream. It included performance curves showing the pressure loss of various gutter type baffles (of the type shown in Figure 6.10) as a function of blockage area and gutter included angle. Altogether it gave a most interesting account of the problems, both performance and mechanical, which were encountered in developing a satisfactory reheat system.

Ignition of a reheat system may be effected by electrical means on the same general principle as that described above for the main engine combustion chamber. Normally the igniter will only have to light up the pilot flame but it will need special protection as it would be exposed to turbine outlet temperatures throughout its life. An alternative means of ignition which has been used successfully on a number of engines is known as the "hot shot" ignition system. In this, when reheat is required, the reheat fuel is turned on and immediately after a small quantity of fuel is injected directly into the main engine combustion chamber. The temperature there is sufficiently high to give spontaneous ignition of the fuel shot and a long flame will pass right through the turbine to ignite the reheat fuel. Despite the very high temperatures which act momentarily on the turbine, the duration of the pulse is so short that the metal does not have time to heat up to any extent so no damage results. The hot-shot system is reliable, easy to install and does away with the problem of fitting electrical equipment in a very high temperature zone.

One afterburning system which, although never particularly successful, deserves special mention, is the Schmidt Reheat System. In this, fuel is injected immediately upstream of the turbine and the ignition delay is such that combustion does not take place until the reactants have passed through the turbine. As originally proposed no special baffles were to be used to stabilize the flame and it was calculated that with the high gas temperatures the reaction would take place within the length of the tailpipe. Practical testing showed a tendency for the flame to stabilize on the tail-cone or on the tail-cone support vanes. If this happens damage can result very quickly. In practice too the fuel cannot readily find the oxygen it needs for burning and the combustion efficiency is low.

Little can be said about the mechanical troubles of designing a reheat system as each particular engine unit will bring in its own problems. The problems that arise stem from the very high temperature gases which must be contained in a relatively cramped region of an airframe. Good thermal insulation is needed to prevent adjacent parts of the airframe overheating but then, when the heat loss is reduced to a low value, the jet pipe metal temperatures will approach the gas static temperature. The paper by Edwards referred to previously mentions a heat shield in the form of a cylinder of metal which is placed in the reheat pipe as shown in Figure 6.10.

This heat shield is cooled by the relatively cool air which passes between it and the jet pipe and it effectively blocks radiation between the reheat flame and the jet pipe.

6.8 FUNDAMENTAL STUDIES APPLIED TO THE COMBUSTION PROCESS

It must be stressed that combustion studies which are directly applicable to aero-engine design and development, at least at the present time, are essentially empirical in nature. The bulk of the work is carried out by actually building combustion chambers and testing them either in engines or in rigs which aim to simulate engine conditions as closely as possible. In parallel with this work—which is really applied research carried out by manufacturers with fairly limited objectives—extensive fundamental studies into different aspects of combustion are under way in various research laboratories throughout the world. Despite the great amount of information flowing from these fundamental studies, there is comparatively little at present which can help in the actual design of combustion systems, although each year sees more progress towards a real understanding.

Why should this be? The main reason would seem to be that aero-engine combustion systems are at present limited in the main by *mixing rate* considerations rather than by chemical *reaction rate* effects. Although it is possible to set up elegant chemical experiments to discover the ultimate rates at which reactions take place, this is not particularly useful if the real barrier to the speed at which fuel burns is how quickly fuel and oxygen come into contact with each other. Mixing rate phenomena involve the penetration and break-up of intermingling gas streams and it is difficult to lay down more than the most elementary of rules. The rules which *can* be stated were originally defined as a result of practical testing and the development of workable systems—rather like the process described in Section 6.3. It should be appreciated though that the possibility of a breakthrough at some stage in the future cannot be ruled out. New techniques and new methods of analysis are continually being developed and very promising work on the computer analysis of flow is being carried out by a number of workers in the field. The new big computers have the capacity to handle the myriad complex equations for the calculations of turbulent and mixing flows and their use may allow the exact prediction of turbulent flow phenomena.

A proviso must be made that the above remarks apply mainly to the operation of combustion chambers at low altitude conditions and that, at high altitude, reaction rates might in fact become limiting. The reason for this is that the rates at which reactions take place depend on the number of collisions between fuel and oxygen molecules as well as in the energies of the individual molecules. From simple kinetic theory considerations (as quoted for example in ref. 4) it can be shown that the number of collisions will be dependent on the absolute pressure so that the reaction rate will fall at high altitude where pressures are low. In fact the controlling parameter will vary inversely with the absolute pressure to the power of n where n is about 1·8–2·0, so clearly this will be the dominant factor. Also at high altitude, the level of temperatures will be lower throughout the engine again making the achievement of satisfactory stability limits more difficult.

With these considerations in mind, it is relevant to ask what the results would be if combustion systems *were* to become reaction rate limited at design conditions, rather than mixing rate limited. Clearly there would be little chance of improving significantly on the existing 98–100% combustion efficiency! Neither would there be much chance of reducing pressure losses or improving stability limits from present day values. The big breakthrough would come in combustion chamber bulk and weight. Theoretically it should be possible to increase the combustion intensity to several times its present value—the actual increase would be determined by the acceptable level of stability limits—which, for a given engine, would mean a corresponding reduction in reaction space volume. This in turn would give a major reduction in overall engine size.

An interesting example of one particular fundamental combustion study with valuable practical applications is that described by A. E. Clarke and J. Odgers in ref. 4. Their work was carried out using a spherical reaction chamber into which was fed a pre-mixed stream of propane fuel and air. Figure 6.11 shows the experimental set-up. By using high velocities from the injector tube, very intense mixing is obtained so that the behaviour of the combustion approximates to that of a homogeneous mixture within a reaction zone. The controlling rates are therefore the reaction rates rather than those of the mixing. Very high combustion intensities are possible with this set-up but only at the expense of low combustion efficiency and poor stability limits. For example, although an intensity five times that of a conventional aero gas turbine was achieved the efficiency was less than 85% and the state of the reaction corresponded to that of the limiting point of the stability curve of Figure 6.6. At the peak intensity the chamber would operate at only one point and any change in fuel/air ratio caused extinction.

Dr. Clarke points out that this type of spherical combustion chamber makes no pretence of being a practical combustor but instead must be thought of as a standard of comparison for other systems and as a tool for combustion studies. The paper deals with this role of the combustor as a research instrument and a number of

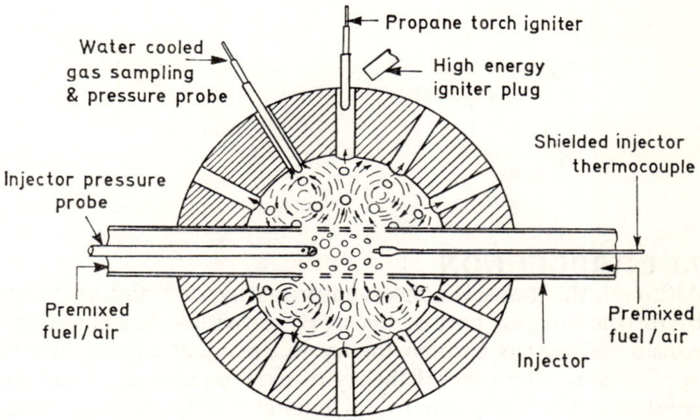

FIGURE 6.11 *The spherical reaction chamber*
(Courtesy J. Lucas Ltd.)

experiments which have been carried out are discussed including the question of simulating altitude conditions by diluting the air in a chamber with inert gases. This can be a cheap way of simulating altitude conditions without the complexity of operating a complete chamber at sub-atmospheric pressure. Test results using steam to dilute the air are shown which demonstrate the validity of this approach.

7 The Installation of Aero-Engines

7.1 INTRODUCTION

Although this book is concerned primarily with aircraft propulsion, by far the greatest part of the work of the last few chapters can be considered as gas turbine theory independent of any aeronautical applications. The general theory of axial compressors, turbines and combustion is applicable to any engine whether for use on the land or in the air, and aero-engines differ only in their emphasis on low bulk and weight. In this chapter work is to be done of relevance to aero-engines only, and the engine will be considered as part of a propulsion unit in a complete flight vehicle.

The important question of the design of intakes for turbine engines will first be dealt with followed by a consideration of some of the problems arising from the power planting of engines and from the need to meet certain aircraft services. This will lead on to a discussion of the interaction of aircraft and engine and to the question of how far the two can and should be integrated.

7.2 DESIGN OF INTAKES FOR TURBINE ENGINES

The function of the intake for a turbine engine is to deliver the required air mass flow at the compressor face with the highest possible stagnation pressure and with the smoothest possible velocity distribution. Any loss in stagnation pressure represents a performance loss for the engine while variations in velocity round an intake are liable to cause compressor surging or blade breakages. These requirements are to be met over a wide range of flight conditions without introducing too much complexity or weight. Quite clearly, it will be necessary to design for one case and accept performance penalties at off-design conditions which must not become unacceptably high. Many modern engine installations will be found to have variable geometry in which duct shapes are varied with flight conditions in order to give good off-design performance.

Consider first the intake problem for a turbojet designed to operate at high subsonic speeds. At the design conditions of flight speed and engine mass flow the velocity at the compressor face will almost certainly be lower than the aircraft velocity so the intake duct must take the form of a diffuser. The area variation will be chosen so that at this design condition the engine air is contained in a stream tube of the same cross-section as the front of the intake as shown in Figure 7.1(a). Figures 7.1(b) and (c) illustrate other flight cases.

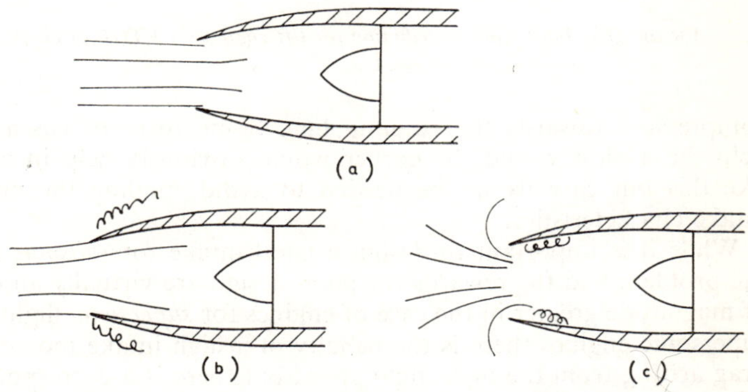

FIGURE 7.1 *The air flow pattern into intakes at subsonic flight conditions*

In Figure 7.1(b) a reduction in engine flow has occurred and some of the air reaching the intake through a parallel-sided stream tube must be spilled around the outside of the engine. When this occurs, the flow within the intake duct will probably not be greatly affected but there may be difficulties with the aircraft aerodynamics.

The static case with the aircraft stationary and the engine demanding the maximum airflow shown in Figure 7.1 (c) is very much more serious. Here there is a very grave danger of severe breakaway occurring as shown at the lip of the intake. Just at the time when the engine is required to give its highest possible thrust the stagnation pressure loss in the intake will be at its worst. To overcome this problem some aircraft have spring-loaded doors in the intake duct walls. When the pressure loss into the intake is high, the differential pressure between duct and atmosphere causes the doors to open automatically and increase the air flow capacity.

A particularly severe maldistribution problem arose with early VTOL aircraft using jet lift such as the Short SC.1. Hewson showed (ref. 1) that when the SC.1 started to move forwards there was a cross-flow across the intakes of the lift engines (as shown in Figure 7.2) and that this had the effect of dropping the surge line of the

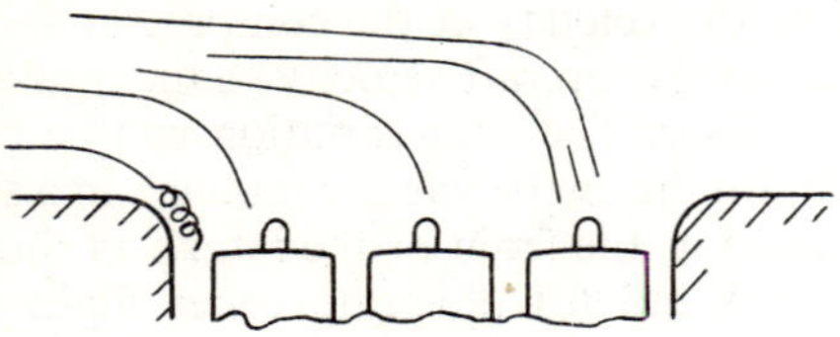

FIGURE 7.2 *Intake maldistribution for lift engines in VTOL aircraft*
(Courtesy Rolls-Royce Ltd.)

compressors towards the running line. Some form of cascade to help the airflow round the corner would obviously help in a case like this but care would be needed to avoid spoiling the normal intake characteristics.

While it is important to design a good intake for *subsonic* flight the problem and the rewards for good design are virtually an order of magnitude greater in the case of engines for *supersonic* flight. For supersonic engines there is the penalty of a high intake momentum drag arising from the high flight speed but there is a corresponding advantage in the high ram pressure ratio available. An ideal isentropic intake would give a ram pressure ratio of

$$p_0/p = [1 + \tfrac{1}{2}(\gamma - 1)M^2]^{\gamma/\gamma-1}$$

Putting in the value of $\gamma = 1{\cdot}4$ shows that at Mach 2 the ideal ram pressure ratio is 7·8 while at Mach 6 it rises to 1 600! This "free" pressure ratio is of course paid for by the momentum drag.

At the beginning of this section it was said that the intake must present the air at the compressor face at the correct velocity. If the free stream conditions are supersonic, then in decelerating the air to subsonic conditions it must pass through *at least one* shock. As discussed in Chapter 1, deceleration through a shock is always accompanied by a loss in stagnation pressure and this loss will vary with the intensity of the shock. An efficient intake will use multiple shocks to keep the intensities low and so keep the loss in stagnation pressure to a minimum.

Supersonic intakes as currently used can be classified under the headings of *external* compression, *internal* compression, and *mixed* external–internal type. These classifications will be adhered to in the general descriptions that follow. In each case the ideal flow pattern is considered and subsequently the real effects which complicate the design of supersonic intakes are discussed.

External Compression Supersonic Intake

External compression intakes are those in which a shock pattern is set up ahead of the outer casing of the intake. As is found so often in engineering, an improved performance can be obtainable at the expense of increased complexity. This in turn means that as flight Mach number goes up, it is worthwhile to make intakes of more

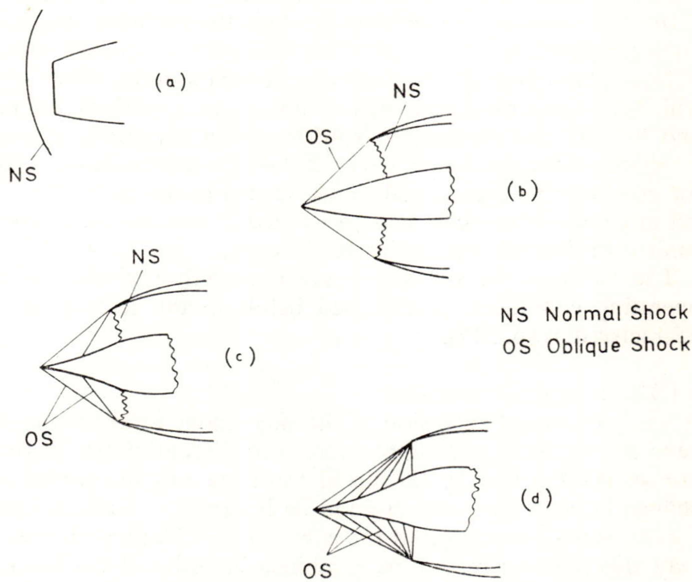

FIGURE 7.3 *Types of external compression intake in supersonic flight*

complicated shape. Four representative intakes with their design shock configurations are shown in Figure 7.3. Referring to this diagram the following points can be made:

(a) The Pitot Intake

The Pitot type intake gives a satisfactory performance up to a flight Mach number of about 1·5. At supersonic conditions a bow wave forms which is effectively a plane normal shock ahead of the intake. Consequently the flow goes in one single step from supersonic to subsonic conditions. The changes in static pressure, stagnation pressure and stream Mach number will be unique functions of the flight Mach number and will be found from Table 1.3. Once inside the intake cowl, the flow will diffuse subsonically to the compressor face.

(b) The Single Cone Intake

As flight Mach number increases, the loss in stagnation pressure for a normal plane shock becomes prohibitively high and it is worthwhile to use a single cone intake as seen, for example, in the BAC Lightning. In this case the cone sets up a conical shock whose intensity will be a function of flight Mach number and the included angle of the cone (Figure 1.19 showed the characteristics of the corresponding oblique plane shock.) The gas stream Mach number after the shock will still be supersonic but will be lower—as explained on p. 51—and will be deflected away from the axial direction. By suitable selection of the intake geometry the shock can be made to fall on the cowl lip and the normal shock needed to decelerate the flow to subsonic conditions can also be located at the lip. As before, subsonic diffusion takes place inside the cowl.

The variation of the shock position with variation in engine operating conditions is discussed below in the section on intake characteristics (p. 217).

(c) The Double Cone Intake

A straightforward extension of the single cone intake is the double cone shown in Figure 7.3(c). Here, two discontinuities of the flow are set up, the first by the tip of the cone and the second by the change in cone angle. With good design (and some luck!) the two conical shocks may meet at the cowl lip and the final normal shock may also settle there. As the combined intensity of the two conical shocks is less than that of a corresponding single one the intake performance is improved.

(d) The Isentropic or Oswatitsch Spike

It is possible to envisage the ultimate extension of the multi-cone principle wherein an infinite number of infinitely weak shock waves—they will actually be Mach waves—are set up by a contoured spike. Such an arrangement is named after its inventor, the German, Oswatitsch, but it is largely of academic interest. Although ideally it would be possible to obtain a compression which was virtually isentropic, design of the spike would be very complicated due to the difficulty of estimating boundary layer effects and in addition the spike would only be of the correct contour at one particular flight conditions. For these reasons the isentropic spike has not been used often so far in aero-engines.

The last three intakes have been described in terms of three dimensions with a cone being used to generate the first part of the shock system. In principle each intake has a two dimensional counterpart in which the cones are replaced by wedges. For these the Mach number relationships across the shocks are of the same general form as for the conical shocks and the characteristics can be presented in the same way with upstream Mach number and wedge angle as the prime variables. This was in fact done in Figure 1.19.

To turn now to the *characteristics* of external compression intakes, we will take as a typical example the single cone type using one conical shock and one plane normal shock. As the intake is acting as part of the compression process, it is convenient to plot its characteristics in the same form as those for a compressor. They will be

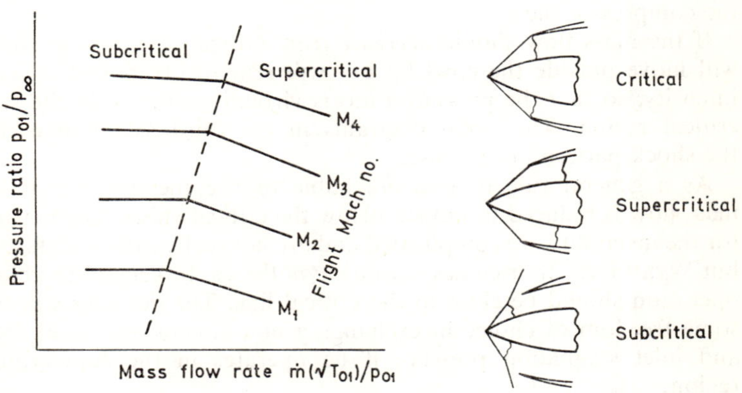

FIGURE 7.4 *Characteristics of the external compression intake*

drawn to the axes of pressure ratio and "non-dimensional" engine intake mass flow. Flight Mach number will be the other main variable so lines of constant flight Mach number will be drawn on the characteristics. Figure 7.4 shows the typical form of intake characteristic and it will be seen that, as would be expected, the pressure ratio goes up with increasing Mach number.

On the characteristic a "critical" line is drawn joining the points of maximum pressure ratio and the regions to left and right are labelled "subcritical" and "supercritical" respectively. To see what is happening, it is instructive to study the variation of the shock pattern with changes in engine operating conditions. It must be realized that the position and intensity of the conical shock, for a given intake cone angle, depends *only* on the flight Mach number. But the position—and so the intensity—of the normal shock will depend on the mass flow being swallowed by the engine. At critical conditions the shock configuration will be that for design conditions with shock-on-lip conditions.

If now the engine air demand increases so that a higher $\dot{m}\sqrt{T_{01}}/p_{01}$ is called for, the oblique shock will be unaffected but the normal shock will move down into the intake as shown in the second sketch. That portion of the intake is a diverging duct so it will cause an *acceleration* of the supersonic flow. Therefore, when the normal shock *does* occur it will be more intense and will result in a larger loss of stagnation pressure. It must be appreciated that the physical mass flow and the stagnation temperatures will be unchanged (they depend on the flight Mach number and the flow area at the cowl lip) so the increase in $\dot{m}\sqrt{T_{01}}/p_{01}$ must arise as a result of a loss in p_0 at the compressor face.

If the mass flow should *decrease* from critical, the normal shock will move outside the cowl lip but will remain of about the same intensity, so that the pressure ratio is virtually constant in the subcritical region. The small diagrams in the third sketch illustrate the shock pattern in this case.

As a general rule intakes are liable to become unstable if the mass flow is reduced too much below the critical value. The reasons for the instability are complicated and are not really fully understood but Wyatt (ref. 2) discusses various hypotheses. In general therefore operation should be close to the critical line. The designer still has some freedom of choice in exchanging non-dimensional mass flow and inlet stagnation pressure if he operates in the supercritical region.

Internal Compression Supersonic Intake

From the first discussion on the flow of gases through nozzles in Chapter 1 it has been said that the flow is ideally adiabatic and *reversible* with, in a nozzle, a static pressure gradient giving an increase in velocity. Because of the reversibility it should be possible to design an intake which is effectively a reversed nozzle. If a supersonic gas stream is fed into a convergent–divergent duct it will decelerate with a consequent rise in static pressure.

Figure 7.5 shows the layout of an internal compression intake which is like a reversed Laval nozzle. In theory it is possible to have

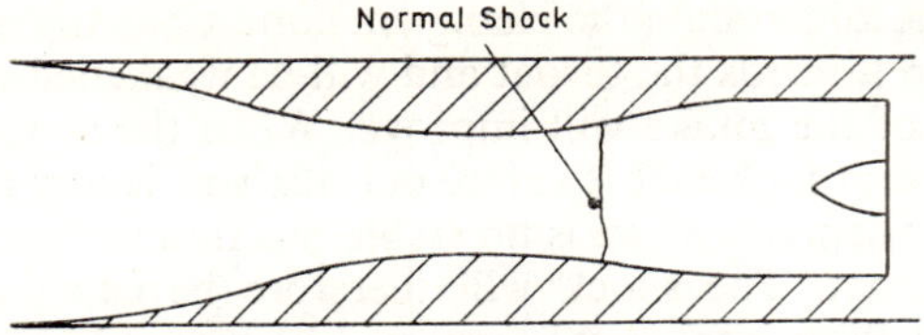

FIGURE 7.5 *The ideal internal compression intake*

a very sharp cowl lip and to have streamlines entering the intake exactly and decelerating to the throat. If the velocity at the throat is still supersonic, then on going into the diverging portion the gas will accelerate. The flow into the compressor will almost always be subsonic so somewhere in the divergent portion there must be a normal shock. The intensity of this shock will decide the efficiency of the intake as the rest of the flow is ideally isentropic (this ignores any shocks which might be set up outside the intake and only considers the internal flow).

The characteristics of an internal compression intake can be plotted in the same form as those of the external compression, that is, with intake pressure ratio being plotted against non-dimensional mass flow for lines of constant flight Mach number. Looking at the characteristics of Figure 7.6 it is seen however that there is a

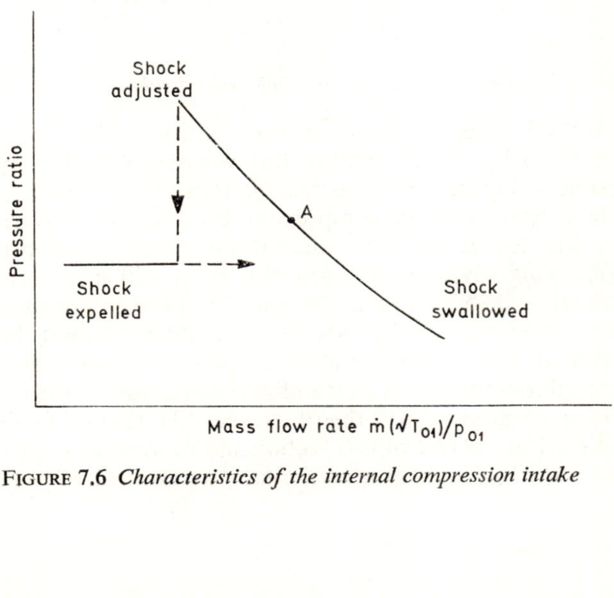

FIGURE 7.6 *Characteristics of the internal compression intake*

fundamental difference which can be understood if the position of the normal shock is considered. Suppose that we start at some point A on the characteristic and reduce the non-dimensional air flow demand of the intake. To accommodate the change in conditions the normal shock will move up towards the throat and will be weakened so that the pressure ratio of the intake will improve. When the shock is at the throat it will be at its weakest value, but, if there is any further decrease in air flow demand, there is no stable position for the shock forward of the throat. The shock will therefore be expelled, the intake will act as a Pitot intake and there will be a very big drop in the pressure ratio. What is more, the shock expulsion is irreversible and the shock can only be swallowed by a considerable increase in the intake mass flow. Clearly any practical internal compression intake would have to operate a safe distance to the right of the "shock adjusted" position.

Combined External and Internal Compression Intakes

Intakes may be designed which combine the features of both external and internal compression types. A typical example is shown with its idealized shock pattern in Figure 7.7. The figure shows that a wedge

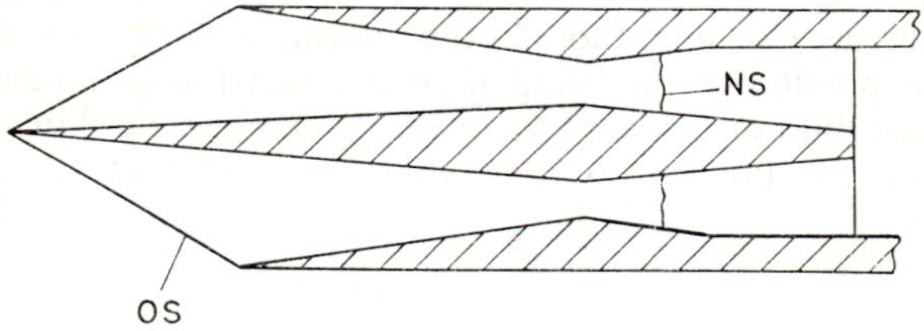

FIGURE 7.7 *Ideal external/internal compression intake*

or cone is used to set up an initial shock which lies, at design conditions, on the lip of the intake. But whereas with the external ompression of Figure 7.3(b) the passage after the lip was diverging, that of the external/internal compression continues to converge. As, therefore, the flow is still supersonic it will decelerate to a throat. In the following divergent portion the gases will accelerate to a normal shock and then diffuse subsonically to the compressor inlet. When operating the intake, the normal shock should be held downstream of the throat for stability purposes, but if it should move through the throat it will stabilize again ahead of the cowl lip. The shock configuration will then be rather like that of Figure 7.4.

Example 7.1 at the end of this section shows some of the methods that can be used in making a preliminary assessment of the design of an intake. Although the theoretical design is fairly straightforward, the practical design of supersonic intakes and their development to a suitable state for service is very much more difficult—the intake engine matching problem has probably been the most serious single difficulty with most high performance aircraft. A full discussion of this matter is beyond the scope of this book but the following notes highlight some of the main factors which must be borne in mind.

(*a*) *Boundary Layer Growth*

The discussion so far has considered flow conditions which are essentially two dimensional and has ignored the effects of boundary layer growth. However, the whole flow pattern in an intake, with its decelerations and increases in static pressure, is of a form most liable to cause boundary layer growth and separation. To overcome this problem it may be necessary to provide facilities for bleeding off the low energy boundary air at points (such as discontinuities) in the intake duct. The trouble then is that the air bled off is virtually useless for propulsion purposes—its pressure is far too low—but it has been accelerated to flight speed so has made a contribution to the intake momentum drag. Luckily the bled air might be suitable for cooling or ventilation so it may be possible to take some advantage of it.

(The boundary layer air which grows on fuselage surfaces ahead of side intakes of a turbojet is always a problem. On many installations, such as those of the Gnat and Starfighter, the actual air intake stands proud of the fuselage side and the fuselage boundary layer is taken off by a separate small duct.)

(*b*) *Shock Interactions*

The ideal shock patterns drawn are in reality a considerable simplification of the true picture. An important aspect which has been ignored for example is that of shock interactions. By this is meant the effects that arise when a shock strikes a solid wall or meets another shock in the gas stream. Figure 7.8 illustrates that when a shock strikes a wall a second shock is *reflected* from the wall and that there is a danger of separation occurring at the wall. Figure 7.9 shows two shocks intersecting with the formation of a vortex sheet separating two parts of the flow. As the upper part has only been through the oblique shock while the lower has been through both oblique and normal, the static pressures in the two parts will obviously be appreciably different.

For a fuller discussion of shock interactions the reader is referred to standard textbooks on gas dynamics such as Shapiro (ref. 3).

(*c*) *Requirements of Variable Geometry*

When an intake is required to operate efficiently over a range of flight conditions it may be necessary to introduce variable geometry. For example it may be necessary to position the oblique shock on the

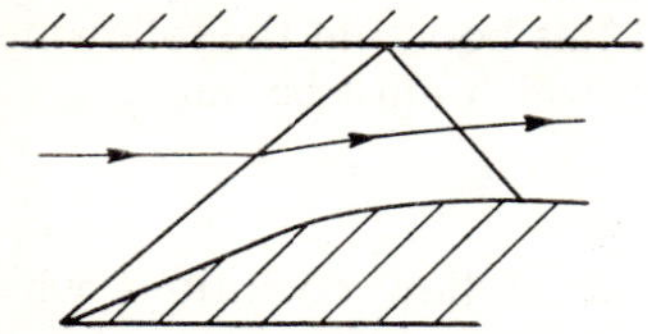

FIGURE 7.8 *Shock interaction at a wall*

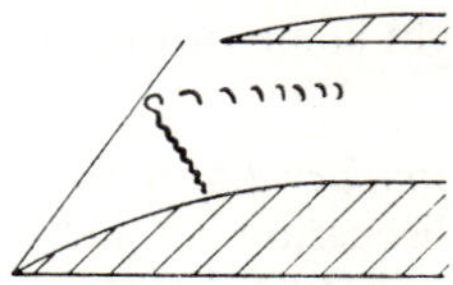

FIGURE 7.9 *Intersection of two shocks*

cowl lip and to do this at varying flight Mach numbers will involve axial movement of the centre cone. The important aspect of this is not so much that of moving the cone—powerful hydraulic jacks to overcome the air loads are readily available—but of *controlling* the movement. Some signal must be chosen to sense the required and actual positions of the cone and the dynamic response of the control system used must not introduce instabilities into the airflow.

Before leaving the problem of supersonic intakes, it must be noted that the intake must be able to accept air entering in a non-axial direction. If the intake takes up some angle of attack the shock pattern will go awry and gross distortions of the air flow at the compressor face may result. As noted before that this might make an intake unusable as it would be liable to cause compressor surge.

Example 7.1 The intake for a turbojet is of the mixed external/internal compression type and is rectangular in section with a

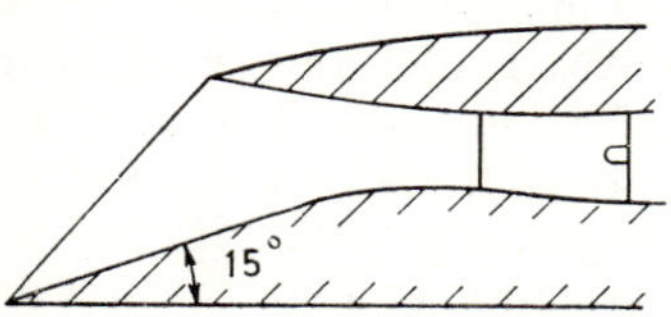

FIGURE 7.10

constant width of 1m. The general arrangement is as shown in Figure 7.10. From this it will be seen that a 15° wedge sets up a plane oblique shock, which, at the design flight Mach number of 2·4 just impinges on the cowl lip. Downstream of the oblique shock the air decelerates to a throat where the Mach number is 1·2. It then accelerates to a Mach number of 1·4 before passing through a plane normal shock and decelerating subsonically to the engine intake.

When flying in conditions where the atmospheric pressure is 6 kN/m^2 and the atmospheric temperature is 217 K the engine mass flow rate is 100 kg/s. Calculate the throat dimensions and the stagnation pressure and temperature at the compressor face assuming that the flow is isentropic except through the shocks.

Solution. In solving this problem use is made of the oblique shock characteristics given in Figure 1.19.

For an upstream Mach number M_1 of 2·4 and a wedge angle of 15° we see that the Mach number downstream of the oblique shock M_2 is 1·8 and that the static pressure ratio across the shock is 2.4.

Now calculate the stagnation temperature and pressure relative to the intake of the approaching air.

$$T_{01} = T_1(1 + \tfrac{1}{2}(\gamma - 1)M_1^2) = 217(1 + 0{\cdot}2 \times 2{\cdot}4^2) = 466 \text{ K}$$

$$p_{01} = p_1(1 + \tfrac{1}{2}(\gamma - 1)M_1^2)^{\gamma/\gamma-1} = 6(1 + 0{\cdot}2 \times 2{\cdot}4^2)^{3\cdot5}$$
$$= 87{\cdot}5 \text{ kN/m}^2$$

Through the oblique shock, stagnation temperature is conserved but there is a loss in stagnation pressure.

$$T_{02} = T_{01} = 466 \text{ K}$$

$$p_{02} = p_1 \times \frac{p_2}{p_1}(1 + \tfrac{1}{2}(\gamma - 1)M_2^2)^{\gamma/\gamma-1}$$
$$= 6{\cdot}0 \times 2{\cdot}4(1 + 0{\cdot}2 \times 1{\cdot}8^2)^{3\cdot5}$$
$$= 82{\cdot}6 \text{ kN/m}^2$$

If station 3 is at the throat we know that $M_3 = 1{\cdot}2$. We also know that, as the flow is isentropic between shocks,

$$p_{03} = p_{02} = 82{\cdot}6 \text{ kN/m}^2$$

As always, stagnation temperature is conserved so that $T_{03} = 466$ K.

At the throat

$$T_{03}/T_3 = (1 + 0{\cdot}2 \times 1{\cdot}2^2) = 1{\cdot}288$$

and

$$p_{03}/p_3 = (1{\cdot}288)^{3\cdot5} = 2{\cdot}425$$

Static conditions at the throat are therefore $T_3 = 362$ K and

$$p_3 = 34{\cdot}0 \text{ kN/m}^2$$

Writing the continuity equation

$$\dot{m} = \rho A_3 V_3 = p_3 A_3 V_3 / RT_3$$

we get

$$A_3 = \frac{\dot{m}RT_3}{p_3V_3} = \frac{\dot{m}RT_3}{p_3M\sqrt{(\gamma RT_3)}}$$

$$= \frac{100 \times 0{\cdot}287 \times 362}{34{\cdot}0 \times 1{\cdot}2\sqrt{(1{\cdot}4 \times 287 \times 362)}} = 0{\cdot}670 \text{ m}^2$$

(R in the expression for the speed of sound must be in J/kg–°C, i.e. $kJ \times 1\,000$.)

Hence the throat dimensions are 1 m × 0·670 m.

After passing through a normal shock with the upstream Mach number M_4 of 1·4 the stagnation temperature will still be 466 K but there will be a loss in stagnation pressure which can be found from Table 1.3. This shows that with $M_4 = 1{\cdot}4$,

$$p_{05}/p_{04} = 0{\cdot}9582$$

At the compressor face

$$T_0 = 466 \text{ K and } p_0 = 0{\cdot}9582 \times 82{\cdot}6$$
$$= 79{\cdot}0 \text{ kN/m}^2$$

7.3 DEFINITION OF INTAKE EFFICIENCY

The question of the ways of defining the intake efficiency of an engine was touched on briefly in Chapter 3 when dealing with propulsion cycles. It was said that the efficiency could be defined in a number of ways which would each have particular merits in certain cases. Two of the most commonly used definitions are the *Stagnation Pressure Recovery* and the *Kinetic Efficiency* which are defined as follows.

Stagnation pressure recovery is the ratio of the actual stagnation pressure at the compressor face p_{01} to the value which would have resulted if the compression had taken place isentropically from the ambient pressure of p_∞. For a flight Mach number of M the efficiency is defined as

$$\text{Stagnation pressure recovery} = \frac{p_{01}}{p_\infty[1 + \frac{1}{2}(\gamma - 1)M^2]^{\gamma/\gamma-1}} \tag{7.1}$$

Kinetic efficiency is the ratio of the kinetic energy which would be obtained by expanding the gas isentropically from compressor face stagnation pressure to atmospheric pressure, compared with the kinetic energy corresponding to the flight speed. (This efficiency has already been defined in Chapter 3 with reference to Figure 3.5, p. 86.) An equation can be written for the efficiency, namely

$$\eta_{KE} = \frac{c_p T_\infty}{c_p T_\infty}\left(\frac{1 - (p_{01}/p_\infty)^{(\gamma-1)/\gamma}}{1 - (p_{0\infty}/p_\infty)^{(\gamma-1)/\gamma}}\right) \tag{7.2}$$

From the relationship

$$p_{0\infty}/p_\infty = [1 + \tfrac{1}{2}(\gamma - 1)M^2]^{\gamma/\gamma-1}$$

the above two equations can be manipulated to show that the two expressions for efficiency can be related by the equation

$$(p_{01}/p_{0\infty})^{(\gamma-1)/\gamma} = \frac{1}{[1 + \tfrac{1}{2}(\gamma - 1)M^2](1 - \eta_{KE})} \tag{7.3}$$

Equation 7.3 has been used to calculate the curves of Figure 7.11

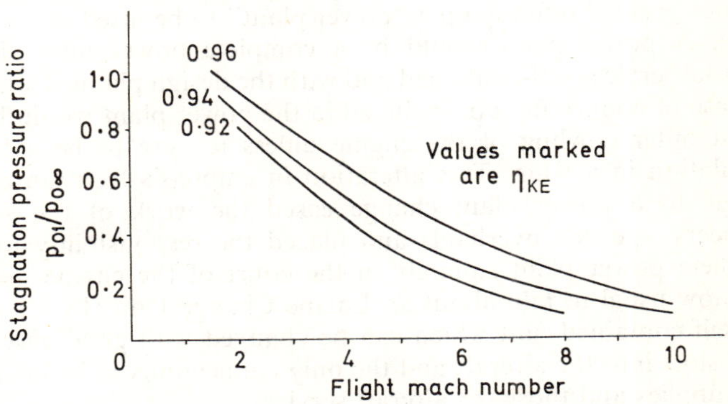

FIGURE 7.11 *Variation of stagnation pressure recovery with kinetic efficiency*

which shows the stagnation pressure recovery against flight Mach number for a number of values of kinetic efficiency.

Coming now to the merits of the two ways of defining efficiency, the stagnation pressure recovery is particularly useful when dealing with the calculation of propulsion cycles. Any ram compression in the intake will contribute to the expansion ratio available across the final nozzle and this will have a direct effect on specific thrust and specific fuel consumption. The kinetic efficiency on the other hand, gives a measure of how good the intake is from the "state-of-the-art" point of view and as such it is much more meaningful when comparing intakes which are to operate over a range of flight conditions. Naturally this is not the whole story as there may be special restrictions on a particular design which will make it difficult to obtain both an efficient intake and a compact design. One with an efficiency of say 0·92 may be preferable overall to a larger one with a slightly higher efficiency. On the whole though, kinetic efficiency gives a reasonable measure of the refinement of the design.

(A useful analogy can be drawn with the delivery of hydraulic power at a power unit at the end of a long pipe. In this case the losses in kinetic efficiency are analogous to the friction factor in the pipe. Clearly if power is to be delivered a long distance—corresponding to decelerating air from a high flight speed—there must be losses but the overall efficiency of power transmission can be improved by using a smooth pipe with a low friction factor.)

7.4 THE POWER PLANTING OF ENGINES

In the earlier days of aviation, it was usual to install an engine on its own as a unit in an aircraft. As things began to get more complicated the idea grew of offering up a "power plant" to be fitted in the aeroplane. A power plant would be a complete power unit with its essential services self-contained and with the design planned specially for ease of connecting up. Included in the power plant would be the actual outer cowling of the engine unless it were to be a buried installation in a wing. This alteration in emphasis from an engine change to a power plant change eased the work of the service engineers, speeded overhauls and placed the responsibility for the complete power plant squarely in the court of the engine makers. It is now usual to talk about an Engine Change Unit (ECU) which is a self-contained unit which can be changed with great ease. The ECU slots into the aircraft and the only connections to be made are fuel supplies and those for aircraft services.

Certain principles must be observed in the design of power plants for engines. The most important are probably those related to fire precautions and regulations on this matter are laid down by the Air Registration Board. In general, a fire risk is held to exist if fuel or inflammable liquids are present and if there are high gas or metal temperatures. To keep down the risk of these coming together, a power plant is split up into a number of bays or zones which are separated from each other by fire-proof bulkheads. If there is a fire risk in a zone then special provision must be made for fire detectors and extinguishers. Any essential electrical services passing through a fire zone must be specially protected by enclosing them in armoured ducts.

Turbine engines are essentially high temperature machines and precautions must be taken to see power plant temperatures do not rise to excessive values. This can best be done by ensuring that each zone of the power plant is well ventilated—remembering of course, that ventilation air represents a drag penalty. Normally it will not be difficult to get adequate ventilation in flight and the most severe condition will be ground running in the tropics. It may be necessary for this condition to take some power from the engine to give forced ventilation of the bays, just at the time when the take-off power is most limited.

One of the gravest hazards which has existed from the earliest days of flying is that of icing up of the airframe or engines. Whenever an aircraft has flown for a time in sub-zero atmospheric conditions and then enters a cloud or rainstorm there is a danger of ice building up on the surfaces of the intakes and slowly strangling the engines. As with the fire risk, therefore, the Air Registration Board lays down stringent regulations on anti-icing and insists that the manufacturers demonstrate that their installations are safe. The most common means of protecting against ice is to provide heating for the vulnerable surfaces, either by a hot gas bleed from the engine or by electrical means. There will always be a performance penalty when anti-icing is in operation and there will also be weight penalties in the power plant if double skinning is needed to form hot gas ducts. With high speed aircraft the icing danger is greatly reduced as firstly the surface temperature is raised by ram effects and secondly there is less moisture in the air at the altitudes where they normally operate.

At the start of route proving of the Proteus in the Britannia in 1956 came the first recognition of a new icing danger—that of dry ice. At Britannia flight conditions over Africa, the aircraft were flying through ice crystal clouds and the crystals collected on various warm parts of the intake subsequently shedding like snowballs into the intake. The problem was one of very great complexity and the solution came eventually when it was realized that *warm* surfaces were the danger. If everything could be kept cold, the crystals just bounced off. As a result of Bristol's experience, the Air Registration Board's requirements were extended to cover the new cases.

7.5 PROVISION OF AIRCRAFT SERVICES

Besides being the means of propulsion of an aircraft, the engines must also generally act as its power station, and must provide the energy for operating the essential services on the aircraft. The word "generally" is used because some aircraft carry small power units specifically to meet the energy requirements. Although fitting an auxiliary power unit (APU) means that the operator can be independent of power supplies and can meet such needs as supplying cabin cooling air before take-off without having to start the main engine, an APU inevitably means increased weight and complexity which is not always justified.

The most important aircraft requirement is probably for electrical power which is used to operate radio and many other services. At one time almost all aircraft electrics were on a d.c. system at 28 V nominal but, as the systems developed and got more complex, the advantages of distributing by a.c. became apparent. In addition, the performance of alternators at low speed is very much better than that of d.c. generators and reliable transformer-rectifier units could be used to allow batteries to be charged by alternators. Once a.c. comes into use, however, there is the problem of maintaining constant frequency from an engine drive which is fundamentally varying speed. To get over this problem a number of constant speed drives have been developed which act like variable speed gear boxes to give a steady output speed despite variations in input rev/min.

The other direct power supplies used in an aircraft are pneumatic and hydraulic. Although each has advantages from the control point of view, pneumatics are especially convenient if the turbine engine is used as the source of air. However, the fact that engine air is essentially low pressure means that high pressure hydraulics can score on the grounds of compactness and low weight.

An important service in the modern aircraft is that of cabin pressurization and air conditioning. The air requirement is of the order of 0·5 kg/min per passenger and it must be supplied at a pressure of about 70 kN/m^2 and a temperature of about 15°C. Air at the requisite pressure can be obtained if necessary by bleeding air from the compressors but this system, though light and compact, has some objections when compared with the use of separate cabin blowers. An important snag is that it is necessary to ensure oil-free air and if this is laid down as an absolute requirement design penalties may be imposed on the engine—no one particularly *likes* oil leaks on an engine but if the air is *not* going to the passengers, some leakage on occasions may be acceptable. The performance picture must also be considered as a whole as engine bleed may be inefficient under some flight conditions but the best solution in others. Finally there is the need to maintain cabin air supplies during a descent when the engines are throttled back and are working at reduced pressure ratios. All in all, it is not surprising to see that the majority of modern airliners use separate engine-driven blowers for cabin air supply.

In Section 7.4 the problem of power plant anti-icing was touched on and it was said that the protection of the airframe was also needed. As a general rule, heating will be needed mainly for the leading edge of the wing and this may well be made of double skin construction to allow hot gases to be passed under the surface. Alternatively, electrically heated pads may be used with the electrical power being supplied from the engines. For the fin and tail plane leading edges electric heating will often be more convenient unless the engines are tail mounted.

7.6 ENGINE/AIRFRAME INTEGRATION

From the overall performance aspect it is important that engine and airframe should always be considered as items in a complete vehicle because they will inevitably interact with each other. So far we have looked at the installation problems but have not dealt with the aerodynamic interaction to any extent. The problems of how engine flows affect the pressure pattern around an aircraft, and conversely how the aircraft pressure field influences engine flows, are exceedingly complex ones which are not really suitable for discussion here. However, a number of methods have been proposed and tried for using power inputs from an engine to influence the detailed airflow around aircraft and some of the main ones are discussed below. Special rigs such as the Rolls-Royce Thrust Measuring Rig—the Flying Bedstead—which were built for one specific research purpose are excluded from the discussion.

(i) Boundary Layer Control

As the angle of attack between a wing and the airflow increases the lift coefficient of the wing will increase steadily until separation of the boundary layer occurs when the wing stalls with a drastic loss in lift and increase in drag coefficient. Boundary Layer Control (BLC) techniques aim to delay this separation and so to increase the usable lift coefficient of the wing and allow a reduced minimum speed for flight. Two methods are possible, either by applying suction at the separation point or by injecting air over the wing surface and "revitalizing" the boundary layer. In either case appreciable volume flows of air must be handled; with a piston engine aircraft separate compressors will be needed but with gas turbine engines it may be feasible to use main engine bleeds.

A review of work carried out on BLC was made by Williams and Butler (ref. 4) and their paper detailed the considerable performance improvements which could be achieved by suction or blowing. Despite these possible gains, the use of BLC has never progressed beyond the purely experimental stage except for the case of flap blowing which is discussed in greater detail below. The reasons for this lack of application are partly that the power requirements can be quite appreciable and the provision of compressors can involve weight and cost penalties. In addition there are structural problems in providing the blowing or suction slots with their associated ducting. But one of the biggest difficulties lies in ensuring absolute reliability. When an aircraft is flying below its normal stalling speed and is relying on air blowing to keep it in the air, it is imperative that the air supply should not fail. This is probably the major factor which has held back the application of BLC to civil aircraft.

(ii) Flap Blowing

A particular case of BLC which *has* been used on production aircraft such as the Hawker Siddeley Buccaneer is known as "flap-blowing". (The Buccaneer also uses a measure of air blowing over the wings.) In a flap blowing installation (see Figure 7.12) air is blown across

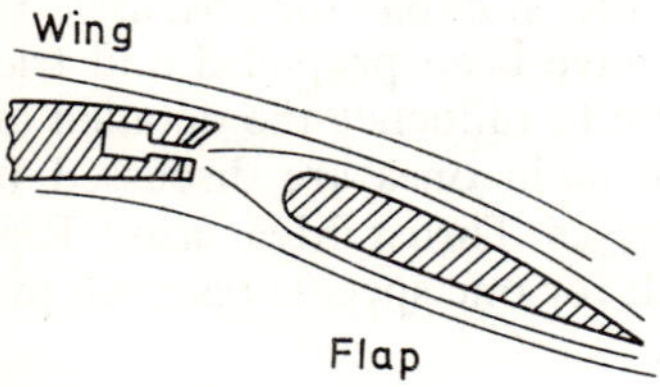

FIGURE 7.12 *Flap blowing*

the nose of the flaps which are used to increase the wing lift coefficient at take-off and landing. Use is made of what is known as the "Coanda effect" which says that air injected over a surface will cause the main flow to stay close to the surface even if this involves the flow following a curved path. By using flap blowing, the flap performance can be improved appreciably especially at higher angles of flap deflection such as are used at take-off and landing.

It is of interest to note that the Buccaneer is a conventional aircraft which is required to operate from an aircraft carrier at the lowest possible speed. The turbojet engines would be at very much reduced power during the approach so there would be ample bleed air available from the engines to carry out the flap blowing.

(iii) The Jet Flap

A paper by Davidson (ref. 5) described work carried out at the National Gas Turbine Establishment on a remarkable new approach to engine/airframe integration, the Jet Flap. In a conventional aircraft the low speed performance is improved by lowering the flaps—mechanical surfaces which are hinged at or near the trailing edge of the wing. Lowering a flap on an aerofoil increases the effective camber of the wing and so converts it into an aerofoil section with an increased lift coefficient. The work described by Davidson had sought to replace a mechanical flap by a sheet of air issuing from a slot at the trailing edge of the wing (Figure 7.13).

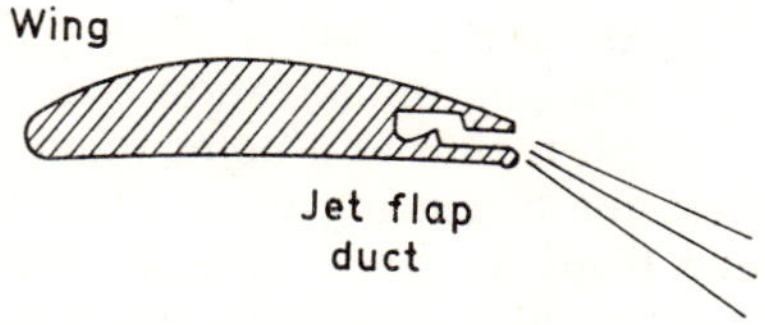

FIGURE 7.13 *The jet flap principle*

Variations in the amount and velocity of the air leaving the slot will alter the effective equivalent flap but with quite modest air flow amounts very large flap areas can be simulated. There is a further advantage that the air leaving the jet flap slot will have a vertical component of velocity and so will give some direct upward thrust.

Against these advantages of the jet flap must be set a number of disadvantages. One of the most important is that large movements of the centre of pressure on a wing occur as the jet flap is brought into operation and this will affect the control of the aircraft. There is also the problem which came up before in connection with BLC of ensuring a high degree of reliability. Loss of power with a jet flapped aircraft while flying at low speed would result in a serious loss in lift. Many of the problems which arise can only be investigated with a full scale aircraft and to gain more experience an experimental aeroplane, the BAC 126 has been built and flown at the Royal Aircraft Establishment, Bedford. Air for the jet flap in this case is bled from the single Rolls-Royce Bristol Orpheus turbojet which provides the propulsive thrust. If and when the jet flap is brought into use on commercial aircraft it is more probable that a number of small engines will be used with the bleed air being taken to a common duct. In this way the risks arising from failure of an engine would be reduced and the ducting problems might be simplified.

(iv) Control of Jet Lift Aircraft

A problem which arises with vertical take-off aircraft which use jet lift methods (that is, as opposed to helicopters) is that of aircraft attitude control at hovering and low speed conditions. An aircraft such as the Hawker Siddeley Harrier, with a single vectored thrust turbofan engine, has conventional aerodynamic controls for use during wing-borne flight but these will clearly be useless at low air speeds. To supply the necessary control for this type of aeroplane it is common to fit "puff pipes", small downward-facing nozzles at wing tips, nose and tail through which compressed air can be passed to give the necessary thrusts. This control air will normally be provided from tappings off the main engine compressor and so a thrust penalty will result whenever they are in use. Design of the puff pipe system brings in a number of interesting optimization exercises. It is important, for example, to tap off the air at a point where it will have the maximum effectiveness for producing the control thrusts but the thrust penalty resulting from the bleed must be kept to a minimum. Having chosen the air bleed-off point the size of ducting will be decided by the allowable pressure losses and the acceptable weight of the complete system. In choosing the run of the ducting it may be worthwhile to use a longer path and avoid regions of the air frame where extensive thermal lagging of the pipes would be needed. Altogether it is a very complex exercise.

8 Engine Dynamics and Control

8.1 INTRODUCTION

The task of building a turbojet which would work even inefficiently was initially so difficult that the provision of a sophisticated control system was at one time given relatively low priority. Gradually, though, it was realized that the fuel system was the heart of the engine and that great care was needed to ensure that the controls were matched correctly to the engine characteristics. When an engine starts its test flying, as opposed to being run only on the test bed, the problems of dynamic behaviour can become the major items in the test programme. It has been said by a successful test engineer that ground testing is concerned very largely with steady state engine performance while flight test deals mainly with engine transient problems. This statement gives a measure of the importance of understanding the dynamic behaviour of engines.

In this chapter the general requirements of a control system are discussed and the basic control philosophies are considered. Following from this, the dynamics of the turbojet are analysed before going on to the characteristics of some representative fuel control systems. Much of this work is applicable to turbofans since the gas generator section of this type of engine is very similar in behaviour to a turbojet. A turboprop, on the other hand, introduces special problems which need particular mention. In many ways these turboprop problems are related to those of advanced turbojets which use variable geometry and reheat.

8.2 GENERAL REQUIREMENTS OF A CONTROL SYSTEM

The requirement, above all others, for the control system of an aero-engine is that the pilot should not have to worry about the engine—it must do exactly as he wishes it to. From this basic need the requirements for three régimes of operation of the control system can be distinguished. There is the requirement of *steady control* which stipulates that when the pilot selects a certain engine rating, this is held under all flight conditions. There is the requirement of *transient control*, which says that when the pilot changes the rating selected, the engine will change its output smoothly and safely and as quickly as possible. Finally there is the requirement of *emergency control* which calls for the engine to be safe in operation despite any malfunction of engine or control system.

To these ideal performance characteristics must be added the general design requirements for aero-engine work, that a component must be of acceptable bulk, weight and price. These are the real limiting factors which make every control system less than perfect—a control which was required to cover every possible eventuality would be so complex as to be impracticable. So, as always, design involves a balancing up of the factors to achieve an acceptable compromise. It may also be necessary to accept a reduction in the transient performance of an engine in order to avoid limiting the engine's steady state performance. (This topic has already been touched on when considering the compressor characteristics in Chapter 4.)

8.3 THE RATING OF ENGINES

Mention has been made of the fact that a pilot will normally set his control lever so that a certain "engine rating" is maintained. To explain this more fully it should be said that the life of an engine is tied in closely with the operating conditions to which the engine is subjected. It is very easy to wear out an engine quickly by working it at excessively severe conditions for long periods. On the other hand, by the very nature of an aircraft and its flight, it is possible to arrange the operation of the engine so that long periods of steady operation are used and that high power is only used for short times. (There is, of course, the obvious exception for fighter and trainer aircraft where frequent power changes will be the order of the day.)

Normally the test-bed proving of an engine will establish a number of ratings which have limits of duration in their use. The ratings will generally include a take-off rating, with a five-minute limitation, a maximum cruise rating with a half-hour limit and a continuous cruise rating which can be used indefinitely. There may in addition be an emergency rating with a one-minute limitation if the engine is used in a multi-engined helicopter where high power is needed from good engines in the event of one engine failing. The permitted life of the engine between overhauls is decided on the basis of the strip condition after running for a period with these limitations.

A rating must be defined in terms of what a pilot can measure easily and must be based on a figure related to engine life. The most convenient parameter is generally the rotational speed of the gas generator as this gives a meaningful figure which can be displayed to the pilot. In theory, it would be better to use the turbine entry temperature as this is the dominant factor in fixing the engine life. Measuring this temperature is however very difficult and it is seldom done. As a compromise the turbine outlet or jet pipe temperature is *always* measured and presented to the pilot. With the two readings of rotational speed and jet pipe temperature available, a rating can be defined as a given rev/min subject to the jet pipe temperature not exceeding some stated value. The pilot will then fly by the rev/min indicator but will reduce power under high intake temperature conditions to avoid exceeding the limiting jet pipe temperature. To relieve the pilot of worry, an automatic temperature control will generally be fitted in the fuel system.

If an engine is of the two-spool type with low-pressure and high-pressure rotating assemblies it will be necessary to state limitations in terms of both shaft speeds, as with varying flight conditions each may in turn be a limiting parameter.

8.4 CONTROL SYSTEM PHILOSOPHIES

For most engines, control is effected primarily by varying the fuel flow. The reason for taking fuel flow as the main variable is that there is a big background of knowledge and experience on fluid control equipment. It is a convenient variable too, as by fixing the fuel flow for an engine of fixed geometry operating at a given flight condition the operating state of that engine is determined completely. For these

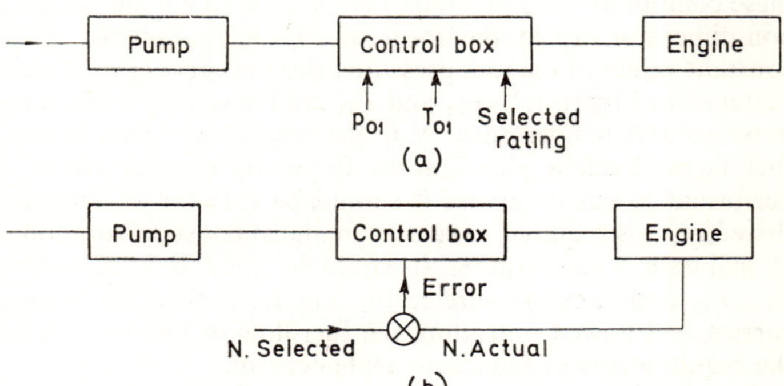

FIGURE 8.1 *Block diagrams of control systems*

reasons, when, in this section, "fuel system" is mentioned, "control system" is implied.

In control work, two main classes of system are used. They are known as the *scheduling* and the *closed loop* types of control. Figure 8.1 shows block diagrams illustrating the two philosophies; these diagrams are drawn to show only the essentials of the system and which signals are fed into the control box, without going into any details on the actual mechanisms. Consider now each system in turn.

(a) Scheduling Control

A scheduling control is based on the principle that the fuel required to run an engine at a given rating for any flight conditions is determined entirely by external conditions and can be calculated from

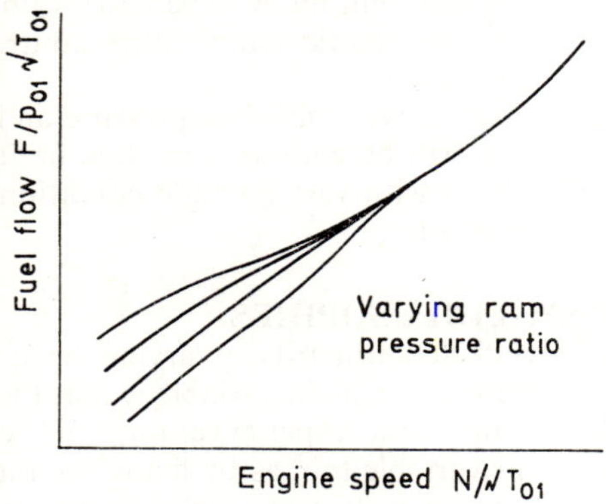

FIGURE 8.2 *Typical non-dimensional fuel flow plot for a turbojet*

these conditions. The fuel requirements for an engine can be plotted non-dimensionally in the form of $F/p_{01}\sqrt{T_{01}}$ against $N/\sqrt{T_{01}}$ at constant values of the ram pressure ratio the curves being of the form sketched in Figure 8.2. (p_{01} and T_{01} are the engine intake stagnation pressure and temperature, N is the rotational speed and F is the fuel flow.) Such a plot ignores Reynolds number effects on the performance but otherwise it should be true for all conditions. If, therefore, it is required to run an engine at constant rotational speed to maintain a given rating, it would be possible in theory to sense p_{01}, T_{01} and ram pressure ratio, and from them to compute the correct fuel flow. Controlling the fuel flow in this way would meet the requirements of the steady state control.

A scheduling control of this type would take no account of the actual behaviour of the engine and would ignore variations of individual engines from a mean performance. It would be expected, therefore, that the control would not be very accurate for any particular engine. With three inputs needed, the computing side of the control would be a complicated mechanism and so would be difficult to design with high accuracy. It might be possible, however, to compromise on the performance of the control and get greater simplicity while retaining a satisfactory accuracy. Often it is found that a satisfactory control is possible using a measure of intake pressure only, provided certain assumptions are made on the flight pattern and climatic conditions where the engine is to work.

(b) Closed Loop Control

Looking at the block diagram of the closed loop control system in Figure 8.1(b), it is apparent that the essential difference between this and the scheduling arrangement is that there is a *feedback* from the engine to the control box. The main item in the system is the "comparator" which senses both the rev/min selected by the pilot and the actual rev/min of the engine. If there is a difference between the selected and actual values, a signal goes to the fuel control valve and the fuel flow rate is altered accordingly.

A closed loop control gets over the big limitations of the scheduling control of limited accuracy and of ignoring the behaviour of an actual engine or the variations between engines. In practice almost any degree of accuracy of control can be achieved by designing a control system of sufficient sophistication. One big difficulty is introduced, though, whenever feedback is introduced and that is that there is a chance of instability of operation. Unless care is taken in matching the transient characteristics of the various components in the control loop they might interact and cause "hunting" of the engine. To take a simple example, suppose that when a selected rotational speed was reached the fuel flow was reduced suddenly to a small percentage of its original value and that it remained at the low value until the speed fell by 500 rev/min, when it as suddenly returned to its full value. With a control having these characteristics steady running would be impossible.

Although a system may be scheduling with respect to the engine, there might be separate feedback loops within the system, for example, between the control unit and the fuel pump. Each loop is a potential source of instability and its characteristics must be checked. Then, too, it is possible to combine the systems and design a system which is basically scheduling but which has closed loop trims. An arrangement like this can give a system which remains basically simple but which will control to a high degree of accuracy.

8.5 THE DYNAMICS OF THE TURBOJET

Before the fuel system engineer can design a control for a turbojet, he must have some knowledge of the engine's steady and dynamic characteristics. This information is needed for the design of an acceleration control as an engine is particularly liable to be damaged during acceleration if too much fuel is added. It is also needed if a closed loop steady state control is fitted. In this case, the engine forms an essential part of the control loop and if its characteristics are not matched properly there is a danger that unstable operation might result under some conditions. The work that follows aims to show how the transient performance for a simple turbojet of fixed geometry may be estimated from component characteristics.

Suppose that for a turbojet component characteristics are available of a form as in Figure 8.3. The compressor performance has been

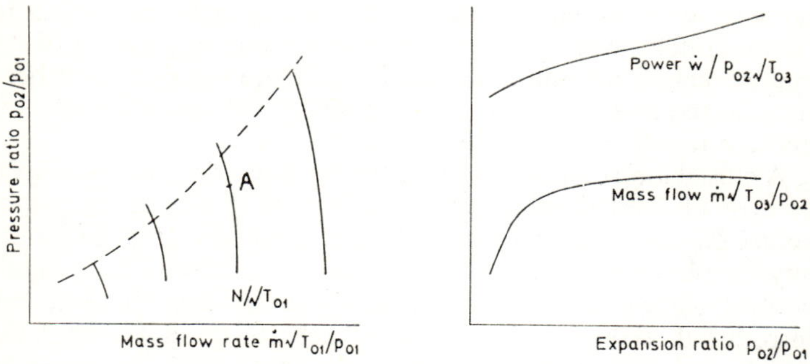

FIGURE 8.3 *Representative turbojet compressor and turbine characteristics*

plotted non-dimensionally (using the methods shown in Chapter 4) as pressure ratio against mass flow at values of constant speed with values of the isentropic efficiency available. For the expansion system, non-dimensional mass flow and power characteristics have been plotted against the overall expansion ratio as was described in Chapter 5. This characteristic has been simplified in that single line curves are drawn and the variations with turbine speed have been neglected. For convenience, too, the pressure loss in the combustion chamber has been debited to the turbine characteristic, so saving a special set of combustion system performance curves. It is for this reason that the non-dimensionals for the expansion system are plotted with the compressor delivery pressure p_{02} rather than the turbine entry pressure p_{03}.

In general, characteristics of the form shown in Figure 8.3 would be obtained from testing of various components under steady conditions. However, the assumption can be made that the characteristics would still hold during engine transients, that is, that each point in a transient could be treated as a quasi-static operating condition. Using this assumption, it was shown by Otto and Taylor (ref. 1) that the engine transient performance could be predicted by the methods given below. Justification for the assumptions comes when it is shown that dynamic characteristics calculated in this way agree to an acceptable degree of accuracy with those found from actual measurements on an engine.

As a start, consider any point A, on the compressor characteristics. At this point the pressure ratio, mass flow and efficiency are known and so for any intake conditions the power absorption by the compressor can be calculated. Now the pressure ratio can be read across to the expansion system characteristics (making due allowance for any intake ram pressure ratio), so allowing the turbine entry temperature and the power output of the turbine to be calculated. This power can be compared with that of the compressor and it will be found that

(*a*) an excess of turbine power over compressor power means that the engine is accelerating;

(*b*) an excess of compressor power over turbine power means that the engine is decelerating;

(*c*) equal compressor and turbine powers mean that the engine is running at steady conditions.

By carrying out these calculations for a large number of points the equilibrium running line for the engine can be defined; above this line lie the points passed through by the engine during acceleration and below lie those for deceleration.

For each of these points the compressor delivery and turbine entry temperature will be known and so the effective fuel flow can be calculated (see equation 3.26). This effective fuel flow is the product of actual fuel flow and combustion efficiency. Also at each point, the engine speed is known, as well as the powers of both compressor and turbine, and so the net out-of-balance torque available to accelerate the engine can be found. A formal analysis to obtain equations for the engine dynamics may then proceed in the following way.

Let the moment of inertia of the compressor/turbine assembly be I, the difference in torque between turbine and compressor be τ, and the acceleration be α. In addition, let N denote the engine speed and F_e the effective fuel flow.

Then

$$\tau = I\alpha \tag{8.1}$$

Also, the out-of-balance torque is a function of engine speed and effective fuel flow, i.e.

$$\tau = \mathrm{f}(F_e N) \tag{8.2}$$

Let this function be expanded as a power series from an initial state denoted by subscript i

$$\tau = \tau_i + \left(\frac{\partial\tau}{\partial F_e}\right)(F_e - F_{e_i}) + \left(\frac{\partial\tau}{\partial N}\right)(N - N_i)$$
$$+ \left(\frac{\partial^2\tau}{\partial F_e{}^2}\right)\frac{(F_e - F_{e_i})^2}{2!} + 2\left(\frac{\partial^2\tau}{\partial F_e \partial N}\right)$$
$$+ \left(\frac{\partial^2\tau}{\partial N^2}\right)\frac{(N - N_i)^2}{2!} + . \; . \; .$$

If the initial state is steady running, $\tau_i = 0$; if in addition the displacement from the initial state is small, the fourth and subsequent terms can be neglected. We get, therefore,

$$\tau = \left(\frac{\partial\tau}{\partial F_e}\right)(F_e - F_e\ ) + \left(\frac{\partial\tau}{\partial N}\right)(N - N_i)$$

This may be written as

$$\tau = \left(\frac{\partial\tau}{\partial F_e}\right)\Delta F_e + \left(\frac{\partial\tau}{\partial N}\right)\Delta N \tag{8.3}$$

Equation 8.1 can be written as

$$\tau = I\alpha = I\frac{d}{dt}(\Delta N) \tag{8.1a}$$

$$I\frac{dt}{d}(\Delta N) = \left(\frac{\partial\tau}{\partial F_e}\right)\Delta F_e + \left(\frac{\partial\tau}{\partial N}\right)\Delta N \tag{8.4}$$

The partial differentials of equation 8.4 can be represented graphically by curves of the form shown in Figure 8.4. Measurement of the slope of the lines at equilibrium running gives the value of the differential. It will be seen that, in fact, $\partial\tau/\partial N$ is negative at constant values of effective fuel flow. A moment's thought should show that this is a not unexpected result.

As has been found previously there are great advantages in plotting engine parameters in non-dimensional or in "corrected" form so that they can be related to operation at a variety of different intake conditions. Using $\delta = p_{01}/14{\cdot}7$ as an intake pressure correction

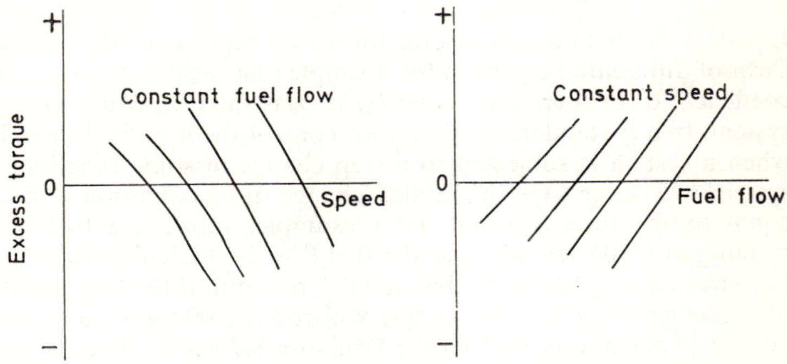

FIGURE 8.4 *Torque versus fuel flow and rotational speed characteristics for a turbojet*

factor and $\theta = T_{01}/288$ as a temperature correction factor, the relevant corrected parameters are

Engine speed $N/\sqrt{\theta}$

Effective fuel flow $F_e/\delta\sqrt{\theta}$

Power $\dot{W}/\delta\sqrt{\theta}$

Also, as torque is the quotient of power and speed the non-dimensional form of torque is τ/σ. These parameters may be substituted in equation 8.1(a) which then becomes

$$\frac{\tau}{\delta} = I\frac{\sqrt{\theta}}{\delta}\cdot\frac{d}{dt}\left(\frac{\Delta N}{\sqrt{\theta}}\right) \tag{8.5}$$

Hence equation 8.4 may be written in terms of the corrected parameters as

$$\frac{I\sqrt{\theta}}{\delta}\frac{d}{dt}\left(\frac{\Delta N}{\sqrt{\theta}}\right) = \frac{\partial(\tau/\delta)}{\partial F_e/\delta\sqrt{\theta}}\frac{\Delta F_e}{\delta\sqrt{\theta}} + \frac{\partial(\tau/\delta)}{\partial(N/\sqrt{\theta})}\frac{\Delta N}{\sqrt{\theta}} \tag{8.6}$$

This again can be represented graphically by plots similar to those shown in Figure 8.4. For convenience a subscript c will be used to denote corrected quantities (e.g. ΔN_c is written for $\Delta N/\sqrt{\theta}$) and the partial differential with respect to fuel flow and engine speed will be written a_c and b_c respectively. Equation 8.6 then becomes

$$\frac{I\sqrt{\theta}}{\delta} \cdot \frac{1}{b_c} \frac{d}{dt} (\Delta N)_c + \Delta N = \frac{a_c}{b_c} (\Delta F_{e\,i})_c \tag{8.7}$$

Equation 8.7 is in a very useful form as it represents the standard form of differential equation for a simple first-order lag system. The coefficient of the derivative term $I\sqrt{\theta}/\delta b_c$ is the time constant of the system. In any standard textbook on control theory it is shown that, when a system is subjected to a step change in some function, the system will cover 63% of the step change in *its* condition in a time equal to the time constant. (For example, suppose a turbojet is running at 8 000 rev/min and the fuel flow is suddenly increased to the value corresponding to that at 9 000 rev/min. If the time constant of the engine is 1 sec, the engine will reach 8 000 + 0·63 × 1 000, i.e., 8 630 rev/min in that 1 sec.) Equation 8.7 shows that, as would be expected, the response time of the engine varies directly with the moment of inertia of the compressor/turbine system. It is less clear from the equation that the time constant *increases* at lower rotational speeds of the engine where the levels of power of both compressor and turbine are lower—the difference in power, which is what goes to accelerate the engine, will also be reduced. In the same way, as altitude increases the levels of power will also fall due to the decreased air density and the engine will be more sluggish. This result has important bearings on the design of the fuel system. Although the *engine* time constant increases at altitude, the time constant of the control system will be more or less the same so the characteristics of the response loop will be greatly changed.

8.6 CONTROL OF TURBOPROP ENGINES

The simple, single shaft, fixed geometry turbojet is the easiest turbine engine to control as, at a given flight condition, the output is a unique function of one control variable, the fuel flow. For the turboprop, an additional variable, the propeller speed or pitch setting, comes into play. At any given flight speed, for a selected fuel flow, wide variations in power are possible by altering the settings of the propeller blades. Great care must be taken at the design stage to ensure that the two control loops are compatible with each other and are matched correctly. The work has relevance to turbojets with variable geometry or reheat as these also have two interacting variables.

It will be recalled from Chapter 3 that there are two main classes of turboprop and that they differ greatly in their control requirements. For the "free turbine" engine (Figure 8.5(a)) there is no mechanical connection between compressor and propeller—the Rolls-Royce Proteus and Gnome are typical examples of free turbine engines. The other class, the fixed shaft engines (Figure 8.5(b))

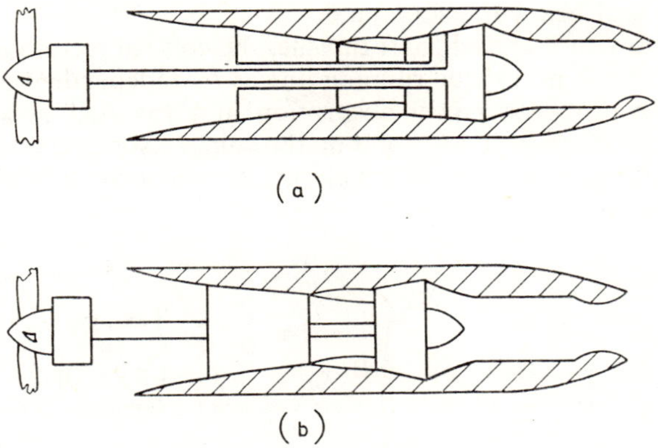

FIGURE 8.5 *Arrangements of turboprop engines*

exemplified by the Rolls-Royce Dart, *do* have a mechanical linkage. From the control point of view, the Rolls-Royce Tyne type layout should be treated as a fixed shaft engine. Dealing first with the free turbine engine, control is quite straightforward. Basically, a fuel control system must be provided which will ensure that the gas generator runs at a fixed rating for any position of the pilot's control lever setting. The problem is very similar to that of choosing a control system for a turbojet and a given control lever setting will give a certain amount of *available* power for the power turbine. In addition to his fuel control, the pilot will be able to choose the setting of a governor which will control the propeller speed. This governor will act by varying the propeller blade settings in order to alter the power absorption characteristics and so keep the speed constant. Normally the engine will be rated so that a certain propeller speed is used for take-off and a lower value for cruise. The settings chosen are not critical and appreciable speed variations from the normal values are possible without incurring heavy power loss penalties (see also Chapter 3, p. 102). This flexibility of operation can be useful as it allows lower propeller speeds to be selected at high flight speeds on cold days and so reduces the tip Mach number of the propeller and cuts down the noise level.

The main control of power output with a free turbine engine is achieved by altering the gas generator speed and an equilibrium running line can be drawn on the compressor characteristics using the methods of section 8.5. This fact, that varying power implies varying gas generator speed imposes restrictions on the acceleration times achievable with free turbines and means that the running line must be set well below the surge line.

Turning now to the fixed shaft turbine, the control picture is very different. There is no unique running line as by independent adjustments of the propeller power absorption and the fuel flow it is possible to run steadily at any point on the compressor characteristic. For example, Figure 8.6 shows a running line which could be chosen

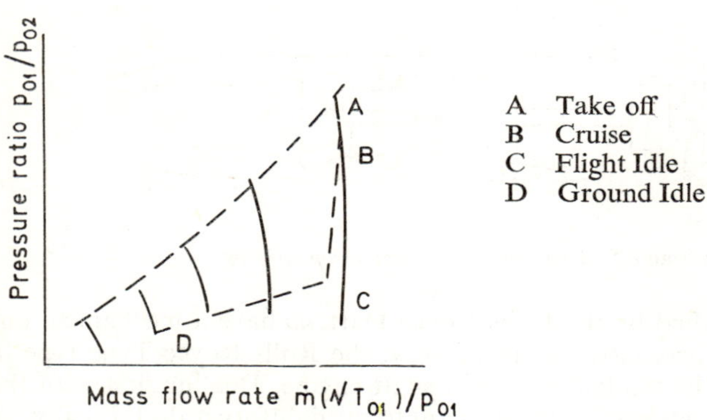

FIGURE 8.6 *Typical operating line for a fixed shaft turboprop*

with the power settings marked at a number of points on this figure. It will be seen that a wide range of power outputs are possible at a constant value of engine speed and so rapid changes in power are possible—there is no need to accelerate a rotating system but only to alter the setting of a fuel pump and the blade angles of a propeller. With this advantage of being able to select any running line, though, is associated the biggest problem of control for fixed shaft engines. Interlocking is needed between the speed control and the fuel flow if the engine is to be safe at all times from danger of surge or over-temperature. It would be easy to reduce engine speed before cutting back the fuel flow and run into surge.

To understand some of the control problems for fixed shaft turboprops under flight conditions it is useful to know something of the characteristics of a propeller. Figure 8.7 shows an element of a propeller in flight with pitch angle β, and forward speed V_a. Briefly the characteristics of greatest interest, the power absorption and the thrust given by the propeller, can be expressed as coefficients involving propeller speed N and air density ρ. These coefficients can then be plotted in terms of the propeller blade angle or "pitch" and the "advance ratio" or ratio of forward motion of a blade element to

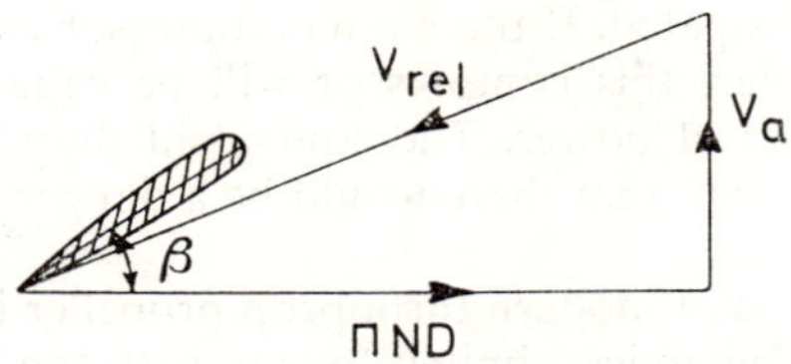

FIGURE 8.7 *Velocity diagram for blade element of a propeller*

rotational motion. The advance ratio V/ND is usually written J and Figure 8.7 shows that, for a given pitch, J effectively defines the angle of attack of the blade element.

Characteristics for a typical propeller are given in Figure 8.8 which

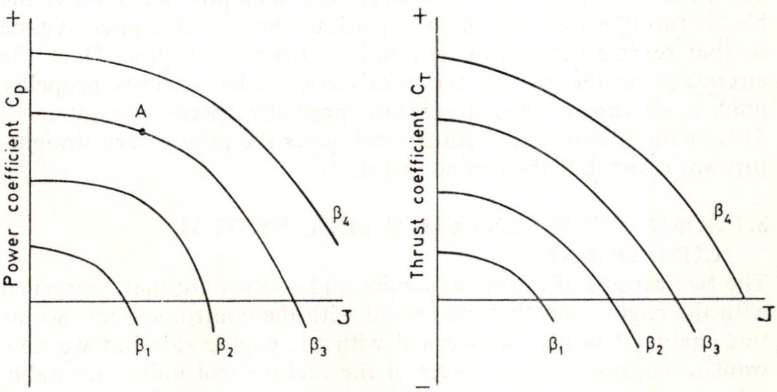

FIGURE 8.8 *Propeller characteristics*

shows that both power and thrust coefficients pass from positive to negative values at certain points. In the negative region, the propeller is acting as a brake absorbing power from the airstream and causing a drag rather than providing a useful thrust.

Suppose that a turboprop engine is operating at a point corresponding to point A on the characteristics and that the control unit is selecting a constant speed of operation. For a given flight speed,

V_a and hence J will be fixed and the propeller will take up a pitch angle of β_3 to absorb the power of the engine. If for some reason the engine fails the constant speed unit would at once try to reduce the blade angle in order to maintain speed and, if this should happen, the propeller would pass very rapidly into the negative thrust region. In doing this it would absorb power from the air and this power would have to be dissipated. If there is a compressor coupled mechanically to the propeller, this compressor will be capable of absorbing the large amounts of power. The equivalent drag forces on the aircraft would be so high that there would be a danger of the pilot losing control completely.

A representative example of a modern turboprop propeller is that built by Hawker Siddeley Dynamics Limited for use with the Rolls-Royce Tyne engine. Every step is taken to ensure that the propeller cannot go into low pitch in flight; mechanical and hydraulic pitch locks are provided and a system sensitive to the torque output of the engine is used to increase propeller pitch immediately in the event of engine failure. Interlocks on the undercarriage allow the flight fixed pitch stops to be withdrawn so that once the aircraft has touched down the blade angle can be reduced. It is then possible to move the blades through the zero pitch angle into the negative pitch régime so that reverse thrust can be obtained from the engine. Once the aircraft is on the ground the pilot's control level selects propeller pitch angle directly and a constant propeller speed is maintained. This method, known as Beta control, gives the pilot a very straightforward control of the engine thrust.

8.7 SOME REPRESENTATIVE FUEL SYSTEM COMPONENTS

The two aspects of engine dynamics and control are that concerned with the engine and that concerned with the control system. So far this chapter has been concerned with the engine side but we now want to consider some aspects of the fuel control units. Inevitably this section will be dealing with "bits and pieces" of fuel system hardware. The aim, however, is to give a rather more basic discussion of matters than would be given, for example, in the technical literature of a fuel system manufacturer. Points are illustrated, though, by reference to actual components.

Two points should be borne in mind in looking at engine fuel systems. Firstly, they are essentially high pressure systems as, in general, a high fuel pressure is needed at the fuel injection point to ensure good atomization of the fuel. Secondly, the flow through any control orifices will almost always be turbulent so that the flow will vary with the square root of the pressure drop. It is common British practice to define a flow number $\bar{N}$ for an orifice such that the fuel flow F in gallons per hour is found from the equation

$$F = \bar{N}\sqrt{\Delta p} \tag{8.8}$$

Δp is the pressure drop in lbf/in^2. An apology must, therefore, be made for the grossly inconsistent units of flow number! In mitigation it may be said that the numerical value of flow numbers generally encountered fall into a convenient range of about 0·1 to 100.

Any description of fuel system components must of necessity emphasize the products of Lucas (GTE) Ltd., and Dowty Fuel Systems Ltd. These two firms have worked with the engine companies from the start on the specialist job of designing and developing engine controls and have done most of the work on engine fuel system design and manufacture. General papers describing some of the work by the companies have been given by Lawrence of Lucas (ref. 2) and Carey of Dowty (ref. 3). The illustrations of the various components in this section are reproduced by kind permission of the two companies.

Lucas Components

The heart of Lucas fuel systems is the pump. This pump, shown in Figure 8.9, is of the positive displacement swash-plate type and the delivery flow is varied by altering the angle of the swash-plate. The figure shows that if the swash-plate is perpendicular to the axis of the pump, the plungers will not move in and out and the flow will be zero. As the inclination is increased, so the delivery flow will increase. There are normally seven plungers in a pump and the drillings in the rotating barrel are inclined to the axis of the pump so that centrifugal forces keep the plungers in contact with the swash-plate.

To control the angle of the swash-plate, a servo-piston is used. This piston is subjected on one side to the full pump delivery pressure while the other side feels a servo pressure, which is the pump delivery pressure after passing through a restriction. The whole essence of control of the pump is that the flow in the servo line is controlled. If the servo line is blanked off, there will be no flow across the servo restriction so the servo pressure will be full pump pressure; then, due to the differential areas and the spring in the servo piston, the pump will go to maximum delivery. If the servo line is opened wide, servo pressure will fall and the pump will go to zero delivery. Under normal operating conditions a control unit somewhere in the system will be partly open so that the servo piston takes up a position to give the required fuel flow.

A fuel line carrying servo pressure must go from the pump to each control unit in a system and it is important to design these connections for complete integrity—a leak in a servo line would cause

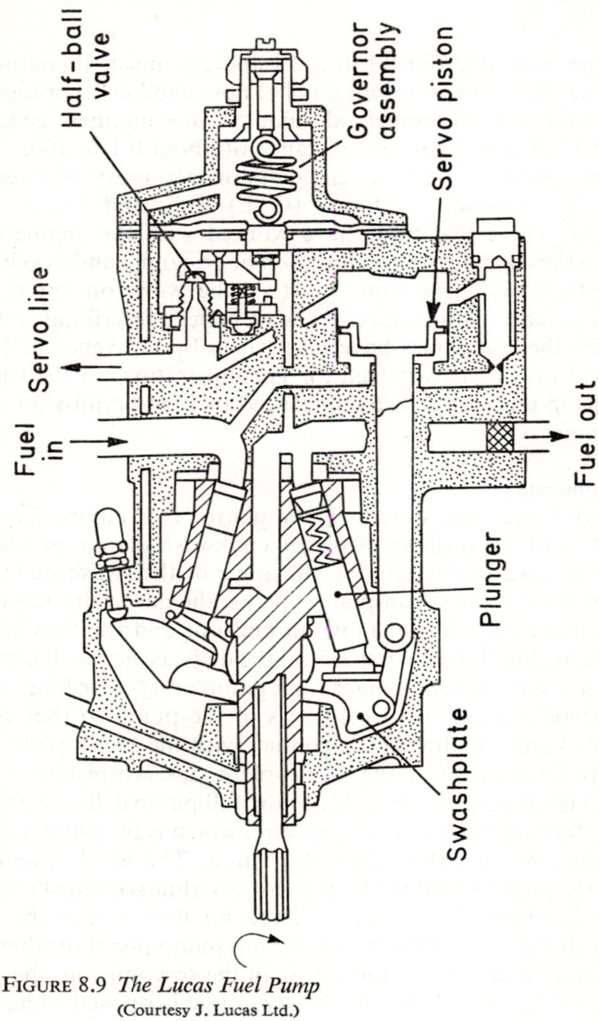

FIGURE 8.9 *The Lucas Fuel Pump*
(Courtesy J. Lucas Ltd.)

the pump to go to zero stroke. The actual control valve in the servo line may be of the half-ball valve type shown in Figure 8.9, or of a more recent design known as the kinetic valve. Each type fulfils the same basic function of controlling the servo pressure but the kinetic valve has the advantage that it is less sensitive to dirt particles in the fuel so needs less thorough filtration in the servo line.

One of the earliest and most successful Lucas units used in steady state control of engines with scheduling type control systems was the Barometric Pressure Control or BPC (Figure 8.10). The BPC senses

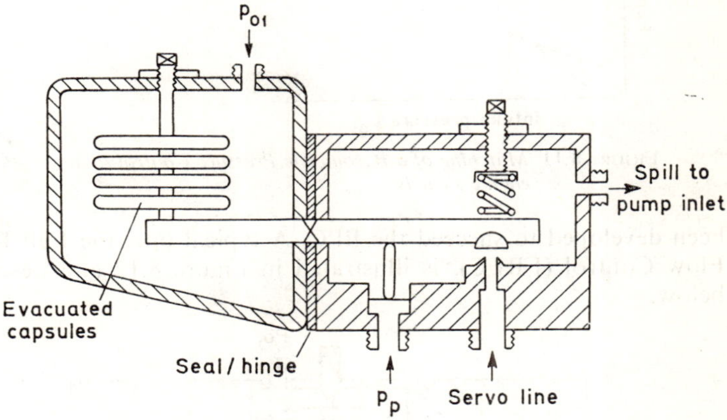

FIGURE 8.10 *The Lucas Barometric Pressure Control*
(Courtesy J. Lucas Ltd.)

intake pressure p_{01} by means of a stack of evacuated capsules and the pump delivery pressure p_p by means of a diaphragm and push rod. A force balance is then set up by applying these to a lever arm so that the BPC gives a law of the form

$$p_p = Ap_{01} + B \tag{8.9}$$

where A and B are constants. From the diagram it will be seen that if the pump pressure is too high the half-ball valve will be lifted so bleeding the servo line, dropping servo pressure and reducing the pump stroke. Figure 8.11 shows the general form of matching diagram for BPC law to engine needs.

The BPC does not match the engine requirements exactly except at two points so the pilot must keep adjusting his throttle lever setting as the aircraft climbs if the engine rating is to be held constant. Accepting this disadvantage, its very simplicity makes the BPC an attractive control unit. However, engineers are almost always prepared to sacrifice simplicity to the cause of improved performance and many more sophisticated scheduling controls have

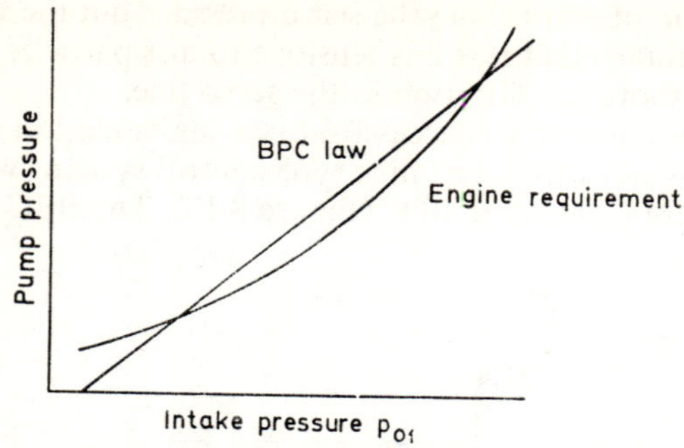

FIGURE 8.11 *Matching of a Barometric Pressure Control to an engine's needs*

been developed to succeed the BPC. A typical one, the Full Range Flow Control (FRFC), is illustrated in Figure 8.12 and described below.

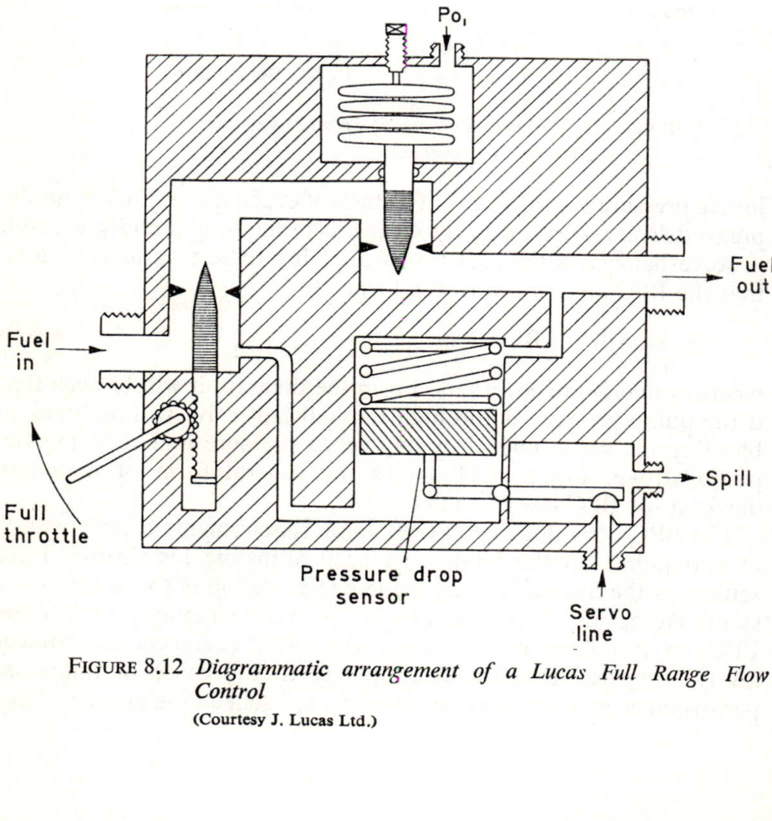

FIGURE 8.12 *Diagrammatic arrangement of a Lucas Full Range Flow Control*
(Courtesy J. Lucas Ltd.)

Like the BPC and throttle valve system discussed above, the FRFC is a control unit which meters the engine fuel as a function of intake pressure p_{01} and pilot's throttle lever position. The principle is that two variable orifices are provided in series, the size of one being a function of intake pressure and of the other, pilot's throttle setting. A pressure drop sensing device is fitted which acts back on the pump servo to maintain a constant pressure drop across the complete unit. There is no restriction on the shape of curves obtained (within reason, of course!) and a typical characteristic will be of the form in Figure 8.13.

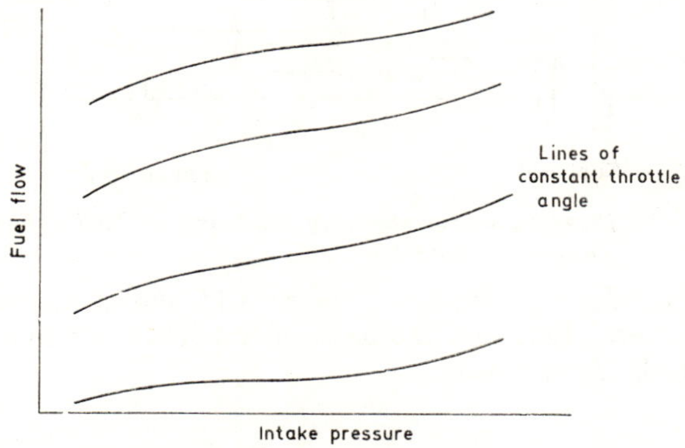

FIGURE 8.13 *The form of the Full Range Flow Control characteristics*

A flow type control unit like the FRFC has an important advantage over a pressure type like the BPC in that its characteristics are independent of the downstream flow number. If for any reason an injector nozzle should get blocked, for example, the flow with a BPC system would fall but the FRFC would pass the same fuel flow and would merely call for a higher pump delivery pressure. The same flow principle is used in the Lucas acceleration control, the Air Fuel Ratio Control or AFRC.

During acceleration it is necessary to supply an engine with more fuel than for the steady running case but the actual quantities provided must not be so great as to endanger the engine. It is found that for very many types of engine there is a simple relationship between the maximum safe fuel flow and the compressor delivery pressure. The AFRC illustrated in Figure 8.14 makes use of this fact, and contains a capsule stack which senses the compressor delivery pressure. Through a servo system the capsule positions a restricting "carrot" in the control orifice, so that the orifice flow number is a function of delivery pressure. If now the fuel flow is above the safe value corresponding to any delivery pressure, the excess fuel pressure drop across the unit will act on a piston and spring and open a half

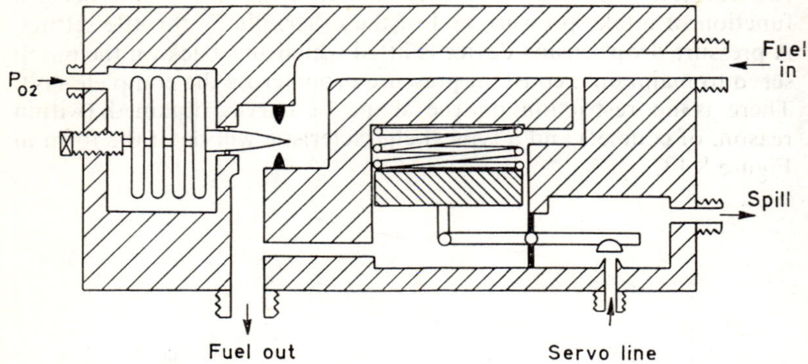

FIGURE 8.14 *Diagrammatic arrangement of a Lucas Air Fuel Ratio Control* (Courtesy J. Lucas Ltd.)

ball valve so reducing fuel flow. Figure 8.15 shows how engine characteristics are plotted on the basis of test results to enable the AFRC law to be determined.

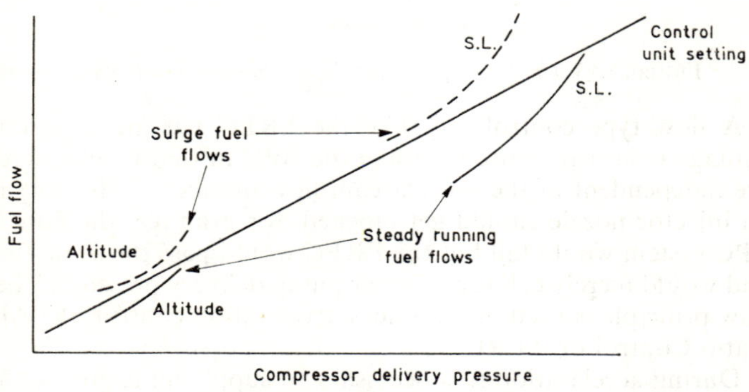

FIGURE 8.15 *Matching of an Air Fuel Ratio Control to an engine's needs*

As a final Lucas component, consider an emergency control, the overspeed limiter which is built into every pump. Although the overspeed governor shown in the pump of Figure 8.9 has now been succeeded by a more sophisticated (and complicated) type the earlier one shows the principle of operation. In this pump a number of radial holes are drilled in the body of the pump so that as it rotates a centrifugal pressure is set up. The pressure rise is fed to a spring-loaded diaphragm and when maximum speed is reached the diaphragm moves opening a half ball valve and so reducing pump delivery flow.

Obviously a governor acting on this principle will be sensitive to fuel density changes. One of the main improvements of newer governors is that they use metal blocks to give an excess speed signal and so get over this disadvantage. Governors also suffer from a problem known as "governor creep" which can be understood by looking at the characteristic in Figure 8.16.

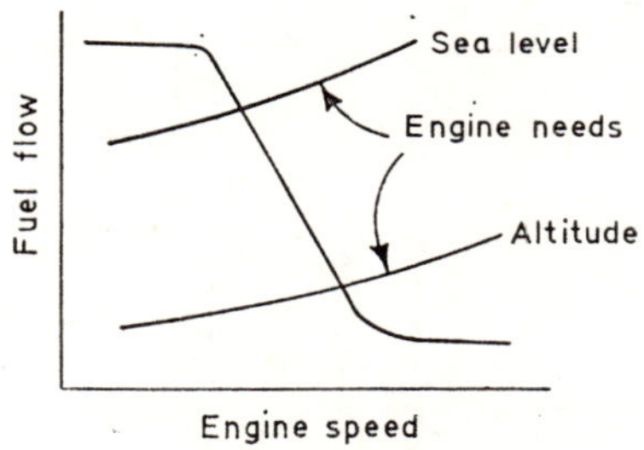

FIGURE 8.16 *Governor characteristics*

For stability purposes it is necessary to build in a certain amount of "rate" into the governor—if the fuel flow fell to zero exactly at the governed speed and then came back to its normal value as soon as the speed fell, the control would not be very satisfactory. Now the fuel flow versus engine speed characteristic will vary with flight conditions and will fall with increasing altitude. Curves for two typical cases have been plotted on the governor characteristic in Figure 8.16. The actual governed speed will be where the requirement meets the governor line and so it will be seen that as the fuel requirement falls off, the governed speed will rise. Again, however, newer designs of governor can be made which do not suffer from this trouble.

Dowty Components

Before making a start on describing some typical Dowty fuel system components it must be emphasized that only simple examples are taken; they are chosen to demonstrate the principles involved. Actual units as made by any company are highly complex devices which are designed with a view to compactness and freedom from faults so that their construction and detailed operation is very difficult to understand.

In Chapter 6, when dealing with fuel atomization, it was said that the Dowty Company have always specialized in the spill-type burner as by this means good atomization is achieved over a wide range of fuel flows. Use of the spill burner has had a considerable influence on the overall design of the engine fuel system as it is possible, for example, to use a fixed capacity, positive displacement pump and control the flow by altering a flow number in a spill line. Figure 8.17

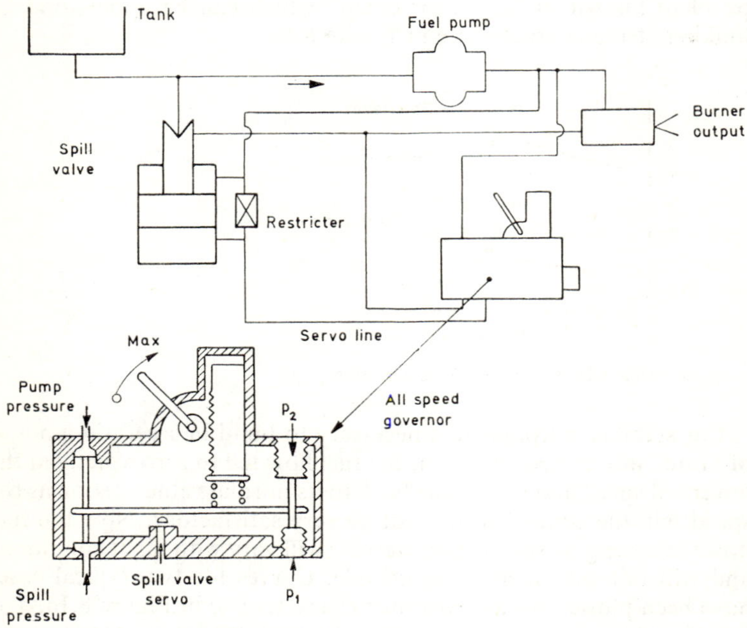

FIGURE 8.17 *A typical Dowty Fuel System*
(Courtesy Messrs. Dowty Fuel Systems Ltd.)

shows a flow diagram for a simple single circuit system and the governor in rather more detail.

Looking at the spill valve it will be seen that this is mounted on a differential area piston which is exposed to high pressure on the smaller side and pressure bled through a restricter on the other. If the servo line is completely blanked then there is no pressure drop across the restricter so the spill valve closes completely. On opening a half-ball valve in the servo line the pressure behind the piston falls and the spill valve opens thus reducing the flow to the engine. During equilibrium running there is always one half-ball valve open and the control is such that, as long as there is an error between a selected parameter and its actual value, the spill valve will move.

One very important characteristic of the spill burner used in the all speed governor shown in Figure 8.17 is that the pressure difference between burner supply and spill line depends only on the pump rotational speed—which will of course be a constant fraction of engine speed. This pressure difference can therefore be applied in opposition to a spring whose force is set by the pilot's lever, with a servo-line half-ball valve then being mounted on the pivot arm. Any fall in speed will close the valve so selecting more fuel until the spill valve has reached its new position. Capsules sensing the compressor pressure rise serve to re-set the governor as altitude increases and so to reduce its effective rate. Clearly other control units could be fitted to this simple system.

For an acceleration control, use might again be made of a compressor delivery pressure signal which would operate on the spill valve servo through an evacuated capsule. A minimum flow control could be fitted to avoid weak extinction on deceleration by setting the lowest flow as a function of intake pressure. Various overriding monitoring signals, such as for excessive jet pipe temperature, can be put on to a further half-ball valve which is in parallel with the others.

A disadvantage of the spill-type fuel control is that at low flows, when a high percentage of the pump output is just being circulated around the pump, the fuel gets hot. There is an energy input from the pump which is proportional to the flow and the pressure rise and is inversely proportional to the pump efficiency. For the high spill case the same physical amount of fuel is being passed repeatedly through the pump so it will get progressively hotter until some equilibrium high temperature is reached. Dowty have accumulated a lot of experience on this problem and have developed fuel systems, which are rather more complex than the one shown, where the difficulty has been overcome.

9 The Ramjet Engine

9.1 INTRODUCTION

The ramjet engine has appeared briefly once or twice before in this book but it is sufficiently distinct and important to merit a chapter of its own. It differs in two main respects from all the other reaction propulsion engines dealt with so far, firstly in that no rotating compressors or turbines are used in its operating cycle and, secondly, in that a ramjet cannot operate until it is moving at some fairly high forward speed. Having no rotating parts (apart from accessory drives) the ramjet can be made light and compact and can avoid the materials limitations encountered at high flying speeds by turbine engines. On the other hand, the necessity to provide some means of attaining the high initial speed required for operation is a severe disadvantage and limits the possible future of application of the ramjet.

Historically the ramjet story is fairly long as the principles of operation were set out in 1913 by the Frenchman Lorin. Early work in France associated with the engineer Leduc was limited in its scope and the ramjet did not come into extensive use until the design of supersonic guided missiles. These aircraft with their restricted duration and their high speeds formed a natural application of ramjet propulsion. In Britain the BAC/Ferranti Bloodhound, powered by the Rolls-Royce Bristol Thor ramjet is the most important type while in America the Boeing and Marquart Companies worked in this field for propulsion of unmanned aircraft at supersonic speeds. Although the main applications have inevitably been for supersonic propulsion, the subsonic ramjet has found some use for tip drive of helicopter rotors and P. A. Hilton & Co. of Southampton have made a useful little subsonic unit for educational purposes.

In this chapter we will first review the general performance of ramjets and see how the thrust appears. Each of the main components —intake, combustion and expansion systems—will then be looked at in greater detail and particular note will be made of the possible operational instabilities which may be encountered. The special variant, the supersonic combustion ramjet or scramjet, which holds great promise for propulsion at very high speeds will then be considered.

9.2 RAMJET PRINCIPLES OF OPERATION

Although the ramjet operating cycle was described in Chapter 3, the work is repeated here for completeness.

Figure 9.1 shows the layout of a ramjet engine and Figure 9.2,

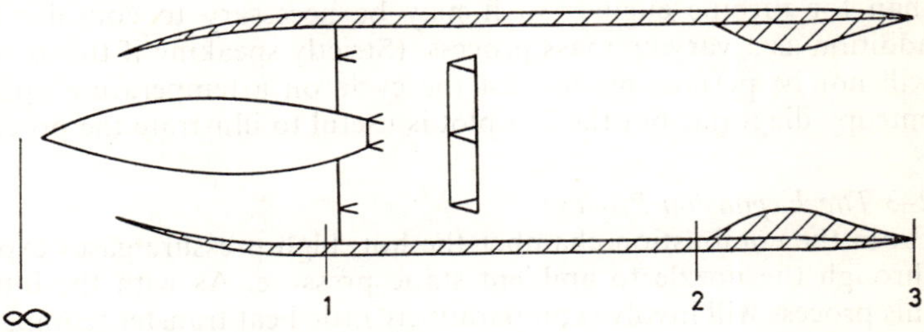

FIGURE 9.1 *Layout of a ramjet engine*

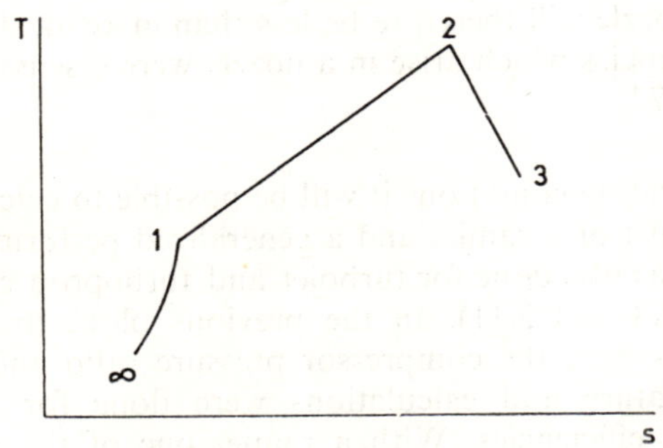

FIGURE 9.2 *The ramjet operating cycle*

which is numbered correspondingly, shows the cycle of operation on temperature–entropy coordinates. The cycle is as follows:

∞–1 Intake Section

Ideally the intake section operates to decelerate the air entering the intake isentropically (i.e. reversibly and adiabatically) from free steam velcoity to a low velocity for combustion. In practice the process is virtually adiabatic but irreversibilities must be present so there will be an increase in entropy. After the real process ∞–1 the stagnation pressure p_{01} will be less than the ideal but, of course, the stagnation temperature T_{02} will equal ambient plus the ram temperature rise.

1–2 *Heat Addition*
Heat is added in the combustion chamber by burning fuel. The real combustion process will involve losses in pressure due to Rayleigh flow effects and due to the need to provide turbulence in order to stabilize the flame. Also, the fuel/air ratio will be appreciably higher than for turbine engines so it may be necessary to consider heat addition as a varying mass process. (Strictly speaking if this is so, it will not be permissible to plot the cycle on a temperature–specific entropy diagrams but the *T–s* plot is useful to illustrate the process.)

2–3 *The Expansion Process*
From the combustion chamber the hot, high pressure gases expand through the nozzle to ambient static pressure. As with the intake, this process will involve comparatively little heat transfer (compared, that is, with the nozzle enthalpy drop) and internal irreversibilities will lead to an increase in entropy. The magnitude of acceleration of the gas in a real nozzle will therefore be less than in an ideal one. Details of the inefficiencies which arise in a nozzle were discussed in greater detail on p. 177.

For any set of operating conditions it will be possible to calculate the fuel flow and thrust of a ramjet and a generalized performance plot can be prepared as was done for turbojet and turboprop cycles previously (Figures 3.8 and 3.11). In the previous plots, the two main design variables were the compressor pressure ratio and the turbine inlet temperature and calculations were done for fixed values of component efficiencies. With a ramjet one of the main variables is no longer relevant as there *is* no compressor to give a pressure ratio. A generalized ramjet performance map must use different variables and useful ones are the flight Mach number number (which determines largely the intake pressure ratio) and the gas stagnation temperature at entry to the propelling nozzle. Figure 9.3 shows the form of such a plot with constant values of intake efficiency, expansion nozzle efficiency and combustion pressure loss being used in the calculations. Of these assumptions that of constant intake efficiency over a wide range of flight Mach numbers is difficult to justify in practice.

Two features are immediately apparent if ramjet figures are compared with those of turbojets. Firstly the temperatures used in ramjets are very considerably higher than those considered for turbojets—this reflecting the fact that there is no highly stressed

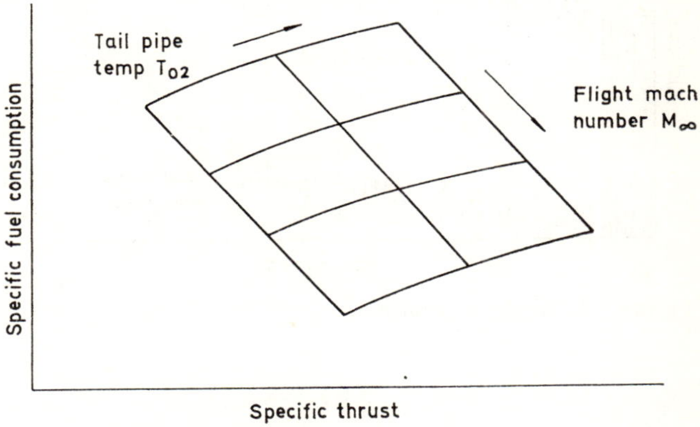

FIGURE 9.3 *Generalized ramjet performance*

machinery in the ramjet being subjected to high temperatures. Secondly, it is only at very high flight Mach numbers that the ramjet approaches the performance figures of the turbojet. The lesson this teaches is that performance is not everything and the ramjets attributes of light weight and simplicity may make it the optimum power unit in many cases despite poorer performance.

While dealing with the general performance of ramjets it is useful to consider one other plot, that showing the possible flight regions where engines can operate. This "flight corridor" as it is called is plotted on coordinates of altitude and flight Mach number as shown in Figure 9.4. To the right of the flight corridor, at high forward speed and low altitude, stress considerations prevent operations as the pressures generated within the ramjet body would be so high that the weight would become prohibitive. Above and to the left on the other hand, a combustion limit is reached where the combustion zone pressure is so low that stable burning is not possible. It is probably obvious that the limits of the flight corridor are somewhat indeterminate and that they can be pushed back by advances in the state of the art of materials and combustion.

It is of interest to ask the question "where do the forces act in a

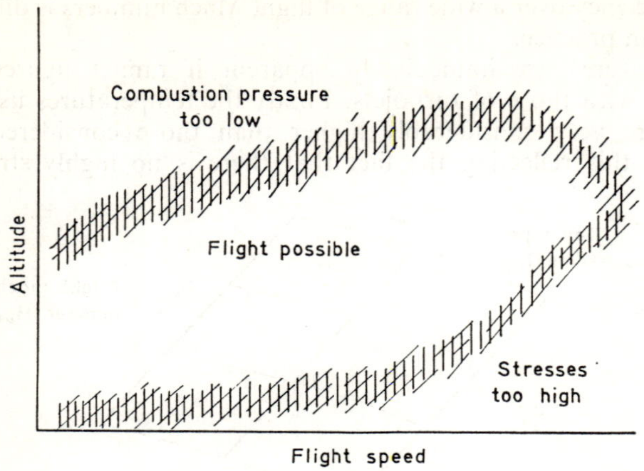

FIGURE 9.4 *The flight corridor*

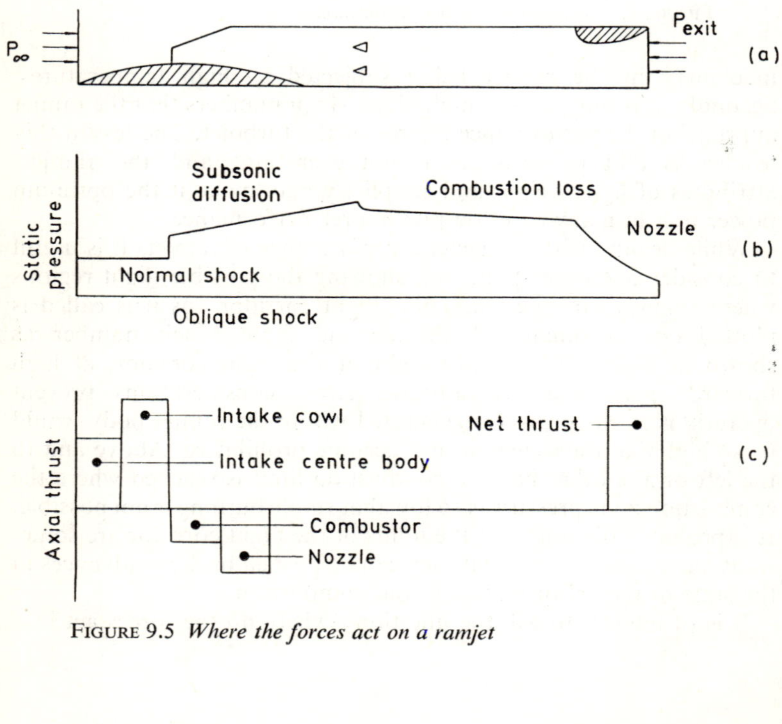

FIGURE 9.5 *Where the forces act on a ramjet*

ramjet which finally appear as thrust?" Newton's laws of motion give us the *overall* answer that the net thrust is equal to the rate of change of momentum of the gases passing through the ramjet but, as the engine seems to be basically a pipe, it is not immediately clear just where the forces act. To resolve this question consider the supersonicram jet shown in Figure 9.5(a) and the staticpres sure distribution through the engine as in Figure9.5 (b). Remembering that an axial thrust results whenever a static pressure difference exists across a projected area, we see that the thrust can be built up from a number of components as shown in Figure 9.5(c). These are

(i) A thrust on the centre body resulting from the large static pressure rise across the shocks and a further rise due to the subsonic diffusion.
(ii) A large thrust on the projected area of the cowl due to the higher than atmosphere pressure within it.
(iii) A negative thrust or drag on the combustion due to the aerodynamic and thermodynamic (Rayleigh flow) pressure drops.
(iv) A drag on the propelling nozzle due to the static pressure drop across it.

The net thrust is the algebraic sum of these forces so it will be seen that really a ramjet is pulled through the air by its nose! The first reaction would be to consider eliminating the combustion system and final nozzle until it is realized that they, in fact, cause the air to flow through the intake.

9.3 DESIGN AND PERFORMANCE OF RAMJET COMPONENTS

We will now look in detail at each of the three main components of the ramjet engine. For convenience only "conventional" ramjets operating at supersonic conditions but with subsonic combustion will be considered; the field of application for subsonic engines is small and the design problems are less severe than for the supersonic application while the supersonic combustion type has a section to itself below.

The Intake Section

In Chapter 7 (pp. 216–18) the problems of design and performance of engine intakes for supersonic flight were considered in some detail and it will be recalled that intakes were classified under two main headings, external and internal compression types. The various intakes, with the single exception of the pitot type which is only really suitable in transonic flight, all featured some form of spike or wedge ahead of the cowl leading edge, the purpose of this spike being to set up an initial oblique shock. In every case the final deceleration to subsonic conditions took place through the inevitable plane, normal shock.

All this work is applicable to a greater or a lesser degree of ramjet intakes. From the published literature it would appear that two dimensional intakes have seldom if ever been used in ramjet installations although there is no reason why non-axisymmetrical intakes should not appear in the future. The literature also seems to indicate that most intakes used so far in practical engines have been of the external compression form with single or double cone centre bodies.

One very important difference between the inlet for a ramjet as opposed to that for a turbojet is in the degree of subsonic diffusion required. Flow velocities which are suitable at the compressor inlet face of an axial flow compressor—say 150 m/s—are far too high for entry to a combustion section. (A considerable degree of diffusion is provided between the outlet of an axial compressor and the inlet of a turbojet combustion system.) Luckily for a ramjet there is no need to accommodate a centre shaft so the full cross-sectional area

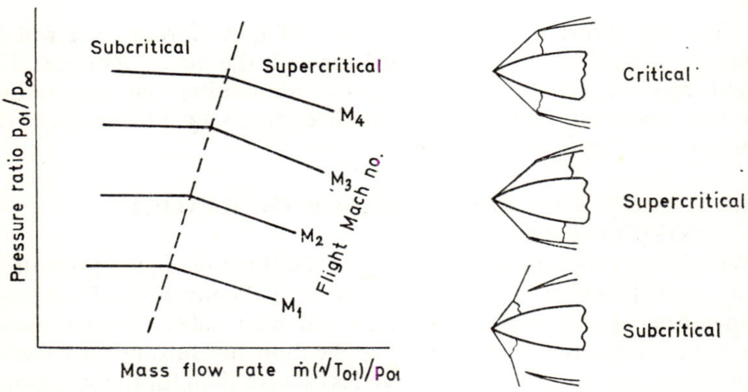

FIGURE 9.6 *Single cone intake characteristics*

of the engine is available for flow but even so the high degree of diffusion automatically implies a long diffusion if the intake is to be efficient.

As before the intake characteristic can best be plotted in terms of the intake pressure recovery and the non-dimensional air flow for lines of constant flight Mach numbers. Figure 9.6 shows a typical characteristic and the sketches give an idea of the intake shock configuration for the subcritical and supercritical régimes of operation at constant Mach number. (For convenience a single cone centre body intake has been shown.) On this characteristic may be superimposed operating lines which, in the case of a ramjet, can be expressed either in terms of the combustion outlet temperature or the air/fuel ratio. The argument for taking combustion temperature as the main variable is as follows.

Let station 1 be the inlet to combustion and station 2 be at entry to the final nozzle and let A be the nozzle throat area. Now if the nozzle is choked (as it probably will be in supersonic flights) then the mass flow parameter $\dot{m}\sqrt{T_{02}}/A_2p_{02}$ will be constant. The non-dimensional mass flow at combustion entry can be expressed therefore as

$$\frac{\dot{m}_1\sqrt{T_{01}}}{A_1p_{01}} = \frac{\dot{m}_2\sqrt{T_{02}}}{A_2p_{02}} \cdot \frac{\dot{m}_1}{\dot{m}_2} \cdot \frac{p_{02}}{p_{01}} \cdot \sqrt{\frac{T_{02}}{T_{01}}} \cdot \frac{A_2}{A_1}$$

Of the terms on the right-hand side $\dot{m}_2\sqrt{T_{02}}/A_2p_{02}$ and A_2/A_1 will be constant for a fixed geometry engine. The combustion pressure loss p_{02}/p_{01} will be related to the temperature rise as will be the mass flow ratio $\dot{m}_1/\dot{m}_2$. (The difference in flows will be accounted for by the fuel added.) If variations in these two parameters are neglected, it will be seen that at a given value of T_{02} the inlet mass flow function $\dot{m}\sqrt{T_{01}}/A_1p_{01}$ will vary with T_{01}, that is with $T_\infty(1 + \frac{1}{2}(\gamma - 1)M_0^2)$ where M_0 is the flight Mach number. (It is assumed that operation is in the stratosphere where the ambient temperature is constant.) A line of constant combustion outlet temperature on the intake characteristic will therefore have a positive slope.

Similar arguments may be applied for operation at a constant value of fuel/air ratio, only in this case the combustion temperature *rise* will be nearly constant and T_{02} will increase with increasing M_0. Allowing for these factors the slope of the constant fuel/air ratio line on the intake characteristic will again be positive and will have a slightly higher gradient.

It may be recalled that when discussing supersonic intakes for turbojets it was explained that a line on the characteristic at *negative* slope was undesirable as the aim was to operate near the critical line and to avoid moving into the subcritical region where there was a risk of instability. With this in mind it is clear that the operating lines considered for ramjets, with constant tailpipe temperature or constant fuel/air ratio, will be quite satisfactory. Of the two methods, that of operating at constant fuel/air ratio will be better in that it is close to the critical and will also be easier to arrange from the control system point of view.

The Combustion Zone

Turning now to the design and characteristic of the combustion system the problems encountered with ramjets are generally similar to those of gas turbine engines except that things are somewhat more difficult in many ways. At entry to a combustion system it is desirable that the pressure and temperature should be as high as possible. (This is true until one gets to the hypersonic flight regime where pressure and temperature get embarrassingly high from the stress point of view. See the section on the supersonic combustion ramjet below.) Now, because of the absence of a mechanical compressor, the pressure and temperature for a ramjet are bound to be less than for a turbine engine at a given speed. Not only does this make the stabilization problem more acute but in addition the temperature *ratio* across the combustion chamber will be higher. For this reason the stagnation pressure losses associated with Rayleigh flow conditions may be significant unless the entry Mach number is kept to a very low value.

Details of the actual hardware of ramjet combustion systems is difficult to come by for security reasons but the general form shown in Figure 9.7 may be taken as being representative. A baffled region

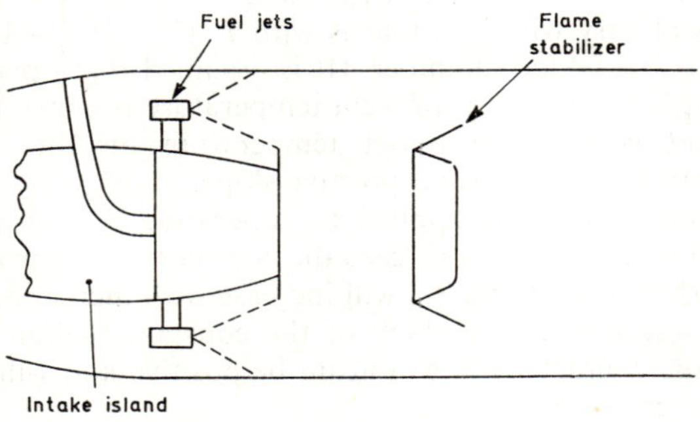

FIGURE 9.7 *A ramjet combustion zone*

downstream of the intake island is used to stabilize the flame and hold it in position and air circulates around the gutter strips until combustion is complete. Some means such as a pyrotechnic flare may be used to provide the initial ignition and this may be allowed to continue burning even after the ramjet has reached its operating conditions. By this means the flame holding can be improved without having to provide excessive blockage of the free stream.

Although the ramjet suffers from lower combustion entry temperature and pressure, it does score in that it is seldom necessary to operate over a very wide range of powers at given flight speed and altitude. In most of the applications to date, in unmanned missiles there is no need to run at idling conditions and the short life means that the engine will be at maximum thrust for most of its operating time. There is of course the need to operate over a range of altitudes and with the high rates of climb encountered by a ramjet the absolute pressure level at combustion will change rapidly. A good control system is needed to ensure that the required fuel condition (e.g. constant air/fuel) ratio is maintained.

While on the question of combustion, mention must be made of two important types of instability which are likely to arise during operation and which can wreck an engine in a very short space of time. They are known as "howl" and "screech". Howling is basically a fairly low frequency, high amplitude fluctuation in pressure which is similar to an organ pipe phenomenon. The frequency will depend on the physical dimensions of the engine and may be of the order of 25–100 Hz. Pressure fluctuations of $\pm$ 100% have been measured so there are clearly large destructive forces present.

Screech is a very much more complex problem as it is essentially a feature of the combustion process. Although the level of pressure fluctuations is lower the frequency (1–2 kHz) is so much higher that components may suffer fatigue failure in a very short time. In his review entitled *Expensive Noise* (ref. 1) S. L. Bragg explained how screech could arise from *circumferential* fluctuations of conditions in the combustion space. As the rate of burning is varying rapidly and the inertia and damping of the oscillating system are very small, screech can build up rapidly. Incidentally, in his paper on ramjets (ref. 2) R. R. Jamison states that at one stage screech was tolerated as under screeching conditions combustion efficiency was found to be improved, but eventually when damage was caused it was necessary to avoid this state.

The Final Nozzle

On the question of the design of the final nozzle for the ramjet little need be added to the work on nozzles for supersonic aircraft covered in Chapter 5. The same problems are encountered of choosing a shape which will be the optimum at some critical operating condition for the whole vehicle but which will have acceptable losses away from this design point. Where the ramjet scores is that it is unnecessary to cater for the stand-off case where good efficiency is needed at low flight speed and low nozzle pressure ratio. On the other hand the ramjet is fundamentally a compact, light-weight engine and any weight or bulk increases which would result from making a variable geometry nozzle would have proportionally larger effects on the overall engine weight than they would in the case of a turbine engine.

All our consideration so far has been on the gas dynamic aspects of ramjet design and performance but there is, of course, the vital problem of ensuring that all the ancillary items of equipment perform adequately. From the development point of view, a fuel system for example, may involve as much effort to get right as an intake—and if it is *not* right, the whole performance of the vehicle can be impaired. Unfortunately, it is not really feasible to discuss ancillary equipment in general terms and security restrictions are such that it is not possible to write about specific equipment in any detail. However a few general principles can be stated.

Firstly, underlying all ramjet design work is the supposition that it is not economic to design a system which will have 100% probability of success under all conceivable operating conditions. To take a specific example, a defensive missile would ideally have 100% chance of destroying its target if fired anywhere in the world under any climatic conditions. It would in theory be possible to approach this state of reliability and readiness but only at the extent of great complexity and cost—and no one would guarantee the last fraction of the 100% under all conditions. Instead it would be preferable to offer 90% probability of success on 99% of the climatic conditions which could possibly be encountered. This could be achieved for a fraction of the cost and it would be cheaper overall to buy, say, three missiles of the poorer quality than one which was well-nigh perfect. If the designer is to adopt this philosophy of deliberately accepting less than the best, he must understand the theory underlying reliability of complex systems and must be in a position to quote results of rig testing to back up his reliability claims.

These ideas can be seen in action in the design of a ramjet fuel system. As was pointed out in Chapter 8, the two main types of control system possible are the scheduling and the closed loop with the closed loop being more accurate but more expensive and more liable to give stability troubles. A typical ramjet control system would be one which was designed to give a constant air/fuel ratio. This could be done by sensing some stagnation pressure in the intake which, for a given flight Mach number in the stratosphere, would give a measure of the air mass flow, and scheduling the fuel flow as a function of this pressure. There would be the inevitable limitations of a scheduling system that no allowance was made for variations between engines or for different climatic conditions but these disadvantages would not outweigh the advantages of simplicity and light weight. Some form of overriding control such as a flight Mach number limitation would normally be fitted to correspond to the top-speed governor of the turbine engine.

The actual mechanical details of a fuel system for a ramjet are influenced by the fact that, as there are no rotating parts, it is difficult to get a drive from the engine to power a fuel pump. Some form of energy storage such as an electrical battery or a compressed gas cylinder might be used to provide the energy for the accessories. Of these two the compressed gas storage has advantages as its "shelf-life" is indefinite (it can stand unattended for years without deteriorating) and it provides its energy in a suitable form for driving a small turbine which can be directly coupled to a centrifugal type fuel pump. This particular form of pump is able to handle the large flows needed by a ramjet but does not need to give the very high pressures such as are wanted by a turbine engine. From the pump, fuel will pass through the metering unit and on to the injectors. The main function of the injectors will be to ensure adequate atomization and stable combustion over the whole range of flight conditions. Their size and disposition will be determined on the basis of development testing of the complete combustion.

Although the primary control of ramjets will be effected through variations in the fuel flow there may be secondary controls acting on the position of the intake centre-body or on the geometry of the final nozzle. If these are required the controls would usually be of the closed loop variety. An intake control for example might sense the position of a normal shock (by detecting the sharp static pressure rise) and altering the axial position of the centre-body so that this shock always lay in the same plane. As always with closed loop controls, testing would be needed to show that the system was always stable. Each additional component which requires to be moved in a vehicle will need some supply of energy and so the demands for energy storage will be increased.

The main line of thinking on ramjets and their mechanical problems is coloured by the knowledge that all practical engines produced to date have essentially been for short-life application. This has always been one of the key considerations in design as it has meant that structures can be lightened and simplified and that comparatively crude power systems—such as gas bottle storage—may be acceptable. Looking to the future, there is a possibility that the ramjet may turn out to be the best engine for some long-range applications. In these cases considerable design changes would be needed to give long life and economic ancillary drive. A promising approach would be to use a ram-air turbine for power generation as in this case power would be available whenever the ramjet was capable of working and the basic engine simplicity would be unimpaired. In fact, some ramjet powered vehicles already use ram-air driven power units to meet their energy needs.

9.4 RAMJET TESTING

With any engine which is expensive to make and which is designed to operate close to its limits of strength it is important to carry out extensive testing before fitting it into a costly vehicle and risking the loss of the vehicle due to malfunction. Ground testing of an aero-engine can never be a proper substitute for flight testing but it is the next best thing and it will generally be possible to start ground testing before test aircraft are available. In the work that follows, we wish to consider some of the more important aspects of testing ramjets both on sea-level testings and in flight.

In the testing of a ramjet at sea-level it is important to reproduce as closely as possible the conditions which the engine will encounter in flight. Two main methods of test are used, the "free jet" method and the "connected test" method. Free jet testing involves producing a stream of air at the pressure, temperature and Mach number corresponding to the flight condition and of sufficient dimensions to encompass the whole engine. Only in this way is it possible to reproduce the flow conditions around the engine and determine the overall engine performance. Connected testing is adequate for trying the combustion system, as the combustion process is not affected by flow outside the engine. In this case air at conditions corresponding to those leaving the intake diffuser is fed directly to the combustion system.

Ground installations suitable for testing ramjets are liable to be very massive pieces of equipment. One of the best known in Britain is the High Altitude Test Plant (HATP) at the Bristol works of Rolls-Royce. (The altitude facilities at the National Gas Turbine Establishment at Pyestock are used very much more for turbo engine testing.) The Bristol HATP shown diagrammatically in Figure 9.8 is able to test complete ramjets at conditions equivalent to 2 000 miles/h (1 250 km/h) flight speed at 100 000 ft (30 000 m). The air supply for reproducing the flight conditions is 75 kg/s at six atmospheres pressure, the air supply compressor requiring a power input of about 18 000 kW. If necessary, the air going to the cell may be heated or cooled. A ramjet on test is mounted in a large pressure vessel which is evacuated to a pressure corresponding to the atmospheric pressure to be simulated. When the test starts the exhausters which suck down the test cell pressure will be required to get rid of huge quantities of air and combustion products. Mechanical

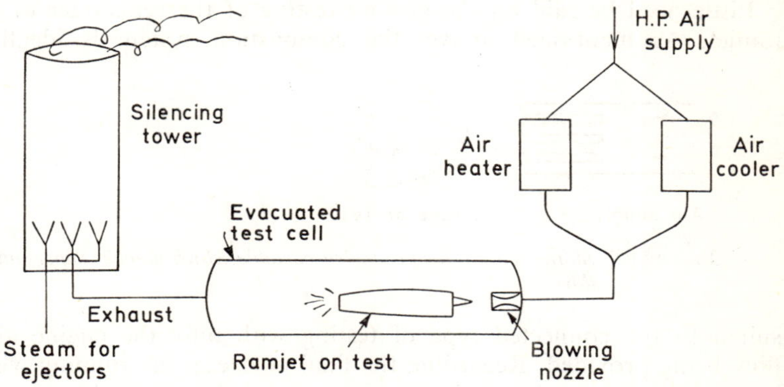

FIGURE 9.8 *Arrangement of the High Altitude Test Plant at Bristol*
(Courtesy Rolls-Royce Ltd., Bristol Engine Division)

exhausters to handle exhaust products are difficult to design and are expensive so ejectors, using steam as the primary pumping fluid, are used in the Bristol plant, the steam being stored in large accumulators which are charged over a period of some hours. Overshadowing the whole plant is the silencing tower which pours out a vast plume of steam while the plant is running.

Testing of the intake section of a ramjet *must* be carried out using the free jet method as the whole intake characteristics are affected by the external and internal flows. For an intake being tested on its own it is important to get the correct flow to simulate the engine mass flow as the required pressure may be less than atmospheric. Some form of evacuated test cell may be needed. In this paper on ramjets referred to previously (ref. 2) Dr Jamison describes a slotted nozzle which has proved very useful in free jet testing. This nozzle, sketched in Figure 9.9, will adjust itself until the static pressure inside the slotted portion is equal to atmospheric; after that, any excess will spill through the slots. It acts therefore like a variable area ratio convergent–divergent nozzle with the outlet Mach number being uniquely determined by the ratio of inlet stagnation pressure to ambient static pressure. Rapid changes in outlet Mach number are possible so that transient testing of intakes is feasible.

One essential feature of any intake testing will be the use of

Schlieren apparatus to study the intake shock pattern. This well-known technique (described in ref. 3) makes use of the sudden density changes through a shock wave to allow the shock configuration to be projected onto screens or photographed.

Little need be said on the ground testing of the remainder of a ramjet. As mentioned above, the combustion system is ideally

FIGURE 9.9 *Slotted nozzle for producing variable Mach number supersonic flow*

suited to the connected type of testing with only the engine air flow being provided. Regarding the final nozzle, some running will be needed to prove the mechanical integrity of the unit and to check the characteristics of any variable geometry. When the complete engine is being run in a test cell it may be convenient to replace the proper final nozzle by one which is convergent only. This will be simpler and cheaper and, provided the nozzle is choked, the correct conditions for the intake and combustion systems may be reproduced exactly. A further advantage is that it is not necessary to exhaust the test cell to a pressure as low as the ambient pressure at the altitude being simulated but only to bring it down to such a value that the nozzle is choked. True thrusts cannot be measured (the convergent nozzle will be less efficient than the fully convergent–divergent) but the power requirements for the exhausters can be greatly reduced.

Passing now to the question of flight test, ground testing, no matter how elaborate, can never be a full substitute for getting some experience in the engine's true environment and operating state. A ramjet will be fitted to a supersonic flight vehicle and will usually be working within the shock pattern set up by the remainder of the machine. There will thus be effects on the air flow pattern in the intake—non-uniform inlet profiles, yawing of the intake, etc.—which are bound to affect the intake performance. It will be necessary therefore, to carry out a flight test programme in order to study these problems, and the collecting of information from the engine will be a very difficult one. Even in the ground test, when weight and bulk are of comparatively minor importance, the fact that ramjets only run for short times means that some form of high speed data recording is needed which can be analysed at leisure later. For flight test, weight must be kept to a minimum and in most missile cases the vehicle is lost at the end of its flight. Some form of radio telemetering is needed to pass test-data from the vehicle where it is collected to some ground station for analysis.

The other important factor which arises in flight test which is difficult to simulate on ground test is that for a major part of the flight the vehicle will be accelerating. This means that the ramjet will be operating away from its design conditions and that it will be undergoing a transient operation so that any instabilities which may be latent in the engine or control system are most liable to come out. In addition the acceleration and the very violent flight manoeuvres carried out by missiles impose large fluctuating loads on the structure and on the components. Checks can be done in the laboratory using centrifuges to show that components will still function when subjected to high acceleration loads but the rapid changes in load are very difficult to simulate.

"That last flight was a failure as nothing went wrong!" This remark by a test engineer sums up one important aspect of flight test work, namely that the important thing is to keep finding the weak spots in a system and to cure each problem in turn. A missile test firing where nothing goes wrong only gives the information that nothing was stretched to the limit; if something *does* break then the designers and development engineers have something to work on. Naturally, as the end of the test programme draws near, the failure rate should fall to zero so that the customer can be sure that he is getting a reliable piece of equipment.

Although it is not strictly an aero-engine one particular ramjet deserves special mention and that is the instructional unit made by Messrs. P. A. Hilton of Southampton. Probably more people deal closely with this little engine that ever encounter its bigger or more exotic brothers in the missile field. The Hilton ramjet is a small subsonic unit which is generally supplied as part of a complete educational set. It is free jet tested at atmospheric pressure using an air supply velocity of the order of 130 m/s and is normally designed to burn gaseous propane which is stored in liquid petroleum gas bottles. Even though the thrust is very low and the shock pattern of a supersonic ramjet is missing the little Hilton unit is well instrumented and gives a clear illustration of how this type of engine works. Apart from this example and some possible applications as tip-drive units on helicopters it is difficult to see much future for subsonic ramjets because, due to the low ram pressure ratio, their thermodynamic efficiency is bound to be low.

9.5 THE SUPERSONIC COMBUSTION RAMJET

Every time the design of combustion systems has been discussed, it has been emphasized that it is very important to provide a sheltered zone for combustion where the gas velocities will on average be lower than the flame speeds. There is an engine of the future that seems at first sight to completely contravene this basic requirement, an engine wherein combustion takes place in a supersonic gas stream. The supersonic combustion ramjet or "scramjet" looks to have a distinct place in the future propulsion spectrum and it is of interest to ask the two questions of *why* this design should be attempted and *how* a practicable engine may be made. Although a large amount of preliminary fundamental research has been carried out no scramjet as such has yet been built.

Out of the considerable literature on this topic two main items have been selected for further reading. Ref. 4 is an article by Woods-Ballard which is an intelligent layman's guide to the whole question, while ref. 5 is a review paper by one of the great workers in the field, A. Ferri.

Why should we consider the scramjet? In the history of aviation so far there has been a continuous increase over the years in the flight speed of vehicles. The subsonic and transonic regimes have been followed by the supersonic and the next step will be to tackle hypersonic flight where flight Mach numbers of 5 and more will be used. This is not to say that very many aircraft will fly at these speeds—any more than many at the moment operate supersonically—but for certain limited but important applications, such as the launching of space vehicles by carriers using air breathing engines, there does seem to be a potential market. For these reasons a number of feasibility studies are being carried out by the designers to try and decide what main problems will be encountered by aircraft operating in the range of about Mach 5 to Mach 10.

The dominant factor in any of these studies is that very high free stream stagnation temperatures will result from the high flight speeds. To give some idea of the order, Figure 9.10 plots stagnation temperatures as a function of Mach numbers and altitudes and it will be seen that, at a Mach number of 10, temperatures of the order of 3 000 K will be encountered. These temperatures are calculated taking account of dissociation effects with the air, a factor which increases the effective value of the specific heats at high temperatures. Quite clearly, the structure at the leading edge of aircraft and engine which will be subjected to this temperature will need special materials and/or cooling provision.

Suppose that a conventional subsonic combustion ramjet were

to be designed to operate under hypersonic flight conditions. The air entering with a very high stagnation temperature would be decelerated with hardly any change in stagnation temperature until the *static* temperature nearly reached the stagnation value. Regarding the pressure, with the very high isentropic pressure available even a

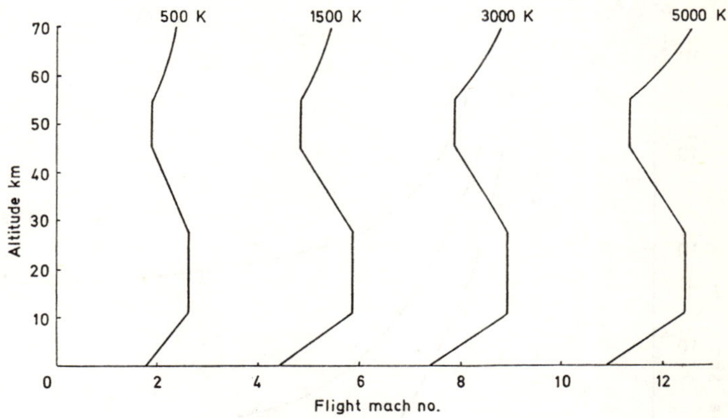

FIGURE 9.10 *The variation of stagnation temperature with altitude and flight speed*
(Courtesy Rolls-Royce Bristol Engine Division)

very inefficient intake would result in very high values of static pressure—of the order of scores of atmospheres. With these conditions of high temperature and pressure the problem of stressing the combustion system would be very severe, and present day materials would not be suitable for this work. In addition, decelerating the air to subsonic flow conditions would require a large cross-sectional area and consequently a bulky, heavy engine.

With these almost intractable mechanical problems it is necessary to see whether supersonic combustion is in any way feasible. Firstly, concerning fuels there are clear advantages if some fuel more exotic than kerosene might be used, An obvious candidate is liquid hydrogen which has great advantages as a cooling agent for structures, etc., besides being a fuel which can be burned very easily. For these reasons, many of the fundamental research studies on topics related to supersonic combustion have made use of hydrogen fuel. Another possibility would be to use methane which could be carried in tanks in liquid form being evaporated before it was fed to the combustion chamber. Although methane has a lower calorific value it has a much higher density and is not at such an embarrassingly low temperature as liquid hydrogen.

The key to the feasibility of supersonic combustion lies in the reduction of the time to complete a chemical reaction as the temperature of the reactants is increased. This can be illustrated by Figure 9.11 which shows the reaction time against initial temperature

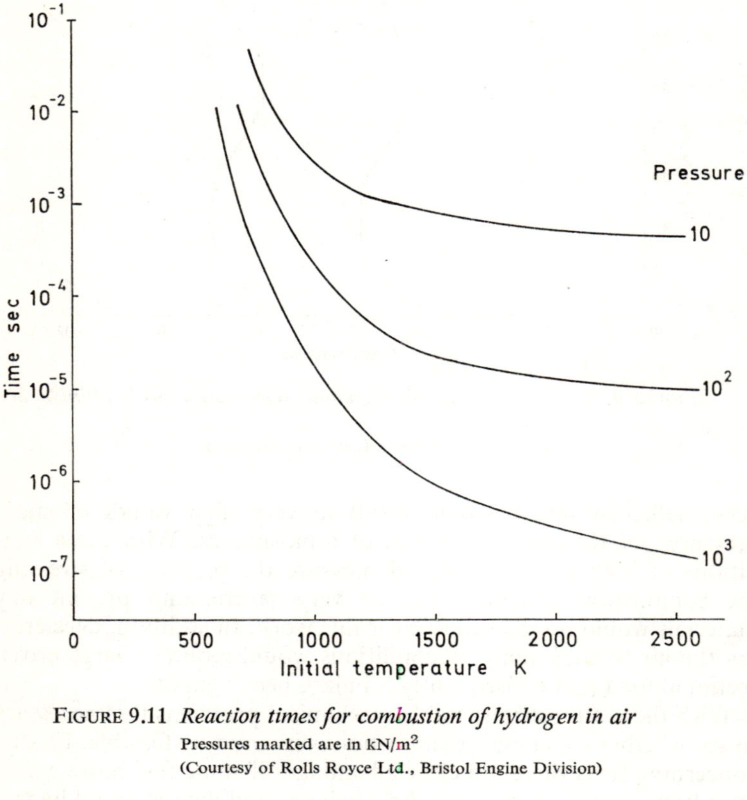

FIGURE 9.11 *Reaction times for combustion of hydrogen in air*
Pressures marked are in kN/m^2
(Courtesy of Rolls Royce Ltd., Bristol Engine Division)

for combustion of hydrogen in air at various pressures. On this graph the time scale is logarithmic and the steepness of the curves in the region below 1 000 K will be noted. Now the *distance* required for a reaction to take place will be the product of the reaction time and the velocity. When the reaction time falls to the order of milliseconds and less, then an acceptable combustion chamber length will be obtained even with mean gas velocities of thousands of metres per second.

The above brief discussion on the scramjet can only hope to introduce the problem by showing that there is a real need to study this type of cycle and that there is no technical barrier to carrying out the work such as would arise if the chemical reaction times were long. Having said this, the question comes down to the technical ones of how to introduce a fuel into a supersonic air stream without setting up a shock pattern, how to get uniform combustion and reasonable combustion efficiency, and how to get over the materials and stressing problems which will be extremely difficult. All these problems are essentially those of the design and development engineer who must work with the experts in each field—gas dynamics, materials science and so on—to produce acceptable engines. It will be a long struggle but, given the requirement, none of the problems would seem to be insurmountable.

10 Rocket Propulsion

10.1 INTRODUCTION

All the types of engine considered so far in this book have been "air breathing engines" which pick up from the atmosphere the oxygen needed for combustion. The rocket differs in this one main respect as it is independent of any outside supplies for its operation. This independence confers great advantages on the rocket, making it simple in construction and allowing it to operate either within or beyond the atmosphere. On the other hand the fact that the oxidizer must be supplied in the form of chemicals carried by the vehicle means inevitably that the rate of consumption of propellants is very high. Operation is therefore costly and the rocket is essentially a heat engine which should only be operated for comparatively short periods at any one time.

Being a heat engine the rocket must obey the laws of thermodynamics and gas dynamics as set out in previous chapters. We will start our work of this chapter by looking at the basic rocket engine and seeing how these laws are applied. One thing that will be immediately apparent is that the high temperatures encountered in most rocket work influence the thermodynamics and fluid properties of the working medium—the phenomenon of dissociation becomes very significant. From this basic work we will pass on to consider in greater detail the main classes of rocket in use today, the liquid fuel, solid fuel and hybrid types. Each of these has advantages and disadvantages which will be discussed in the light of their main applications. Following this comes some elementary work on rocket mechanics which aims to look at the complete vehicle rather than just the engine. Finally brief mention is made of some of the non-chemical rockets which are being investigated for space exploration work.

The interest in rockets in recent years has led to the publication of a number of good textbooks. Of these, that by Sutton (ref. 1) deserves special mention as giving both the fundamental background and a wealth of detail on a wide variety of engines. Although much more concise, the book by S. L. Bragg (ref. 2) is excellent in giving a clear, compact explanation of fundamental principles especially as applied to liquid fuel rockets.

10.2 THE BASIC ROCKET ENGINE

The basic rocket engine consists (Figure 10.1) of a chamber in which

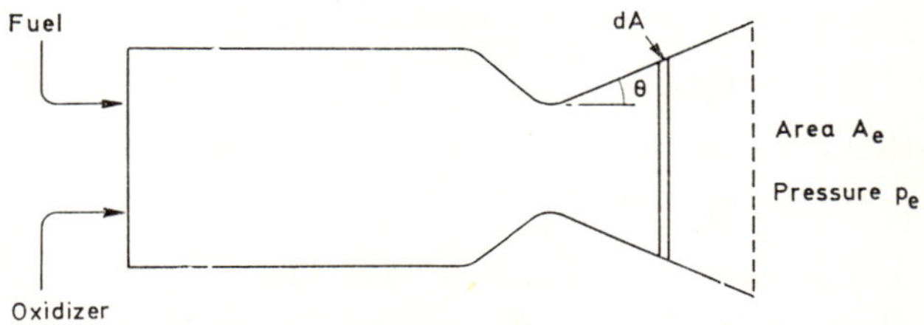

FIGURE 10.1 *The basic rocket engine*

a chemical reaction takes place producing high temperature gases, followed by a nozzle through which the gases expand to the surrounding conditions—it would be wrong to say "expand to atmosphere" as the rocket might be working in the vacuum of empty space. The rate of generation of the gases is matched to the nozzle throat area so that some desired pressure exists in the reaction chamber. Pressures customarily used depend on the type and application of the rocket but range from perhaps 1 000 kN/m^2 to 7 000 kN/m^2 or more. Consequently the nozzle expansion pressure ratios are very much higher than those encountered in turbine and ramjet work.

When the rocket cycle was first mentioned in Chapter 3 the general statement was made that the thrust could be calculated from the overall rate of change of momentum and that only gross thrust was of interest as there could be no intake momentum drag when a vehicle carried all its propellants from the start of the flight. It is now necessary to look at this question of thrust production in rather more detail and to see the effect of variations in surrounding pressure on the overall vehicle performance.

Consider the layout of chamber and nozzle shown in Figure 10.1. Acting on any elementary area dA of the walls there will be a static pressure p and the element of wall will be inclined at some angle θ to the axial direction. The nozzle exit area will be A_e and at the plane of the exit the gas static pressure will be p_e. If the surroundings are at a static pressure p_a then the axial components of the forces on the gases and on the rocket walls can be calculated as follows:

(i) The walls will exert a rearward pressure force on the gas given by

$$F_0 = \int p \cos \theta \, dA$$

There will be an equal and opposite reaction force acting forwards on the walls.

(ii) There will be a forward force on the gases due to exit pressure p_e acting over area A_e given by

$$F_e = A_e p_e$$

(iii) Hence the net rearward force F_N on the gases will be

$$F_N = F_0 - F_e$$

But, by Newton's second law of motion, net force is equal to rate of change of momentum of the gases, that is

$$F_N = \dot{m} V_e$$

Therefore

$$F_0 = F_N + F_e = \dot{m} V_e + A_e p_e$$

F_0 the forward force in the rocket walls is the reaction force produced by the gases on the rocket.

(iv) The surrounding pressure acts over the whole outside surface of the rocket except for the actual nozzle exit area. This pressure will exert a rearward drag force on the rocket given by

$$F_D = \int p_a \cos \theta \, dA = A_e p_a$$

(A moment's thought will show that this is true whether the final nozzle area is the largest cross-sectional area or not.)

(v) Combining all these forces, we get the equation for the thrust of a rocket as

$$X = F_0 - F_D = \dot{m} V_e + A_e(p_e - p_a) \quad (10.1)$$

The units are those used in previous chapters.

Equation 10.1 shows that in the simplest case of a nozzle operating at its design pressure ratio where $p_e = p_a$, the thrust is made up of a purely rate of change of momentum term. It shows also that as the surrounding pressure falls the thrust will rise due to the decreasing value of the $A_e p_a$ term. In the work that follows the simplifying assumption is made that the rocket is operating in an atmosphere with $p_e = p_a$ so that the simpler thrust equation

$$X = \dot{m}V_e \qquad (10.1a)$$

may be used. It should be appreciated however that in the more general case there will also be pressure components of thrust.

V_e, the nozzle exit velocity in this equation, is found by applying the steady flow energy equation (equation 1.12) to the flow; making the usual assumptions that the nozzle flow is horizontal and with negligible heat transfer we get

$$V_e = \sqrt{[2 \times 1\,000c_p(T_{01} - T_e)]} \qquad (10.2)$$

(T_{01} is the nozzle entry stagnation temperature in K).

Assuming for the ideal case that the flow is isentropic the equation becomes

$$V_e = \sqrt{[2 \times 1\,000c_pT_{01}(1 - T_e/T_{01})]}$$
$$= \sqrt{[2 \times 1\,000c_pT_{01}\{1 - (p_e/p_{01})^{(\gamma-1)/\gamma}\}]} \qquad (10.2a)$$

A nozzle efficiency may be introduced to allow for thermodynamic irreversibilities which occur in an actual nozzle.

The most important single parameter defining the performance of a rocket is the *specific impulse I*. The impulse of a force is the product of force and time, and specific impulse is the impulse per unit mass; it has the units of force × time/mass, that is force/rate of mass flow. Specific impulse for a rocket is therefore defined by the equation

$$I = X/\dot{m} \qquad (10.3)$$

where X is the thrust in newtons and $\dot{m}$ is the mass flow rate of propellants in kg/s. Specific impulse, therefore, has the units of N-s/kg.

Going back to the definition of the specific fuel consumption of an airbreathing engine we see that specific impulse is related to the *inverse* of the specific fuel consumption. The differences are that specific fuel consumption is a measured on an hourly basis of time (SFC is in kilogramme of fuel per hour per newton of net thrust) and that the specific impulse is based on the total flow of *propellants* whereas the SFC only takes account of the fuel supplied (which for a normal turbojet will be but 15–20% of the total mass flow rate of propellants).

The units of I, namely N-s/kg, are in fact those of a velocity in m/s and the specific impulse is equal to the ideal nozzle outlet velocity, i.e.

$$I = V_e \tag{10.4}$$

Consider now the factors influencing the specific impulse of a rocket. Writing equation 10.4 in equation 10.3 gives

$$I = \sqrt{[2c_pT_{01}\{1 - (p_e/p_{01})^{\gamma/\gamma-1}\}]}$$

From equation 1.11 we can put $c_p = R\gamma/(\gamma - 1)$; also if the products of combustion have a mean molecular weight M, R is calculated from the universal gas constant $\bar{R}$ using equation 1.19, that is $R = \bar{R}/M$. Thus

$$I = \sqrt{\left[\frac{2T_{01}\bar{R}\gamma}{M(\gamma - 1)}\left\{1 - \left(p_e/p_{01}\right)^{(\gamma-1)/\gamma}\right\}\right]} \tag{10.5}$$

In the limiting case where the rocket nozzle is exhausting to a vacuum $p_e/p_{01} = 0$, and even when exhausting to atmospheric pressure this term is small. Combining the constant terms in this equation into a single constant K gives for the limiting case

$$I_{lim} = \sqrt{\left[\frac{K_1\gamma T_{01}}{M(\gamma - 1)}\right]} \tag{10.6}$$

From this equation it is clear that to get a high specific impulse

(i) The reaction temperature T_{01} should be as high as possible

(ii) The mean molecular weight M of the products should be as low as possible

(iii) Low values of γ for the products are advantageous.

(These three qualitative statements will still hold true for the cases where the exhaust pressure has a finite value although in this case the actual specific impulse will be lower than the limiting value given by equation 10.6.)

Later in this chapter, when looking at the propellants for liquid fuel rockets the specific impulse on a *volumetric* basis will also be quoted. Some propellants which look promising with high values of I may be of such low density that the tankage needed to carry them would impose excessive penalties of structural weight. Liquid hydrogen is a fuel which tends to fall into this category. In this case too, there is the further complication that liquid hydrogen can only be stored in well insulated tanks which again add to the bulk and structural weight problem.

When designing a nozzle to expand the gases correctly over the whole expansion pressure ratio, use can be made of the variable area isentropic flow curves to find the area-ratio at the different stations. Allowance must be made for the various irreversibilities which occur in the non-ideal case. Inevitably with the pressure range to be covered, the nozzle will be of convergent–divergent form and care must be taken that the divergence from the throat is not too abrupt. If too large an angle is chosen then there will be a strong tendency for the flow to leave the nozzle non-axially with a resultant loss in thrust. A thrust coefficient can be defined to allow for these non-uniformities in flow, and tests show that this coefficient is primarily a function of the equivalent cone angle of the divergent portion. Use of a small cone angle is clearly desirable but this adds to the length, weight and bulk of the rocket.

These statements on how to calculate the specific impulse and nozzle flow areas for a rocket have been set down in their simplest form but, like this, they give little idea of the true complexity of the problem. In using the equations, the first thing to do is to find the composition of the reaction products so that the mean molecular weight and the mean value of the specific heats ratio γ can be calculated. But finding the products of the reaction is itself a most difficult problem. It is all very well to write the equation for the reaction as was done in Chapter 1 (p. 20) but is there any certainty that combustion will be complete? The fuel and oxidizer will be supplied at nearly their stoichiometric ratio in order to get a high combustion temperature and it will be difficult to ensure adequate mixing and complete combustion. Any uncombined reactants will mix with the products and alter the mean thermodynamic properties.

Dissociation too will make the problem much more complicated. At the high temperatures in the combustion chamber there is bound to be an appreciable breaking down of the normal combustion products so the estimation of the product composition will be more difficult. The effects of dissociation will also reduce the temperature of the gases entering the nozzle as it is really equivalent to a loss in combustion efficiency.

On coming to the final nozzle and the calculation of its profile, a further important problem arises. Assuming for the moment that it is possible to determine correctly the composition of the gases entering the nozzle and that their temperature has been found, as the gases expand down the nozzle their static temperature will drop so the equilibrium composition will vary—compounds which have dissociated at the high temperature will be reformed as the temperature falls. However, the time taken for the gases to pass down the nozzle (it will be measured in milliseconds as the average gas velocity is so high) will be of the same order of magnitude as the reaction time for recombination of the dissociation products. There will be a strong chance therefore that the recombination rate cannot keep up, so that, at any station in the nozzle, the actual composition of the products will not be the equilibrium composition for that static temperature.

An exact solution to the problem of a real expansion, involving as it does the nozzle length as well as the thermodynamic properties of the working fluid, would be extremely difficult. Instead two extreme cases are taken representing the limits of what may happen. The assumptions made are known as "frozen composition" or "shifting equilibrium". In the former case it is assumed that the flow through the nozzle is so rapid that the composition of the gases is constant throughout the nozzle—that no recombination of dissociation products takes place. Shifting equilibrium on the other hand, assumes that at each station in the nozzle the composition of the gases is the equilibrium composition as determined from the static temperature. Naturally in this case, the mean molecular weight and mean γ for the mixture will be varying continuously down the nozzle so the calculation of the flow conditions will be much more difficult. As the temperature of the reaction and the actual composition of the products are different for every combination of fuel and oxidizer each case must be treated on its own in carrying out the calculations; they are however independent of the nozzle dimensions. The frozen equilibrium case assumes a nozzle so short that no recombination takes place whereas the shifting equilibrium implies a nozzle long enough to allow everything to become settled.

The figure of specific impulse gives a measure of the thrust produced by the propellants and is a performance figure applicable to the complete engine. Two other overall performance figures can be defined which are of particular interest. They are the internal or thermodynamic efficiency and the external or propulsive efficiency. These two correspond in general to the parameters defined at the beginning of Chapter 3 as the internal efficiency is largely a function of the engine while the external efficiency depends on how the engine is used in flight. The thermodynamic efficiency is defined as

$$\text{Thermodynamic efficiency} = \frac{\text{Final kinetic energy of propellants}}{\text{Initial total enthalpy of reactants}} \tag{10.7}$$

The propulsive efficiency on the other hand is given by the equation

$$\text{Propulsive efficiency} = \frac{\text{Final kinetic energy of vehicle}}{\text{Total kinetic energy imparted to propellants}} \tag{10.8}$$

10.3 THE LIQUID FUEL ROCKET

When compared with its solid fuel counterpart the liquid fuel rocket has had a very short history but in the time it has been in existence it has been responsible for most of the major advances in rocket technology. This has been true in the fields of both military weapons and space flight although the particular advantages of the solid fuel type have allowed it to make something of a comeback in certain military applications. For the really high performance vehicles needed for space work the liquid fuel engine has proved unbeatable.

Surprisingly it was not until about the beginning of this century that liquid fuels were seriously proposed for rocket use. The Russian pioneer K. E. Tsiolkovski looked forward to the exploration of space using rocket vehicles and was the first to put the mechanics of spaceflight on a rational basis. (Ideas such as Jules Verne's of firing a gun to propel capsules to the Moon were imaginative but completely beyond any bounds of possibility.) Later, Oberth, in 1923, in Germany, wrote a paper on *The Rocket into Interplanetary Space* which showed how space vehicles of multi-stage design would be needed. On the practical side Dr. Goddard made the first successful known flight of a liquid propellant rocket on 12th March, 1926 in Massachusetts. Following this important step slow but steady progress was made in the late 1920s and early 1930s. It was in Germany, with the realization of the military potential of rockets, that the real effort was then made. The research team at Peenemunde including Dr. Wernher von Braun developed within six years the first of the modern ballistic missiles, the V.2. This weapon far outclassed all previous rocket vehicles and laid the foundations for the work that was to follow on intercontinental ballistic missiles and space flight.

The layout of a typical liquid fuel rocket engine is shown in Figure 10.2. Separate tanks are provided for the fuel and the oxidizer and from these the propellants are conveyed to the injector in the combustion chamber. As shown, a pump is used to inject the liquids but an alternative arrangement which is sometimes used is to pressurize the propellant tanks with high pressure gas and so blow out the liquids. Very often the fuel being pumped to the injector passes through a double wall construction of nozzle and combustion chamber so providing some degree of cooling for these very high temperature parts. The gases produced by the reaction which takes place in the combustion chamber expand in a convergent–divergent nozzle and are exhausted at high velocity to the surroundings.

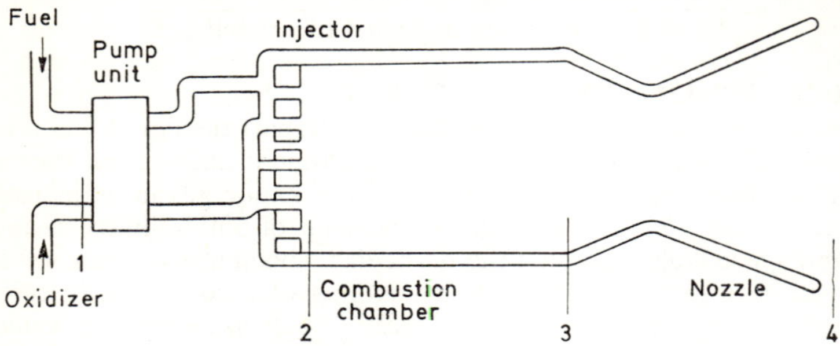

FIGURE 10.2 *Layout of a liquid fuel rocket*

Propellant Supply

Looking at each of the components of this complete engine in turn we will start with the system which pumps the propellants to the injector and controls their rates of flow. The most common method particularly with large engines is to use high speed turbo pumps to handle the liquids. Being high speed machines these pumps are compact and are capable of handling the large volumes of liquid used by big rockets. Direct drive of the pump by a turbine may be used and this will necessitate a supply of high pressure gases. For a long time the turbine supply came from a special gas generator in which small quantities of the propellants reacted together but major simplifications were found to result if a "boot-strap" principle was used in which the gases to drive the turbine were tapped off directly from the main combustion chamber. Clearly in this case some special provision will be needed for starting but the overall system can be made lighter and more compact.

Besides the actual need to *pump* the liquids and get them to the injector at a pressure appreciably above the combustion chamber pressure, the supply system has the functions of controlling the flows and of ensuring safe operation. The control function is very important as the supplies must be closely metered so that the correct fuel/oxidizer ratio is always maintained and that light-up and cut-off occur at precisely controlled times. (An error of one second in thrust cut-off with a long range missile can make a difference of four miles in the impact point.) Regarding safety, the problem here is that two very active liquids are being handled which may in themselves be corrosive or extremely cold and that these liquids must not be allowed to come into contact with each other until the combustion chamber is reached. The problems resolve to those of applying advanced engineering techniques to ensure a high degree of reliability. Over the years it has been found that one potent means of improving reliability is through simplification of systems and there is no doubt that the complexity of the control systems has been reduced greatly since the early days.

The alternative way of getting propellants from the tanks to the injector by using a tank pressurizing system implies that the construction of the tanks is such that they are capable of withstanding the full fuel line pressure. Inevitably they will be heavy and so this type of system is only really feasible for short duration operation such as the rocket assisted take-off of heavily laden aircraft. Gas to expel the liquids from the tanks—normally nitrogen—can be stored in bottles at high pressure and the same pressure can be applied to both fuel and oxidizer. The relative flows will then be a function of the injector geometry and a very simple control system can be used. One further advantage of the gas pressure supply set-up is that when fuel and oxidizer are exhausted the nitrogen gas will purge the pipe lines automatically.

Coming now to the injector (Figure 10.3) this component will

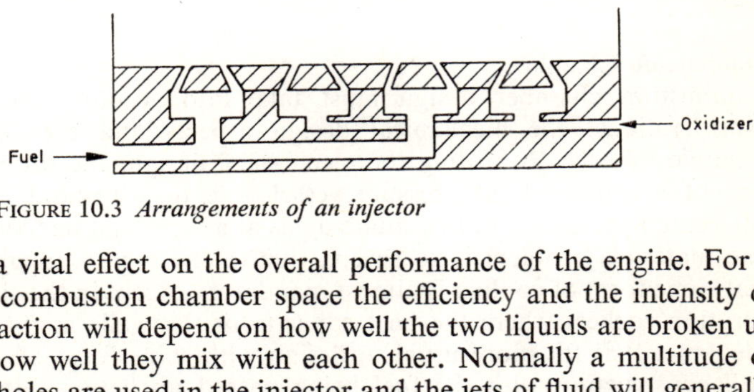

FIGURE 10.3 *Arrangements of an injector*

have a vital effect on the overall performance of the engine. For a given combustion chamber space the efficiency and the intensity of the reaction will depend on how well the two liquids are broken up and how well they mix with each other. Normally a multitude of plain holes are used in the injector and the jets of fluid will generally be arranged to impinge on each other and so break up the flow—this is more economical than using a large number of swirl type atomizers as are used on turbine engines. The jet impingement will normally be of "fuel-on-fuel" and "oxidant-on-oxidant" rather than of "fuel-on-oxidant" as the latter set-up is liable to be unstable under certain conditions, and may give rough combustion.

A type of instability described by S. L. Bragg (ref. 2) as "chugging" may arise as a result of variations in the injector flow. Let us consider a case where the fuel and oxidizer line pressures are being maintained constant at some value not greatly above the combustion chamber pressure. Suppose for some reason that the chamber pressure rises, then the injector pressure drop will fall with an immediate fall in the fluid flow. (The fluids will obey some orifice law with flow $\propto \sqrt{}$pressure drop.) With a reduction in flow the gas pressure in the chamber will start to fall, there being a slight delay due to the reaction time so that, at constant fuel line pressure, the fluid flows will start to rise. Clearly a situation is possible where conditions may become unstable and a steady oscillation may be set up. This instability is particularly liable to arise if the fuel and oxidizer line pressures are fairly low as then relatively small fluctuations in combustion chamber pressure give large percentage variations in injector pressure *drop*.

Some years ago trouble was being experienced with a large rocket engine due to burning of the injector face. A number of attempts were made to cure the trouble but at last it was realized that burning could not be taking place unless oxygen was present. The injector system was modified to ensure that oxidizer penetration was increased and that a fuel-rich mixture would be present at the face of the injector. This simple but ingenious piece of detective work cured the problem and saved a long and expensive series of development tests.

Combustion Chamber

Combustion chamber design must take into account two main considerations, the mechanical design aspects and the thermodynamic aspects. Regarding the mechanical design, the problem is that of making a light pressure vessel able to withstand a large differential pressure and to contain a gas at a very high temperature. The materials problem is so severe, due to the high temperatures, that some form of cooling is almost mandatory except in the shortest duration rockets where the heat capacity of the walls might prove adequate. It is far more usual to design with some form of double wall construction with the fuel being passed up between the two walls. Great care is needed to ensure that the cooling flow is even; if at any point there is insufficient flow then boiling of the fuel might take place which would cause a local hotspot and might lead to burning out of the chamber wall. S. L. Bragg shows (ref. 2) how large chambers are formed out of a number of tubes brazed together.

On the thermodynamic side, the most important point to be observed in chamber design is to ensure that the residence time of the reactants within the chamber is adequate to allow complete combustion. Consider the simple case shown in Figure 10.4 where the velocity at the injector face is low and the gases accelerate up to some high velocity V_T at the throat. For a residence time t_R, assuming that the mean velocity between injector and throat is $V_T/2$, the required chamber *length* will be $t_R V_T/2$. Also, if the mass flow rate is $\dot{m}$, the mass of reactants in the combustion chamber will be $\dot{m}t_R$

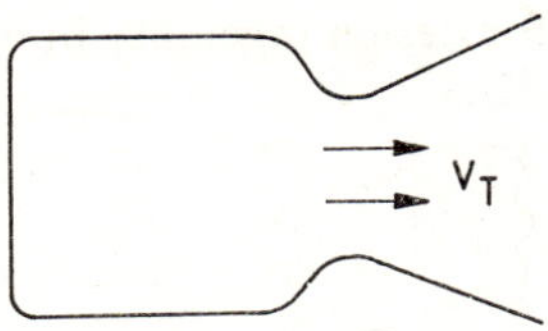

FIGURE 10.4

and the mean temperature, pressure and composition of these reactants can be estimated. By using the characteristic equation of the gases, an approximate value for the chamber *volume* can therefore be found. In this way the main internal dimensions of the reaction chamber might be found.

Although those statements explain what decides the chamber's main dimensions in fact the actual reaction chamber design problem is one which is too complex to analyse theoretically and which can only really be solved on the basis of wide practical experience. For example, only extensive empirical work will show what is an adequate reaction time—it will vary with different reactants, different chamber conditions and different injector design—and only experience will show what form of "average" values should be chosen. Over the years a number of semi-empirical design rules have been formulated which help to give a first idea of combustion chamber dimensions. One of the best known methods is to define a characteristic length L^* which is the volume between injector and nozzle divided by the nozzle throat area. A suggested value of L^* for given chamber conditions can then be used as the starting point for a design. L^* is related to the residence time as would be expected from the discussion in the previous paragraph.

Nozzle

When dealing with the design of propelling nozzles for air breathing engines in earlier chapters it was seen that it was often worthwhile to introduce some measure of variable geometry to cater for the range of different operating conditions encountered by an engine. Although a rocket performance would benefit from the same form of geometrical variation, it is not feasible to provide this as the conditions determining the stressing—pressure, temperature and velocity—are so much more severe. The most that is attempted in the way of mechanical movement is that at times the complete rocket motor may be swung on gimbals as is discussed below when considering vehicle control (p. 289).

As the propelling nozzle forms such a large part of the bulk of a complete rocket engine there is scope for design ingenuity in making

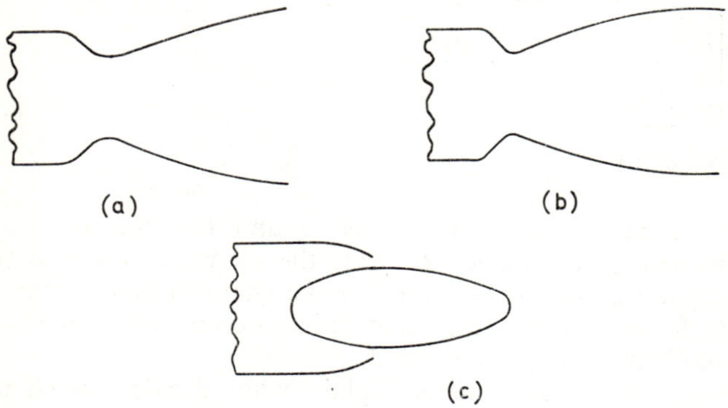

FIGURE 10.5 *Some forms of propelling nozzle*

a more compact nozzle. In a review article Rao (ref. 3) described a number of different final nozzle arrangements which had shown promise. Figure 10.5 shows three of the main designs which ar eas follows:

Figure 10.5(*a*) *The Conical Nozzle*

This is the simplest type to design but, if excessive cone angles are to be avoided, the nozzle will be rather long. There are two important results arising from the fact that the stream-lines of the flow will tend to be straight lines radiating from the throat. Firstly the effective outlet area (i.e. the area perpendicular to the flow lines) will be that of a *spherical* surface based on the outlet rather than being merely the nozzle outlet cross-sectional area. Secondly, as the stream-lines are non-axial there will be a lower axial momentum change, that is a lower useful thrust, than if they were truly axial. (This effect was discussed earlier in this book.) To allow for these factors, a coefficient which is a function of the divergence angle must be applied to both the area ratio and the thrust from the nozzle.

Figure 10.5(*b*) *The Bell Nozzle*

A nozzle shape can be designed which ensures that the stream-lines leaving the nozzle are parallel to the nozzle axis. Such a nozzle turns out to be bell-shaped as shown in the figure. It is more difficult to manufacture than the plain conical but has an appreciably better performance and so is very commonly used in large size rocket engines.

Figure 10.5(*c*) *The Plug Nozzle*
The plug-type nozzle has a throat of annular form and a shaped centre body which constrains the direction of movement of the *inner* surface of the propelling jet. On the outside the only restraint is that a balance must be struck between the gas stream static pressure and the ambient static. If these are not equal then the jet cannot flow axially. The design problem is therefore one of contouring the centre plug to give the desired velocity profile so that pressure balance is achieved. Tests show that the losses resulting from using a truncated plug rather than using one which covers the whole ideal distance are not very great and on balance it is worthwhile to use a shorter nozzle with slightly lower overall performances.

Thrust Vector Control
For manoeuvring the complete rocket vehicle, accurate control of the thrust, both in magnitude and direction, is needed. The thrust of a liquid fuel engine can be reduced by cutting down the rate of flow of propellants but this will generally involve some loss in efficiency as the nozzle will then be operating at less than the design expansion ratio. Also the combustion efficiency will probably decrease due to reduced penetration of the fuel and oxidizer jets leaving the injector. There is an added risk that at lower flows there will be an increased danger of combustion instabilities.

On the control of thrust *direction*, a number of methods have been used in the past. With many of the early rockets such as the V-2, rotatable vanes were fitted at the nozzle exit which deflected the efflux to give control. The life of the vanes was limited due to the high temperature and velocity and, even when they were not being used to deflect the gas, they introduced some loss in thrust. More recently, when multi-engined rocket vehicles have been used, direction control has often been effected by mounting the thrust chambers on gimbals and swinging them as required. Figure 10.6 shows how, with four rockets, the main directions of movement can be obtained by moving the rockets in unison or differentially. When the control forces are not too high, it may be possible to produce them by using small gimballed engines separate from the main

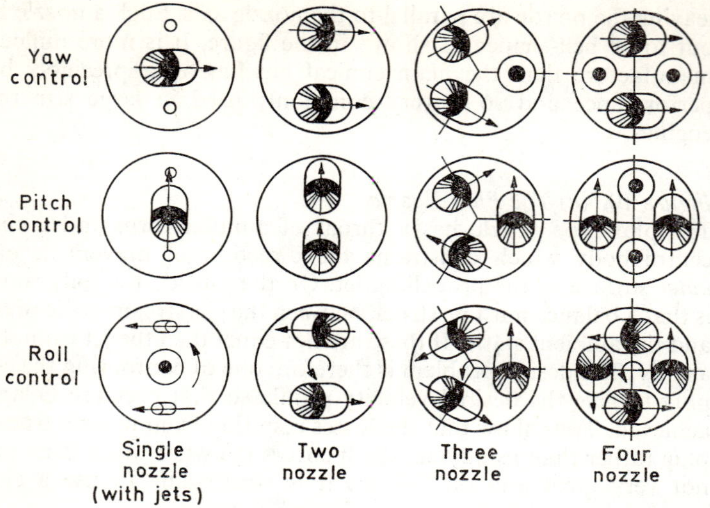

FIGURE 10.6 *Thrust vector control using swinging nozzles*
(Courtesy Rolls-Royce Ltd., Power Division)

propulsion units, or even by tapping gas off the main chamber and passing it through small swinging nozzles. These are sometimes described as vernier engines.

Choice of Propellants

The discussion above of the various components of a liquid fuel rocket has been completely general and has not assumed that any particular fluids are being used. Detail design of the components will however be very largely influenced by the fluid characteristics. It will be necessary to know the flow properties—density, viscosity, etc.—for calculating orifice sizes in injectors and such like quantities; it will be necessary, from the mechanical design point of view, to know whether it is corrosive or very cold. In the work that follows, the aim is to discuss in general terms the types of properties which must be considered and to look in rather more detail at some of the liquids that are used.

Section 10.2 showed that for high performance, the fuel and oxidizer should react rapidly to give a high temperature and that the products should have a low molecular weight. With these conditions being satisfied, the performance, particularly the specific impulse, will be high. In addition the following properties are desirable:

(i) From the overall vehicle point of view it is useful if the fuel and oxidizer have a high density as this means that the size of tanks can be reduced. This in turn will mean a much more compact vehicle with a lower drag, which is especially important if the vehicle is to spend a large part of its operating time in accelerating through the atmosphere.

In the tables on propellant properties below a volumetric specific impulse is quoted which is the product of density and the specific impulse defined by equation 10.3. Some fuels such as liquid hydrogen have a low merit rating on this basis.

(ii) From the operational point of view the liquids should be non-toxic, safe if spilt, and storable under reasonably normal conditions. Liquids which need to be stored at very low temperatures such as liquid oxygen and liquid hydrogen are an embarrassment when compared with others such as kerosene.

(iii) From the economic point of view the liquids should be cheap and readily available in large quantities. Both these aspects are important since it is no use having a cheap fuel if it is not available in the large quantities required by rocket engines.

At the present state of rocket development and application very many of the duties to be carried out by rocket engines are so near the practical limits of engineering ability that the overriding requirement is to achieve the ultimate in performance. For example the cost of a moon vehicle is so high that economically every effort must be made to succeed and this may involve the use of propellants which are expensive or awkward to handle. Liquid hydrogen is an example of a fuel which has come into prominence despite severe drawbacks such as the need to store it at very low temperatures, and its very low density. When it reacts with liquid oxygen high temperatures result and the products have the desirable low molecular weight. As a result liquid hydrogen is used extensively now and a wealth of experience in handling it and producing it in large quantities has been built up. It appears that the exotic fuels of yesterday may become the commonplace ones of tomorrow.

Full details of the properties of a very wide range of possible propellants are given in such standard texts as ref. 4. To give some idea of the figures involved Table 10.1. has been prepared showing the performance obtainable from a few of the more common propellants. Brief notes are then given on the physical properties of the various liquids.

TABLE 10.1 *Performance of Some Liquid Propellants*

Oxidizer	Fuel	Average Density of stoichiometric mixture, ρ kg/m^3	Specific Impulse I N-s/kg	Volumetric Impulse $I\rho$ kN-s/m^3
Oxygen	Ethyl alcohol	990	2 800	2 780
	Kerosene	1 020	2 940	3 000
	Hydrogen	280	3 820	1 070
Hydrogen peroxide 95%	Kerosene	1 300	2 670	3 480
Fluorine	Hydrogen	450	4 000	1 800

Notes. Above figures derived from Sutton (ref. 1). All specific impulses are for shifting equilibrium expansion from 60 bars to 1 bar pressure.

Considering now each of these liquids in turn:

1. *Oxygen.* Liquid oxygen—generally known as Lox—is a very commonly used oxidizer. It boils at 90 K at atmospheric pressure and has a density of 1 140 kg/m^3. Handling and storage are relatively easy if care is taken to maintain clean cold conditions.

2. *Ethyl alcohol.* This was used in the V-2 rocket. It is liquid at normal temperatures with a density of 790 kg/m^3 and can be handled easily provided normal fire precautions are observed.

3. *Kerosene.* One of the most common fuels due to its convenience and low cost. It is liquid at normal conditions with a density of 810 kg/m^3. Like ethyl alcohol it can be handled easily with straightforward precautions against fire.

4. *Hydrogen.* Liquid hydrogen, like Lox, is a cryogenic fluid, that is it must be stored at low temperature. The boiling point at atmospheric pressure is 23 K and the corresponding density is 71 kg/m^3. Clearly the low temperature and low density will introduce storage problems as bulky well-insulated tanks are required but by now handling problems are well understood.

5. *Hydrogen peroxide.* The hydrogen peroxide used as a rocket propellant is in concentrated form and is often known as high test

peroxide (HTP). It is liquid at normal temperature with a density of about 1 300 kg/m³. Storage and handling are not difficult but care must be taken as the liquid is corrosive and is liable to react with many organic substances.

6. *Fluorine.* Although not strictly an oxidizer, fluorine reacts violently with hydrogen to give a very high specific impulse. It is a highly corrosive liquid which boils at 85 K and has a density of 1 660 kg/m². Extreme care is needed in handling and storage due to its great chemical reactivity with most substances.

Example 10.1 following shows how the mass and bulk of a rocket vary with the propellants chosen.

Example 10.1 A rocket engine is required to produce a thrust of 300 kN for 15 sec. Calculate the relative mass and volume of propellants required if the propellants are

(i) Liquid oxygen and kerosene
(ii) Liquid oxygen and liquid hydrogen

Solution. Table 10.1 quotes ideal performance figures for the various propellants. In practice there will be losses due to inefficient combustion, non-ideal expansion, etc., so the actual specific impulse attained by any real rocket engine will be less than the ideal. A typical actual specific impulse would be 80–90% of the ideal. However we have have been asked to calculate *relative* changes so the assumption can be made that the overall efficiency is the same in each case.

Using equation 10.3 the mass flow rate for the kerosene fuel is

$$\dot{m} = X/I = 300 \times 1\,000/2\,940 = 102 \text{ kg/s}$$

For the hydrogen fuel

$$\dot{m} = 300 \times 1\,000/3\,820 = 78{\cdot}5 \text{ kg/s}$$
$$= 77\% \text{ of the kerosene/lox propellant rate}$$

A saving of 23% in propellant mass is therefore possible. For the 15 sec run the total mass of propellants is of the order of 1 500 kg, that is 1½ tons, so 23% saving is appreciable. On the volumetric basis:

$$\text{Volume of lox/kerosene} = 300/3\,000 = 0{\cdot}1 \text{ m}^3/\text{s}$$
$$\text{Volume of lox/hydrogen} = 300/1\,070 = 0{\cdot}28 \text{ m}^3/\text{s}$$

This illustrates the greater bulk involved in using liquid hydrogen.

Mono-propellant Rockets

One particular type of liquid fuel rocket which deserves special mention is the mono-propellant version. This makes use of the fact that certain liquids may, given the correct conditions, decompose spontaneously. Heat is released and the liquids gasify. A very simple rocket can be made using this phenomenon.

The best known mono-propellant rockets are those using hydrogen peroxide as the propellant. In Table 10.1 this was listed as a possible oxidizer and the comment was made that it was liable to react with organic materials with the release of energy. With certain catalysts however, HTP will decompose violently by itself, forming superheated steam at about 990 K and oxygen. Catalysts which may be used include potassium permanganate or, more conveniently, silver gauze.

For many years the former de Havilland Engine Company carried out work on mono-propellant rockets principally for assisting the take-off of heavily laden aircraft. When potassium permanganate was used to catalyse the HTP the rocket would pour forth clouds of heavy smoke but when the change was made to a silver catalyst a clean exhaust resulted. Developments resulted eventually in the Sprite engine which reached the stage that it was test flown in the Comet and other aircraft. Its successor, the Super Sprite, was a conventional bi-propellant rocket in which kerosene was burned with the HTP in the reaction chamber.

10.4 THE SOLID FUEL ROCKET

In dealing with the solid fuel rocket motor we will be concerned firstly with its advantages and disadvantages compared with its liquid fuel counterpart and then with looking in rather more detail at the design and performance characteristics. (The convention of speaking of liquid fuel rocket *engines* but of solid fuel rocket *motors* is established in Britain.) Figure 10.7 shows the layout of a typical motor and it will be seen that the main parts are the charge, the case. the nozzle and the igniter. A bursting diaphragm is fitted to the case which acts primarily as a safety valve but which may also be used for thrust control purposes.

A first impression on looking at Figure 10.7 compared with the liquid fuel engine shown in Figure 10.3 is of the fundamental simplicity of the solid fuel motor. This lack of complexity will give an inherently high reliability, especially as it is the relatively less reliable rotating parts and plumbing which have been eliminated. Since there is no turbo-pump with its associated gas generator and piping, the empty weight of the complete power unit will be less. Consequently it will be possible to design for a higher value of the mass ratio of the vehicle—the ratio of launch mass to mass with all propellants consumed—and this has an important effect on the overall vehicle performance. Economically too a solid fuel rocket is very much cheaper than a liquid fuel. It has been estimated that for long range missile applications, the cost of a solid fuel weapon is about one third of that of its liquid fuel counterpart.

Against these advantages which arise from the simplicity of the solid fuel motor must be set the disadvantage that solid propellants

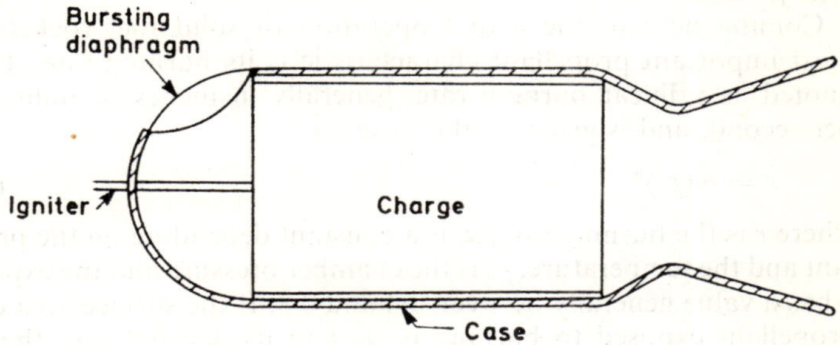

FIGURE 10.7 *Layout of a solid propellant rocket*

generally have a lower performance with maximum specific impulses of the order of 2 500 N-s/kg. In addition, because no cooling of the case is possible, such as is provided by the double wall in the liquid fuel engine, the running time is restricted. A limit is set by the heat capacity of the case until the case reaches a temperature when its strength falls appreciably. Finally on the performance side, control of a solid propellant motor is difficult both from the points of view of thrust control and of thrust vector control.

The most important factors which have led to the increase in interest in solid fuel rockets in recent years have not been related to performance effects but have been concerned with the convenience and ease of storage of the solid propellants. This has been particularly true in the case of military applications as can be illustrated by reference to two advanced solid propellant missiles, Polaris and Minuteman. Polaris, designed to be fired from nuclear submarines and to be fully effective after many years of storage, needed as its prime requirement a fuel which would not deteriorate with time and which could be stored easily and safely. Provided a reasonable amount of care is taken, solid propellants are quite safe to handle and complete rockets can be stored fully charged with their propellants—with liquid fuel engines the tanks are not filled until nearly the time of firing so this means that extra space is needed for rockets plus storage of propellants. For Minuteman the main design stipulation was that the rocket should be easily transportable and always ready for use. The original proposals were that Minuteman missiles were to be mounted on trucks and moved continuously about the railroad network of the United States so that safety and compactness were also major design needs. These advantages of ease of handling and immediate readiness at all times have also led to the use of solid fuel motors for small meteorological rockets such as the Bristol Aerojet Skua.

Coming now to the actual operation of solid fuel rockets, the most important propellant characteristic is its burning rate. This is quoted as a linear burning rate, generally in inches or millimetres per second, and is given by the formula

$$r = a_r(p_c)^n \tag{10.9}$$

where r is the burning rate, a_r is a constant depending on the propellant and the temperature, p_c is the chamber pressure and the exponent n has a value generally between 0·4 and 0·8. If the surface area of the propellant exposed to burning is A_p and its density is ρ_p then the rate of generation of products will be

$$\dot{m}_p = rA_p\rho_p = a_r(p_c)^n A_p\rho_p \tag{10.10}$$

Under steady conditions the nozzle flow must equal the rate of generation of products as given by the above equation. A plot of

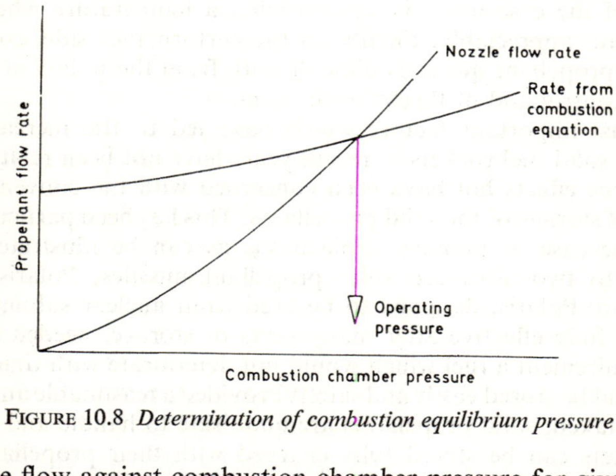

FIGURE 10.8 *Determination of combustion equilibrium pressure*

nozzle flow against combustion chamber pressure for given propellants and ambient pressure (Figure 10.8) will be much steeper than the curve of burning rate against pressure. The equilibrium operating pressure will be determined by where the two curves intersect.

Equation 10.10 shows that the rate of production of propellants depends on the surface area of burning A_p, so that if a steady thrust is required, care must be taken in the shaping of the charge. The diagrams in Figure 10.9 show a few of the possible charge shapes

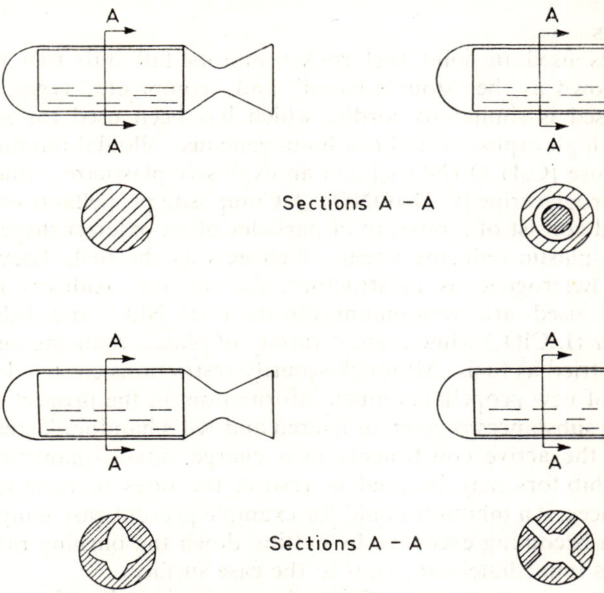

FIGURE 10.9 *Shapes of charge for solid propellant rocket*

and illustrates some of the problems which arise. With the simplest expedient of using end burning, although the area remains constant throughout burning, the surface available is low for a motor of reasonable properties. A charge with a cylindrical hole through the middle will clearly have a burning area which will increase with time. More complicated shapes such as the star section, conduit and the slotted tube are commonly used so that the surface area is constant throughout combustion.

For the charge shapes with internal passages shown in Figure 10.9 a problem arises due to the flow of gases across the surface of the charge. This causes what is known as "erosive burning", an increase in the effective burning rate resulting from the scrubbing action of gases across the surface. Clearly the scrubbing will be more intense at the nozzle end of the rocket as the mass flow across the surface will be highest there. Erosive burning effects can be mitigated to a certain extent by further shaping of the charge so that flow areas increase towards the nozzle and reduce the scrubbing velocities. Whatever shape of charge is used, it is important to keep the "dead-space" around the charge as small as possible as this will have a direct influence on the overall bulk of the rocket motor.

Propellants

Propellants used in solid fuel rocket motors fall into two main classes known as the "double-based" and "composite" types. The double-based is similar to cordite which has been used for many years as a high explosive and is a homogeneous colloidal mixture of nitrocellulose [$C_6H_7O_5(NO_2)_3$] and an explosive plasticizer which is usually nitroglycerine [$C_3H_5(ONO_2)_3$]. Composite propellants on the other hand consist of a mixture of particles of an oxidizer dispersed through a plastic reducing agent which acts as the fuel. They are markedly heterogeneous in structure. Among the oxidizers most commonly used are ammonium nitrate (NH_4NO_3) and lithium perchlorate ($LiClO_4$) while a great variety of plastic reducing agents have been tried as fuels. Although security restrictions cover a lot of the work of new propellants much information on the properties of interesting substances is given in Evered and Kit's handbook (ref. 4).

Besides the active constituents in a charge, small quantities of various inhibitors may be used to restrict the rates of reaction in certain places. An inhibitor could for example prevent case temperatures from becoming excessive by cutting down the burning rate of propellants immediately adjacent to the case surface.

For the performance point of view the specific impulse of a cordite type propellant will be of the order of 2 000 N-s/kg when operating with a chamber pressure of 6 000 kN/m^2 and exhausting to 100 kN/m^2. Composite propellants will, in general, be able to give rather higher values—figures of 2 300 N-s/kg have been quoted—although they are still very much lower than those achieved by good, high-performance liquid propellants.

The physical properties of the charge have a very considerable influence on the mechanical design of a rocket motor and on its overall performance. Double base type propellants are formed as castings or extrusions which are comparatively rigid with low ability to adjust to thermal expansions. They must therefore be loaded as loose charges in the case so provision must be made for removing completely one end of the case. For small size rockets construction as shown in Figure 10.10(a) would be typical. Composite propellants on the other hand are formed in the plastic state and have sufficient flexibility to allow direct bonding to the case. A simpler one-piece construction, of the form shown in Figure 10.10(b), is possible. When being made, the charge is poured directly into the case with the final curing being done *in situ*. By bonding the charge to the case all waste space is cut out and the charge itself, having a low thermal conductivity, acts as a heat insulator to protect the metal of the

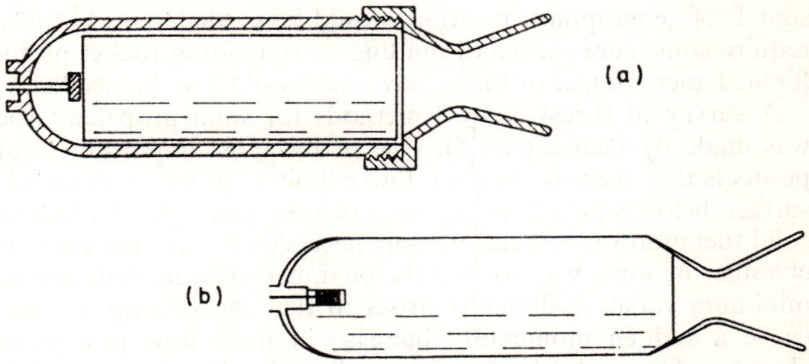

FIGURE 10.10 *Effect of charge type on construction of solid fuel rocket*

casing. It is largely due to these advantages that such spectacular progress has been possible in recent years with small to medium sized solid fuel motors.

While dealing with the mechanical construction aspects of these rockets it is worth mentioning that the throats of the nozzles are commonly made with a lining of some high temperature refractory material. This is desirable because the area at the throat is a very important dimension which must remain constant and the heat transfer there is high because of the high gas velocity and the thinness of the boundary layer.

Thrust Vector Control

To turn now to the problem of controlling solid fuel motors, it should be noted in the first instance that, for very many applications of these rockets, exact control is not called for. If being used for such purposes as launching meteorological instruments, firing life-lines or assisting the take-off of heavily laden aircraft, the requirement is only that the motor should fire reliably and keep burning steadily until all the propellants are consumed. Motors for these applications will generally be of fixed geometry and will be rigidly attached to the vehicle so that control of thrust *direction* will not be a problem.

It is when solid fuel motors attempt to compete with liquid fuel rockets in long range missile type applications that very close control is needed. As was mentioned earlier, with an intercontinental ballistic missile the velocity at burn-out is such that an extra second of burning time would cause the missile to miss the target by four miles or more. Clearly a scheduling type control would be wholly inadequate for these purposes and a closed loop control, working on signals of geographical position, would be needed. This in turn, will require some sure means of igniting or re-igniting rocket motors in flight. Exact control of thrust *direction* would also be needed.

A survey of thrust control methods for solid propellant rockets was made by Pearce (ref. 5). One of the most important technical points is that there is always a lower limit of pressure on the charge surface below which burning will not take place. To "switch off" a solid fuel motor it is therefore only necessary to vent the combustion chamber in some way so that the chamber pressure falls below this minimum value. A difficulty arises in that the venting is bound to cause a sudden momentary increase in mass flow rate from the chamber. Care must be taken regarding the direction in which this discharge takes place as it will produce a thrust which might be an embarrassment. For example, if radial venting was used, it would be necessary to provide two or more vents to avoid large side-loads on the vehicle.

If it is required to achieve thrust control over a fairly limited range rather than accepting "on-off" control, some form of nozzle throat area variation is really needed. Increasing the area will in the first place cause a drop in the chamber pressure and this will result in a fall in the burning rate, that is, in the rate of production of propellants. At the same time the reduced pressure will mean a lower exit velocity from the nozzle. These two effects will combine to give a fall in thrust. Unfortunately, although it is easy enough to say that a variable throat area should be provided, the actual mechanics of the design are much more difficult due to the arduous conditions of temperature and pressure. A translatable bullet has been proposed for this duty, but the materials problems would be severe and it is not clear how the actuating mechanism could be accommodated in a conventionally designed motor.

When dealing with the control of thrust direction for liquid fuel rockets (p. 289), mention was made of various deflectors which could be placed in the efflux and of the alternative of mounting thrust chambers on gimbals so that they could be swung about to give the required transverse thrust components. Although solid fuel motors may make use of the jet deflector devices, it is not feasible, except in the smallest sizes, to mount the complete motor in a pivoted frame. Instead, various swivelling nozzles have been designed and used successfully. The main engineering problem is to devise some form of seal around the nozzle joint which will stop gas leakage and withstand the gas forces while still remaining free to pivot.

One other form of thrust vector control which has become very popular in recent years is that of using liquid injection at a point in the divergent portion of the nozzle. The system is illustrated diagrammatically in Figure 10.11. When liquid is injected, it will

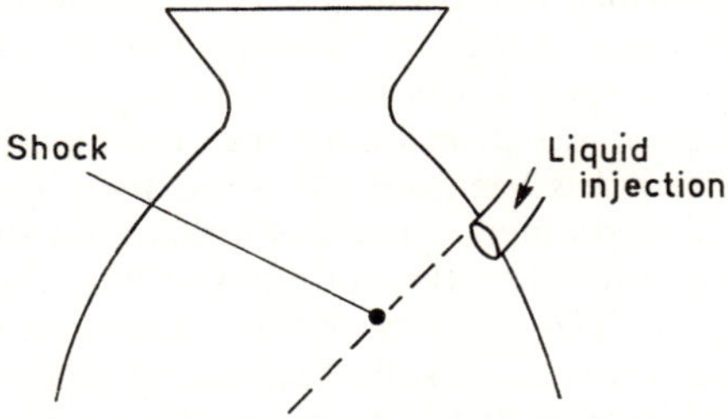

FIGURE 10.11 *Thrust vector control by liquid injection*

cause a blockage and set up a shock pattern in the nozzle so causing the jet to be deflected from the axial direction. Clearly the system will be easy to install and will give a readily controllable means of altering the deflection of the jet. Of great importance too is the fact that when not in use there are no performance penalties. Among the rockets using the liquid injection method of thrust vector control are the later versions of Polaris and Minuteman missiles.

Before leaving the subject of solid fuel rockets, mention must be made of some of the very large American motors. The first of these to be flown, the United Technology Center (UTC) Type 1205, acts as a booster for the Titan 3, a vehicle designed as a standard space launch system. In this role two Type 1205 motors provide a total of over 2 000 000 lbf (9 MN) thrust for two minutes to start the Titan on its way.

A solid fuel booster of this size is much simpler and cheaper than a comparable liquid fuel rocket. As it is only used in the boost phase of the flight, it is unnecessary to have very sophisticated means of controlling the magnitude of the thrust as the vehicle can be controlled more conveniently by means of the later liquid fuel stages. In the Type 1205 the propellant used is of the composite type using ammonium perchlorate (NH_4ClO_4) as an oxidizer dispersed in a fuel of synthetic rubber and aluminium powder. The total charge is cast as five separate cylindrical segments each 120 in. (3 m) diameter and about 10 ft (3 m) thick, the segments then being loaded into a high tensile steel case and locked together. When nozzle, nose facing and control system have been added the complete rocket stands 75 ft (25 m) high and weighs about 500 000 lb (250 000 kg). This motor is only the first of a series and work is being carried out on larger units of up to 260 in. (6·6 m) in diameter.

10.5 THE HYBRID ROCKET

A hybrid rocket is one in which two propellants are used which are initially in different phases, one being a solid and the other a liquid. Although comparatively little work has been done on this type as yet, compared with the effort which has gone into the purely solid and liquid rockets, sufficient *has* been done to show its advantages and to bring out some of its problems. There is little doubt that for certain applications the hybrid offers the optimum solution.

In theory, either fuel or oxidizer may be in the liquid form but so far most practical engines have used a solid fuel charge and a liquid oxidizer which is sprayed in to start the reaction. It is immediately apparent that with this system much simpler control of the thrust magnitude is possible than can be obtained with a wholly solid propellant motor. To switch off, it is only necessary to stop the flow of the oxidizer, a much simpler operation than dropping the whole combustion chamber pressure. The reaction will cease as soon as the fuel loses its supply of oxygen but, with a suitable design, it can be restarted merely by turning on the oxidizer once more. Again compared with a solid propellant motor, it is useful to have a liquid available for cooling purposes. Should the design warrant it, the liquid may first be passed through cooling passages in the nozzle or casing before reaching the injector and this may permit longer duration of runs than would be possible with an uncooled rocket.

When compared with a fully liquid propellant engine, the hybrid scores on the grounds of simplicity in that only one pump is needed compared with two. There is of course the fact that generally with a liquid fuel engine one turbine can be used to power both propellant pumps so there is not a complete halving of hardware but the control side can be simplified greatly as complicated interlocks and safety circuits are no longer needed.

Naturally these advantages of the hybrid rocket are accompanied by some disadvantages—or the hybrid would by now have swept the board for all applications! Probably the most important drawback is that the specific impulse obtained is something between that possible with solid motors and the best that can be achieved with wholly liquid engines. For applications where the highest possible performance is needed, this rules out the hybrid engine. Further, although it is easy enough to *say* that the thrust can be varied by altering the liquid flow, the practical problems of doing this efficiently are considerable. An injector must be used which ensures that the oxidizer reaches the fuel in a suitable form to sustain the reaction throughout the burning—while of course, the dimensions and shape of the solid will be changing continuously. There is the additional problem of providing ignition for a relight should it be required to restart a rocket after shut-down in flight.

A brief description of a Swedish design of hybrid rocket is given by Gudmondson (ref. 6). This engine uses fuming nitric acid as the liquid oxidizer and a solid plastic known as Tagaform for fuel. These two compounds are hypergolic (that is, they ignite spontaneously) and react with a low ignition delay to produce a smokeless exhaust. A specific impulse of 2 600 N-s/kg is obtained with a chamber pressure of 7 000 kN/m^2 when exhausting to atmosphere. To illustrate the controllability it is interesting to note that this particular rocket is required to give a thrust of 1 700 N for 6 sec followed by 350 N for 10 sec.

10.6 ROCKET VEHICLE MECHANICS

For fixed wing aircraft of conventional design, the thrust force produced by the engines is generally small compared with the aerodynamic forces produced by the wings so the flight mechanics of the complete vehicle are determined largely by these aerodynamic forces.

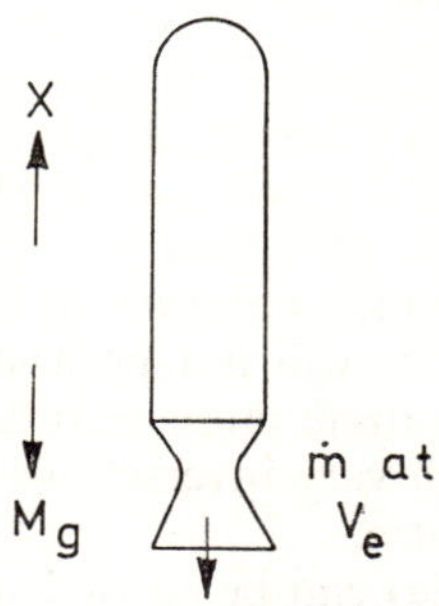

FIGURE 10.12 *Forces acting on a rocket in vertical flight*

With a rocket vehicle on the other hand, if one excludes the relatively minor field of use where rockets are used to boost the power of winged aircraft, the dominant forces are those produced by the propulsion unit. Although large inertia forces might result when sudden manoeuvres are carried out, these will be essentially transient effects with more significance from the structural than from the performance point of view. As the overall vehicle performance is so tied up with the engine performance it is relevant to consider the mechanics of rocket flight and see how they influence the requirements for the engine.

Consider the simplest case of a rocket in vertical flight with aerodynamic drag forces being neglected in the first instance. Figure 10.12 shows a rocket of mass M operating with a constant jet efflux velocity of V_J in a vertical gravitational field of strength g. If the rocket has a velocity V at time t, the equation of motion may be written

$$M\frac{dV}{dt} = \frac{dM}{dt}V_J - Mg \tag{10.11}$$

(The mass change of the whole vehicle results from the expulsion of the propellants.)

This may be rearranged as

$$dV = \frac{dM}{M}V_J - gdt \tag{10.11a}$$

At $t = 0$, the instant of firing the rocket, $V = 0$ and the mass will be the launch mass which we will denote as M_L; let us denote the time when all the propellants have been burnt as t_B with velocity and mass at "all burnt" being V_B and M_B respectively. Equation 10.11(a) can be integrated to give

$$V_B = V_J \ln\left(\frac{M_L}{M_B}\right) - gt_B \tag{10.12}$$

Although V_J and g have both been assumed constant to enable the integration to be carried out easily, this will not be a strictly true assumption. V_J will in fact vary with nozzle pressure ratio and g, the local acceleration due to gravity, will vary inversely with the square of the distance from the Earth's centre.

A further integration of equation 10.11(a) can be carried out to find the altitude attained by the rocket during burning. By putting in the appropriate limits and making the same simplifying assumptions as before, the height at "all burnt" can be shown to be

$$h_B = V_J\left[t_B - \frac{M_B}{\dot{m}}\ln\left(\frac{M_L}{M_B}\right)\right] - \tfrac{1}{2}gt_B^2 \tag{10.13}$$

where $\dot{m}$ is the rate of consuming propellants.

After the propellants have all been used up, the rocket will coast on upwards converting its kinetic energy into potential energy until finally it comes to rest at some maximum altitude h_M. The altitude increment during the coasting period will be given by

$$h_M - h_B = \frac{V_B^2 r_B}{2g_B r_B - V_B^2} \qquad (10.14)$$

This equation *does* allow for the variation of gravitational acceleration; r_B is the distance from the centre of the Earth at all burnt, that is, it is (h_B + 6 400) with h_B measured in kilometres.

In the limit the propulsion problem for every rocket vehicle resolves to that of imparting a certain amount of energy to the vehicle, the required amount being the sum of its potential energy and its kinetic energy when it is in its required flight path. For a vehicle going into orbit* the velocity must be such that the centrifugal force acting upwards exactly balances the gravitational attraction while a rocket which is to leave the earth's gravitational field must attain an escape velocity of about 10 km/s. Normally the potential energy at the moment when all propellants have been consumed will be small compared with the kinetic energy at this point, so to a close approximation we can say that the propulsion problem is to impart a high kinetic energy to the vehicle.

Looking at equation 10.12 which shows the velocity at "all burnt" it will be seen that the variables at the disposal of the designer are the jet efflux velocity V_J, the mass ratio M_L/M_B and the burning time t_B. Of these, the efflux velocity is dependent largely on the choice of propellants (assuming that they can be used efficiently) and there are limits to what is achievable. The burning time will be determined by the mass flow rate and the possible propellant capacity ($M_L - M_B$.) The real skill in design lies in achieving the highest possible mass ratio M_L/M_B, that is in reducing the mass of structure, engines, etc., to the lowest possible value.

Even with the most advanced materials and techniques there is bound to be a practical upper limit to the mass ratio, the value being in the region of 20:1. With this value and with currently available propellants escape velocities cannot be obtained without recourse to multi-staging. When a multi-stage rocket is used the first stage is jettisoned after it has accelerated the vehicle up to high speed and has got it through the denser parts of the atmosphere where most air resistance is encountered. Subsequent stages have greatly reduced

* Although intended mainly as a handbook for observers, Desmond King-Hele's text on earth satellites (ref. 7) gives an excellent and highly readable account of the motion of these vehicles.

structure mass compared with the complete rocket so acceleration to the required high velocities is possible. Multi-staging has the added advantage that peak accelerations are reduced (the maximum acceleration will be the initial acceleration times the mass ratio) and this is particularly useful for manned vehicles.

Besides the purely performance problems of space flight there is a multitude of special difficulties arising from the environment in which space craft work. There are, for example, great difficulties of transporting a tank of liquid hydrogen to the Moon without losing too much by evaporation or by the tank being punctured by micrometeorites. Many special problems too stem from the zero-gravity conditions in space. Equipment cannot be cooled by convective means and the fuel in a tank will take up a spherical shape not necessarily in contact with the tank walls. Auto-welding of surfaces may take place if two metals touch in the clean conditions of the near absolute vacuum in space. There is little doubt that the advances in engineering made necessary by a space programme must lead eventually to improvements in more mundane fields.

10.7 NON-CHEMICAL MEANS OF PROPULSION

Two classes of rocket engine remain to be discussed neither of which relies on chemical reactions as its source of energy. They are the nuclear rocket in which a nuclear reactor acts as the power source, and the ion propulsion devices in which electrostatic means are used to accelerate the particles. Both types offer promise of meeting certain specific needs in the future but, as yet, actual test experience with them is limited.

Taking first the nuclear rocket this is still a comparatively conventional machine. A single propellant is used which is pumped through a nuclear reactor where its temperature is raised to a high value. The hot, high pressure gases so formed are then expanded to atmosphere through a convergent–divergent nozzle in the usual way.

Where the nuclear rocket scores is that the temperature of the propellant entering the nozzle is not limited to that attainable by a chemical reaction process but it may, in theory, be made much higher. A top limit is set by the temperatures which the actual reactor will stand. In addition, if hydrogen is used as the working fluid, the propellant has the desirable feature of having a low molecular weight. As only a single propellant is used, tankage and plumbing are simplified. Against these clear advantages must be set the obvious weight penalty of the reactor (especially when provision is made for shielding) and a rather more difficult power control problem.

Most of the information published on nuclear rockets has been about the American work carried out by the NASA. Finger (ref. 8) reviewed the programme and the results obtained up to the end of 1966. As an indication of what can be done, he quoted a specific impulse of 7 400 N-s/kg as having been demonstrated experimentally by one of the test rockets.

The ion propulsion devices—a large variety are possible—make use of electrostatic devices to accelerate charged particles up to very high velocities. Some thermal means must be provided for ionizing the particles which are to form the working fluid and further power is needed to generate the magnetic fields which carry out the actual acceleration. Theoretically very high velocities can be attained and specific impulses of many thousands have been measured with experimental engines. Ion propulsive devices work with very low mass flow rates however and despite the high specific impulses, the actual thrusts are generally very low. It must be appreciated that the power required to accelerate the ions will be measured by the kinetic energy they are given so the power requirement for a given thrust will be very high.

Both nuclear and ion propulsion rockets come into their own when long voyages into space are to be undertaken. For travel to Mars and Jupiter for example the total journey times will be hundreds of days so even if a small thrust is provided continuously it can cause an appreciable acceleration. The main requirement must be that the propellant should be used as effectively as possible, that is with a high specific impulse. Though the thrusts produced by the ion rockets are so low, it can be useful for trimming purposes and for altering course over an appreciable period of time. One final point in favour of the ion type unit is that it is essentially an electrical device and it might well be feasible to operate it with batteries which can be recharged by solar cells in a long flight.

11 The Piston Aero-Engine

11.1 INTRODUCTION

It may be wondered—particularly by younger readers—why a book on aircraft propulsion published more than twenty-five years after the first flight of a jet aircraft should have a complete chapter devoted to the piston aero-engine. On consideration, it will be seen that there are a number of reasons for giving the piston engine this degree of coverage. For one thing there is a very large number of piston-engined aircraft—light-planes, airliners and military machines—still in use; in terms of total numbers of aircraft they far outnumber those with turbine engines. For another, the history of the design and development of the big piston engines contains a number of interesting technical lessons which are still worth stating. Finally, in the field of personal aircraft no turbine engine has yet offered it serious competition. There are fundamental difficulties in making a small efficient turbine for a reasonable price and despite the turbines admitted advantages of smoothness of operation and reliability the piston type still reigns supreme in this field.

In this chapter we will look first at the design and performance of the larger piston aero-engines. These in their day were at the forefront of propulsion technology and over the years they were developed to give greater and greater powers at higher and higher altitudes with continuous improvements in specific weight. Sometimes it is salutary to think of the effort which went into their production and to realize that if the gas turbine had not come forward still greater advances would have been made. At a fairly late stage in piston engine development the exhaust turbine driven supercharger was brought into extensive use and this led to the development of the turbo-compound engines. The most successful in this field were the American Wright Turbo-compound engines which are still in extensive use but an attempt to extend and develop the concept of the compound engine in the Napier Nomad led to a dead-end. Nonetheless the compound engine story merits a special section in this chapter. Finally we look at the modern low power piston engine which is being produced still in large quantities for light aircraft applications.

11.2 THE HIGHER POWER PISTON AERO-ENGINE

Looking at the history of piston aero-engines from the earliest days it is possible to pick out examples of almost all possible reciprocating i.c. engines. Four stroke and two stroke cycles have been used; at one time compression ignition engines looked like being keen competitors to the petrol ones; air and liquid cooling were tried with at least one steam-cooled engine (the Rolls-Royce Goshawk) and one which ran for short times cooled by solid carbon dioxide; poppet valves have had to compete with sleeve valves; and the number of arrangements of crankshafts and connecting rods has been immense. After all the experimentation, however, the only successful high power engines which have survived have been spark-ignition petrol engines operating on the usual four-stroke cycle of induction, compression, power and exhaust.

A comparison between the thermodynamics of the piston engine and turbine can best be made by reference to the cycle diagram on pressure–volume or temperature–entropy coordinates. Fig. 11.1

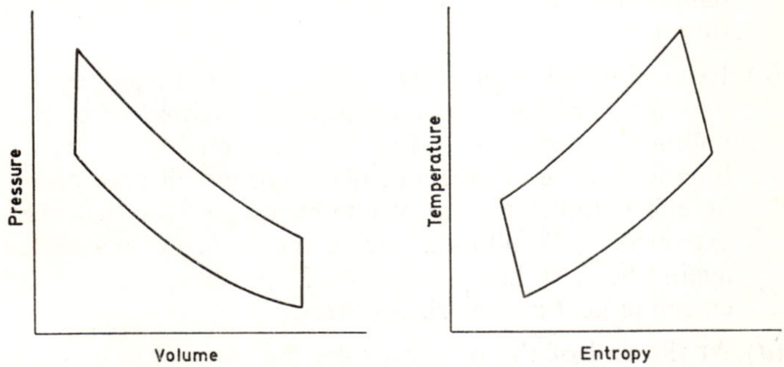

FIGURE 11.1 *The operating cycle of the reciprocating aero-engine*

shows such a diagram for the compression and power strokes of a four stroke cycle engine. This shows for the ideal cycle

(*a*) The charge trapped in the cylinder at the end of the induction stroke is compressed according to a law of the form pv^n = constant. In the case of an unsupercharged engine the pressure and temperature at the start of compression will be close to atmospheric conditions but if the engine is supercharged they will be appreciably higher than atmospheric. During the compression heat will be rejected to the surroundings so there will be a reduction in entropy of the charge. Consequently the index of compression n will have a value less than that of the specific heat ratio γ—a typical value of n would be in the range 1·2–1·35.

The first law of thermodynamics as applied to a closed process may be used to calculate the work and heat transfers during the compression process.

(*b*) Near the end of the compression stroke the charge of petrol/air in the cylinder will be ignited so giving a rapid rise in temperature. The fuel/air ratio will be close to stoichiometric so the temperature rise will be very much greater than in the case of a gas turbine—the temperature reached will be such that appreciable dissociation of the combustion products will take place so increasing the effective specific heat of the gases. As the combustion takes place at more or less constant volume there will be a large rise in pressure.

(The reciprocating engine can withstand the higher gas temperatures because of the intermittent nature of the thermal loading. Cylinder and piston are only subjected to the peak temperature for a small fraction of the total engine running time.)

(*c*) Expansion takes place from the maximum temperature and pressure through a *volume* ratio approximately equal to the volume compression ratio. As the temperature falls recombination of the products of dissociation will take place giving an effect similar to that of late burning of the fuel. Again the expansion will follow a law of the form pv^n = constant and again the first law for a closed process may be used to calculate heat and work transfers.

(*d*) At the end of the power stroke the exhaust valves open and allow the products to blow down into the exhaust system. This process approximates to a constant volume process.

The net work done per cycle can be found by taking the algebraic sum of the works for each of the processes and the power produced in the cylinders can be found by multiplying the power per cycle by the number of working cycles in unit time. This power may be measured experimentally by using an engine indicator so it is known as the *Indicated Power*. All mechanical losses must be debited against this to find the useful or *shaft power* output of the engine. The indicated work per cycle of an engine of given design will depend primarily on the mass of charged trapped in the cylinder before compression. The indicated power will therefore depend on the overall engine mass flow rate.

The most significant differences between piston and gas turbine engines arise from the very much higher level of pressure in the piston engine. Whereas few gas turbines ever exceed combustion pressures of about 1·5 MN/m^2 (say 225 lbf/in.^{2}) cylinder pressures in petrol engines can well exceed 10 MN/m^2 (1 500 lbf/in.^{2}). There are also, as has been mentioned, much higher peak cycle temperatures. The immediate results of these increased conditions are to give the piston engine a much higher specific output and a much better thermal efficiency. On the other hand mechanical losses in piston engines are high, being up to 25% of the indicated power while a turbine would seldom lose more than about 1% in friction losses. Also, the piston engine is not suitable for using high volume flow rates of gas.

Over the years the development pattern of piston aero-engines was to improve the engine air swallowing capacity by supercharging, increasing speed and improving the induction process and at the same time to push up the limits of pressure and temperature in the cylinder. Between 1909 and 1950 it has been shown (F. R. Banks, ref. 1) that the power output per cubic inch of piston displacement rose from 0·12 hp to 1·34 hp (5·5 W/cm^3 to 61 W/cm^3). In the same period the specific weight of engines fell from 3–4 lbf/hp to less than 1 lbf/hp (corresponding to 2 kg/kW to 0·6 kg/kW).

Pushing up the power of a piston engine by increasing the throughput of air causes difficulties in two very important fields. As the power of the engine goes up, the problem of cooling becomes more and more important because the cooling load increases more or less proportionally with the power. The aim will always be to keep engine bulk as low as possible so there was always a great field of development work in improving engine cooling. Fuel limitations were the other great barrier towards increasing output. Over the years engine improvements ran side by side with improvements in fuels which allowed higher pressures to be used in the cylinders before detonation set in. These two main problems will now be considered.

Engine Cooling

The great point of discussion on engine cooling was always on whether to use liquid or air cooling. Looking back over the years it is possible to see an ebb and flow of the tide favouring first one and then the other as various innovations gave each in turn the advantage. By the early 1930s the situation had settled to the extent that the high power radial engine (layout as in Figure 11.2(a)) using air

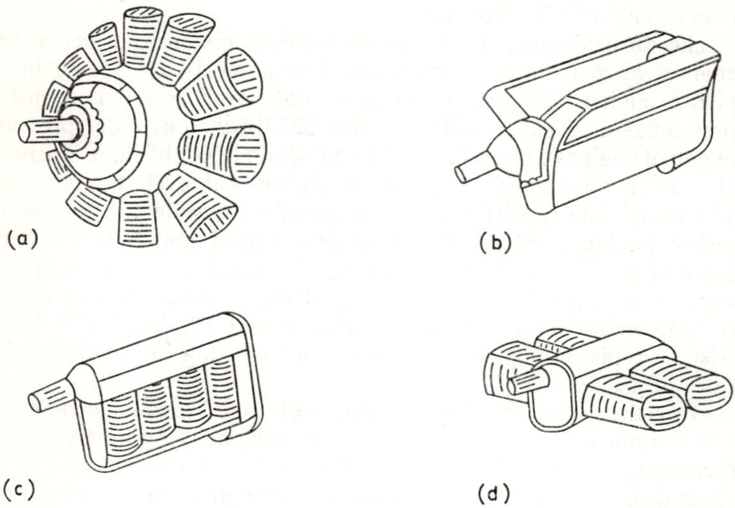

FIGURE 11.2 *Configurations of reciprocating aero-engines*

cooling was the most common type but the inline vee, with liquid cooling (Figure 11.2(b)) was beginning to offer strong opposition. An inline was inevitably of lower frontal area so could give an aeroplane very clean lines, while with liquid cooling the designer had a degree of freedom in positioning the radiator. For example the Supermarine Schneider Trophy racer, the S.6.B., had a radiator which formed the leading edge of the wing so there was virtually no drag penalty for cooling this engine. Even though the radial made a comeback with improvements in cowling design the inline always had this fundamental advantage of a lower frontal area.

By the beginning of the Second World War Rolls-Royce had overcome most of the plumbing problems of liquid cooling and had a highly efficient and reliable system. A secondary-surface radiator had been developed by them in which the cooling surface area on the air side was about five times that on the water side. (This could be done because the film coefficient with water was much higher than with air.) By suitably ducting a radiator it could act rather like a rudimentary ramjet and produce useful thrust instead of giving a drag penalty. Instead of plain water for cooling, a pressurized system using an anti-freezing mixture of methanol and water had been introduced which allowed operation at rather higher temperatures. There was however one serious drawback of the liquid cooled engine from the military point of view, which was that it was very vulnerable to battle damage. Any penetration of the cooling jacket, radiator or piping led to a loss of cooling and rapid engine failure.

In the United States the liquid cooled engine had been largely ignored until in the late 1930s Allison developed a V-12 roughly comparable with the Rolls-Royce Merlin. This was used in fighter aircraft but it was never as successful as its Rolls-Royce counterpart and eventually the Merlin was built under licence in America in very large numbers. Germany entered the War with one highly developed liquid cooled engine, the Mercedes-Benz DB 601, which was used as their main fighter engine until the advent of the Focke-Wulf 190 and which also powered a number of German bombers.

The Merlin and its contemporaries in Britain represented the culmination of design in the field of liquid cooling and though one or two later engines were built there were no significant developments in liquid cooling. The variant of steam cooling, as used in the Rolls-Royce Goshawk, was designed as basically a standard liquid cooled system but with the coolant being allowed to evaporate. The aims of having a more compact system and of allowing a higher engine operating temperatures were commendable but it was found that when violent manoeuvres were carried out by the engine, the water and steam would not stay in their proper places!

At the present time (1970) the only high powered liquid cooled aero-engines which are still in service in appreciable numbers are the Rolls-Royce Merlins and Griffons. Merlins are flying still in a number of military aircraft and are in civil use in Argonauts and Yorks; the main application of the Griffon is in the Hawker Siddeley Shackleton.

Considering now the high power air cooled piston engine, the outstanding advantage has always been its simplicity and lack of plumbing compared with the liquid cooled type. Against this had to be set the need to ensure an adequate airflow past each cylinder at all flight conditions. The logical layout to meet this requirement in the high power range is the radial engine as each cylinder then has its full supply of fresh air. At low forward speed and high power, that is particularly at take-off conditions, the cooling problem is at its most severe but the fact that the engine is in the propeller slipstream helps to keep it cool.

A number of highlights in the development of the radial engine over the years may be recognized. The early engines were of the single-row type generally with seven or nine cylinders but by 1917 Armstrong Siddeley had introduced the Jaguar two-row radial with 14 cylinders. This two-row layout became a classic pattern with the rear bank cylinders staggered behind those of the front so that all got a "clear view" of the air. In about 1927 came the first serious attempts to cowl-in the radials. Until then, to ensure adequate cooling all the cylinders were completely exposed, but in that year H. Townend of the National Physical Laboratory found that if a cowl was placed around the outside, cooling was still adequate and the aerodynamics of the installation were greatly improved. Having found that radials could be enclosed without undue loss in cooling performance the design of cowlings developed rapidly. Outstanding were the NACA types which gave low drag and a measure of control over the cooling through the use of gills at the rear of the cowls which could vary the airflow over the cylinder.

One of the last big engines to be built in quantity was the Pratt & Whitney Wasp Major, a 28-cylinder engine using four banks of seven cylinders. As a measure of the success in power planting this type it can be noted that the rearmost bank runs cooler than the first three. Some of the final developments in the history of the radial in Britain were designs at Bristol which incorporated a cooling fan driven by the propeller shaft (still to be seen in the Varsity aircraft) and a project, which did not reach the production stage, for using cylinders in which the barrels had longitudinal fins instead of the usual circumferential ones. If development of the large piston engine had continued, ideas such as these would have had much wider application in the search for lower drag and a smaller cooling load penalty.

This discussion of the cooling problem of engines has concentrated on the question of liquid and air cooling. One other important aspect of cooling is in the role of the engine oil in removing heat. With a piston engine there are a large number of components in sliding contact with each other and the lubricating oil has a very significant part to play in removing the resulting frictionally generated heat. In addition the piston will be subjected to a high thermal loading with a large amount of heat being carried away by the oil. All this meant that the design of the oil system for a piston engine was much more complex than for a gas turbine where the only lubrication loads are for ball and roller bearings.

Fuels for Reciprocating Engines

Whereas the engine cooling problem is one that does not exercise the minds of motorists unduly, everyone who has purchased petrol knows that a range of different qualities is available and that it is a bad thing for an engine if a grade of fuel is used of lower standard than that really required. Sufficient publicity has been put out by the oil companies for it to be generally known that "knocking" or "detonation" is the thing to be avoided. The history of piston engine fuels for aircraft is tied up completely with the fight against detonation.

What is detonation? Very briefly, when detonation occurs in a cylinder, the combustion takes place at a very high rate and with an explosive violence. Pye (ref. 2) describes the normal combustion process as being a controlled reaction which gives a steady, even rise in pressure through the cylinder. When detonation occurs, the pressure rise rate is very much more rapid and the peak pressures are reached before the piston reaches the top of its stroke. Loss in power results and there is a rise in temperature with a grave danger of mechanical damage to the engine due to the intensity of the pressure waves.

The mechanism causing detonation is very complex and is related to the molecular structure of the fuel. With some fuels the molecules are in the form of long chains and under high loads of pressure and temperature these chains are liable to break up into smaller molecules. It may happen that the spontaneous ignition temperatures of these smaller molecules is less than the temperatures of the charge in the cylinder so that burning takes place throughout the charge instead of spreading as an orderly flame front. The main physical factors which will cause a mixture to detonate are therefore increases in the pressure and temperature of the charge in the cylinder. These unfortunately are just the factors which the engine designer will be trying to push up in his attempt to get more power.

The full significance of detonation as the limiting factor in engine performance was gradually realized and basic combustion studies into the problem were carried out from 1918 by Ricardo, Tizard and Pye. During the 1920s (when so much great work on piston engines seems to have been done) the excellent properties of iso-octane as a fuel were discovered, this then being available as a suitable reference fuel against which others could be compared. Laboratories in different parts of the world set up research engines with variable compression ratio so that fuels could be graded and classified in a scientific manner. Meanwhile in the United States, Midgley had discovered (1922) that the anti-detonation properties of fuels could be improved by the addition of tetra-ethyl-lead (tel) in very small quantities—less than 0·1% by volume gave very significant improvements.

Blessings are seldom unmixed and although the addition of tel improved the knock rating it brought in many difficulties. For example even though the lead was present in such small percentages, during the normal running life of an engine appreciable quantities would be used and would cause unpleasant deposits on the sparking plugs and valves. There was also a distribution problem as there was a danger of different cylinders getting different concentrations of lead. Various additives were discovered by the fuel technologists over the years which served both to keep up the detonation resistance and to avoid harmful deposits.

By definition a fuel composed of 100% iso-octane has an Octane Number (ON) of 100. As better and better fuels became available the term octane number was replaced by a Performance Number (PN) which could be in excess of 100. The performance and octane numbers are related by the equation

$$PN = \frac{2\,800}{128 - ON}$$

so that an octane number of 100 corresponds to a performance number also of 100.

The rating of a higher performance petrol for aviation work is now quoted as a two-figure number such as 100/130 or 115/145 grade fuel. The first figure is the PN with a weak mixture and the second is with a rich mixture. (When running with a high fuel/air ratio, the additional fuel being evaporated will cause a cooling of the charge entering the cylinder and so delay the onset of detonation.) One method of determining the rating of fuel is to carry out tests on a standard single-cylinder engine with a separate supercharger so that the power may be varied at a selected engine speed. Two tests are carried out at varying fuel/air ratio with constant rev/min, firstly one using a standard reference fuel of PN 100 and then one with the fuel to be rated. In each case the intake pressure is increased until the fuel is on the point of detonating and the results are plotted as in Figure 11.3. The PN of the fuel being rated is calculated by finding what percentage of the power with the reference fuel it gives at the weak and rich mixture.

If excess fuel can act as a charge cooler and reduce the chance of detonation, it would be reasonable to expect other liquids to have a similar effect. Tests carried out using water injection show that this is so, although in this case, besides the straightforward cooling effect, the steam in the cylinder acts as a suppressor of detonation. Even better results are obtained using a mixture of methanol and water; the methanol acts as an anti-freeze and also reduces the boiling point compared with that of pure water. Water and methanol injection is frequently used therefore for boosting the take-off power of engines. One of the most beautiful sights on engine test was always that of an engine with stub exhaust pipes running with water/methanol injection as the exhaust flames would be a pale violet colour.

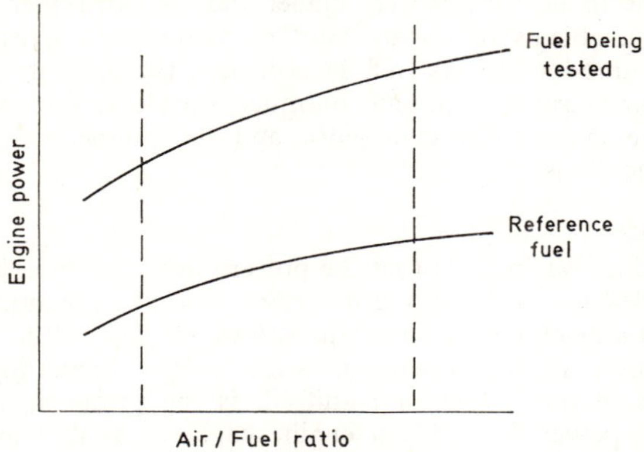

FIGURE 11.3 *Rating of fuels*

The problem of the deposits left by various fuel dopes in exhaust valves was a very real one and to overcome this difficulty the team under Roy Fedden at Bristol decided to concentrate on the sleeve-valve type of engine as opposed to the more common poppet-valve type. In this a sleeve acts as a liner between piston and cylinder and a number of ports are cut in the sleeve. A sleeve drive crank imparts a reciprocating motion to the sleeve the motion having both circumferential and axial components. In this way the sleeve ports cover and uncover the inlet and exhaust ports to the cylinder in any required manner.

Although the sleeve valve was not a new idea when the Bristol engineers took it up (a Burt MacCollum engine using it had been built in 1912) the problems of making it successful had never been overcome properly. With changes in engine temperature between starting and full power the clearances would vary and this introduced materials and sealing problems. However, after many years of hard work the first successful sleeve-valve engine of the Bristol series, the Perseus, was type tested in 1936. This was followed by the larger Hercules and Centaurus. Napier took up the sleeve-valve in their big Sabre engine and Rolls-Royce used this type of valve in the experimental air-cooled Exe and in the last big piston engine they made, the Eagle.

The controversy between sleeve and poppet valves generated almost as much heat as the controversy between air and liquid cooling. Looking back it can be seen that the sleeve valve fulfilled the original design object of getting over the exhaust valve problems and that it allowed a higher compression ratio to be used. Inevitably though, the mechanical losses were higher and the extra liner made heat transfer problems to the cylinder more difficult. Writing in 1950, Air Commodore Banks (ref. 1) expressed the opinion that if big piston engine development had continued, the sleeve valve would have been pre-eminent for civil work and the poppet valve for millitary applications.

Supercharging

It was pointed above (p. 311) that the power developed by a piston engine depended primarily on the *mass flow* rate of air handled by the engine. At a given engine speed the *volumetric flow* will be more or less constant, so supercharging, which means increasing the density of air in the induction manifold, is the prime means of increasing the power for a given engine capacity. Although it is possible to achieve a small amount of supercharge by using resonance effects in a tuned induction pipe, the emphasis for aero work has always been on mechanical compression methods.

One of the great pioneers, Sir Harry Ricardo, wrote in 1950 on the supercharging of internal combustion engines (ref. 4). He pointed out that until about 1920 there were mechanical limitations on the piston engine—thermal loading for example—such that an unsupercharged engine at sea level could develop as much power as could conveniently be handled. On the other hand, as soon as an aircraft started to climb, the air density would fall-off and so the power of a naturally aspirated (i.e. unsupercharged) engine would also fall. Supercharging in this instance might be used to restore the power as the aircraft climbed. From this time however, advances in the general design and the introduction of new fuels meant that engines could stand an increased density at sea level with the accompanying rises in cylinder pressure and temperature. Supercharging then came in with the prime purpose of increasing the take-off power of an engine.

A number of different types of blower have been used at various times as superchargers but the most successful and the widest used has been the centrifugal compressor as described in Chapter 4. This type is compact and flexible and can operate stably over a wide range of mass flow rates at a given speed. Being a high speed machine it is necessary to use a step-up gearing between crankshaft and blower if a mechanical drive is used (the special case of the exhaust turbine driven blower is considered below). The delivery pressure can be controlled quite safely with a centrifugal compressor by throttling the intake to reduce the compressor flow. In this way excessive supercharger delivery pressures, which might lead to detonation or overloading of the engine can be avoided.

The gross increase in power of a supercharged engine due to the greater charge density must be debited with the power required to drive the compressor. It is therefore important that the compressor used should be as efficient as possible. In addition an efficient supercharger will give a lower temperature rise than a less efficient one so the density ratio for a given pressure ratio will be better.

Figure 4.15 showed the typical form of a centrifugal compressor characteristic and it was seen that the region of high efficiency is fairly limited in extent. When the requirement came for aircraft to fly higher it was found that the use of only one gear ratio for the supercharger drive was unacceptable and so multi-speed gear trains were introduced. Another complication came in when it was realized that sufficient compression could not be obtained in one stage of a centrifugal compressor so two stages were fitted in a number of cases. This degree of compression resulted in a very high charge temperature which was liable to cause detonation so it was necessary to fit a cooler after the supercharger. This reduced the charge temperature hence increasing its density, so besides reducing the risk of detonation a higher degree of supercharge was achieved.

Although the great majority of aero-engines use a mechanical drive to the supercharger there are advantages in using a turbo-supercharge arrangement wherein the compressor is driven by a turbine powered from the exhaust gases. As early as 1924 a Napier Lion had been built using this type of drive but it did not come into extensive use until the 1930 when the General Electric Company in the United States commenced development in this field. During the war the GE type supercharger was used widely on the radial engines which powered the majority of American aircraft.

While there are obvious advantages in using the exhaust gases to power a supercharger care must be taken that the engine is not damaged by putting back pressure on the exhaust system. Any restriction to the flow will tend to give the exhaust valves a harder life as the temperature of the gases will be higher. The materials problems of the turbine are quite severe as the exhaust gases are hot and may be corrosive. From the turbo-supercharger, it is a logical step to develop the fully compound engine which is described in greater detail in a later Section (p. 324).

Altitude Performance

For a supercharged engine with a variable speed propeller the pilot will have at his disposal two independent variables by which to control the power output. The engine speed is controlled by varying

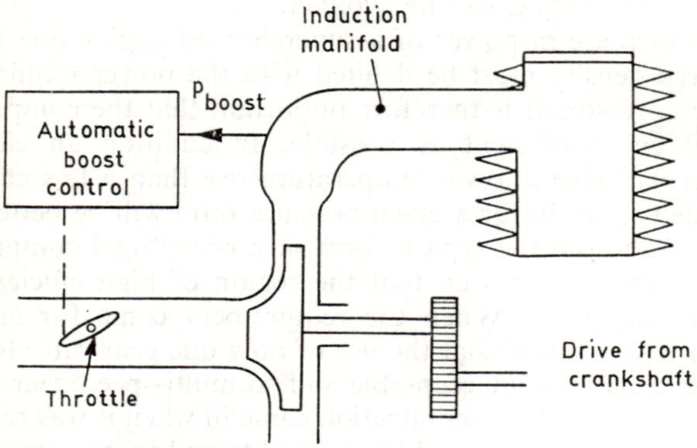

FIGURE 11.4 *The automatic boost control*

the setting of the constant speed unit in the propeller, while at a given speed the power can then be varied by altering the "boost" which is the pressure in the induction manifold downstream of the supercharger. The boost pressure is controlled by means of an automatic boost control device which acts on the throttle valve at the inlet of the supercharger. (See Figure 11.4.)

In British parlance the boost is the absolute pressure in the induction manifold measured above a value of 14·7 lbf/in. so that + 10 lbf/in. boost means a charge pressure of 24·7 lbf/in. American practice is to state the boost in inches of mercury absolute so + 10 lbf/in. boost would be approximately 50 in. Hg in American terms.

The form of the altitude characteristics of a supercharged engine are as in Figure 11.5. This shows power against altitude curves for a typical engine operating at some constant value of rotational speed with representative boost figures of + 10, + 5 and 0 lbf/in.^{2} (+ 70, + 35 and 0 kN/m^2). The description of the curves is as follows:

From A–B, at + 10 lbf/in.^{2} boost the throttle is initially partly closed to avoid excessive boost pressures. As the aircraft climbs the throttle is opened progressively until at B, "full throttle height" is reached; this is also known as the rated altitude.

From B to C the throttle remains fully open but due to the fall in intake pressure the boost pressure cannot be maintained at the selected value. It will fall to + 5 lbf/in.^{2} at E and zero at G. The *rise* in power between A and B is due to the fact that although the

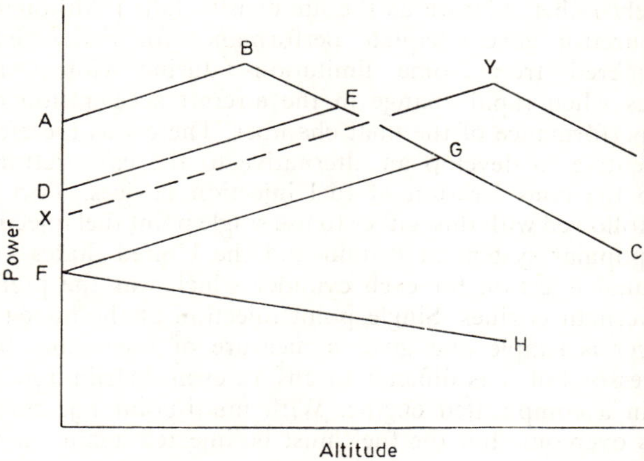

FIGURE 11.5 *Representative altitude characteristics of a piston engine*

induction manifold absolute pressure remains constant the back pressure on the exhaust falls.

D – E and F – G are similar to A – B but with the increased full throttle heights.

F – H shows how the power for a normally aspirated engine would fall with altitude.

Finally, X – Y – Z shows the power for a higher supercharger gear ratio again at + 10 lbf/in.^{2} boost. It would obviously be foolish to work on the dotted part of this line but with the higher gear, power is maintained up to higher altitudes.

Besides power, a pilot will also be interested in fuel consumption variations at various boosts and engine speeds. As a rough and ready rule for long duration, operation should be at the highest possible rev/min with the lowest boost which will give the required power.

So far, all the topics we have discussed in this chapter have been basically thermodynamic ones. Two other questions of a more mechanical nature have been vital to the success of the piston engine and so deserve special mention. They are those of fuel supply to the cylinders and of ignition.

Early aero-engines used conventional-pattern carburettors of the type used in motor cars but with special provision for altitude compensation. (This is necessary since the simple carburettor gives a progressively richer mixture as the air density falls.) Although the basic carburettor gave adequate performance for many years it always suffered from some limitations during violent aerial manoeuvres when rapid change in the aircraft acceleration could upset the performance of the float chamber. There was therefore, a strong incentive to develop an alternative to the carburettor and this led to the consideration of fuel injection devices. Two main lines were followed with this, either to use single point fuel injection—the most popular system in Britain and the United States—or to use individual injection for each cylinder which was the preferred way for German engines. Single point injection at the inlet to the supercharger is simple and gives a measure of precooling before the compressor but it is difficult to ensure even distribution to all cylinders in a complicated engine. With multi-point injection this problem is overcome but the fuel must be injected against a much higher air pressure.

On the question of aero-engine ignition little need be said. Although a low tension system has been used on occasions the use of high tension is more general with the H.T. supply coming from a magneto and supplying two sparking plugs per cylinder. The magneto is preferred to coil ignition as it is lighter and makes the engine independent of batteries during normal running. In addition the ranges of engine speed covered in normal operation are very much less than those encountered in automotive applications so the fall off in the voltage output of the magneto at reduced speeds is not significant. Using two sparking plugs per cylinder each set being supplied by its own magneto improves the reliability of the ignition and gives a useful increase in power compared with having a single plug only. (On test each magneto is switched off in turn and the drop in power is quite noticeable.)

This summary of some of the highlights in the development of the high power piston aero-engine has necessarily been brief and has ignored many points which at various stages seemed to be setting insurmountable difficulties in the way of progress. At one time engines were in severe trouble with torsional vibrations of the crankshaft; at another the problem of surging in valve springs threatened to hold back the design of all engines.

Each problem in turn was overcome by the designers and by test engineers whose job it was to push the engines to the limits of their endurance and beyond.

11.3 PROPELLERS

The development of the propeller has been associated so closely with that of high power engines that it is appropriate at this stage to look briefly at its theory and history.

The function of a propeller is to convert the power output of an engine into useful thrust. To do this it makes use of Newton's laws of motion, that is it accelerates backwards a mass of air. An overall picture can be obtained by treating the propeller as an "actuator disc", a plane through which air flows and which gives the air a pressure rise which can be converted into a velocity increase. Although the actuator disc idea is useful, in some ways a clearer picture of how the propeller produces thrust can be obtained by considering it from the "blade element" point of view.

Figure 8.7 (p. 245) shows the vector diagram for an element of a propeller at diameter D moving at forward speed V with rotational speed N. If the element is at a pitch angle β to the plane of the propeller disc, the air will meet the element at an angle of attack of

$$\left(\beta - \tan^{-1}\frac{V_a}{\pi ND}\right)$$

The aerodynamic force will depend on the angle of attack, the resultant velocity of V_a and πND, the shape of the element and the air density. This overall force may be resolved into two components, a useful thrust perpendicular to the plane of the propeller disc and a drag force at right angles opposing motion which must be overcome by the torque of the engine. To find the total forces acting on the propeller, the forces on each element are integrated from root to tip of the blades. Clearly if all elements of a propeller blade are to be at approximately the same angle of attack, the pitch must vary with radius being coarse at the roots and decreasing towards the tips.

It was shown in Chapter 8 when discussing the control of turboprops that propeller characteristics can conveniently be plotted in terms of thrust and torque coefficients against the advance ratio (V/ND) for constant values of propeller pitch angle. The coefficients incorporate air density and rotational speed factors while the advance ratio brings in forward speed and rotational speed. Figure 8.8 on p. 245 showed the typical form of these overall performance curves. As a general performance figure it can be taken that a modern propeller will produce a static thrust of 15–20 N per kW input (about 2½–3 lbf per hp) and that at cruise conditions the efficiency of converting engine shaft power into propulsive power will be of the order of 70–75%.

Turning now to the history of propellers, those used in the earliest days were of fixed geometry design generally being carved from solid or built up to the required shape from wooden laminations. A propeller of this type absorbs power according to a cube law, that is, for a given air density and flight speed there is a unique relationship between power and speed in which the power varies with $(\text{rev/min})^3$. The unique relationship is liable to be an embarrassment as it makes the engine inflexible to operate—only one engine speed can be used at a given power—so in the 1920s the first attempts were made to design variable pitch propellers. The difficulties of designing a mechanism which would allow the propeller blade to be rotated about its axis while being strong enough to resist the centrifugal loads were considerable but they were overcome eventually. Early variable pitch propellers had two pitch settings, a fine pitch allowing high engine speed for take-off and landing and coarse pitch for higher speed flight. Before long, continuous variation of pitch settings was possible and this was eventually tied in with a governor unit which allowed constant speed operation of the engine over a wide range of power outputs and flight speeds. Provision for feathering the blades—setting them at 90° pitch angle—and for using reverse pitch for braking were also made.

The other main line of development of propellers lay in designing units to absorb the ever increasing engine powers. To absorb more power more blade area is needed but there are limits to what is acceptable by increasing diameter or blade chord. The alternative of fitting more blades to the hub brings in mechanical problems but by careful design it was found possible to fit up to five blades to one hub. After that contra-rotating propellers were designed with two discs, these having the further advantage that the torque reaction of the propeller on the aircraft was largely eliminated. Probably the ultimate in this field is the propeller fitted to the Kuznetsov NK–12M turboprops in the Tupolev TU–114 airliner. These propellers are eight-blade contraprops of 5–6 m (18 ft 4 in.) diameter absorbing approximately 10 MW (12 000 hp).

11.4 THE COMPOUND ENGINE

The *p–V* diagram (Figure 11.1) shows that as engine power is increased by pushing up the peak cycle pressure and temperature there will be a corresponding increase in the pressure at the end of the power stroke. Now the rejection of this gas, which is at well above atmospheric pressure, represents a loss in potential output of the engine. The difficulty arises in that this potential can only be realized by increasing the stroke of the engine, that is by using a larger swept volume. However the law of diminishing returns is coming in and useful power increases can be obtained only with unacceptable increases in engine size. At this point in the cycle gas density is comparatively low and the piston engine is not happy handling large volume of gases. It is here though that the turbine—with its inherently high swallowing capacity—shows promise of coming into its own.

A wide variety of different arrangements of compressor, piston engine and turbine are possible in devising compound cycles.

For aero-engine work only three have reached fruition. They are shown diagrammatically in Figure 11.6 as

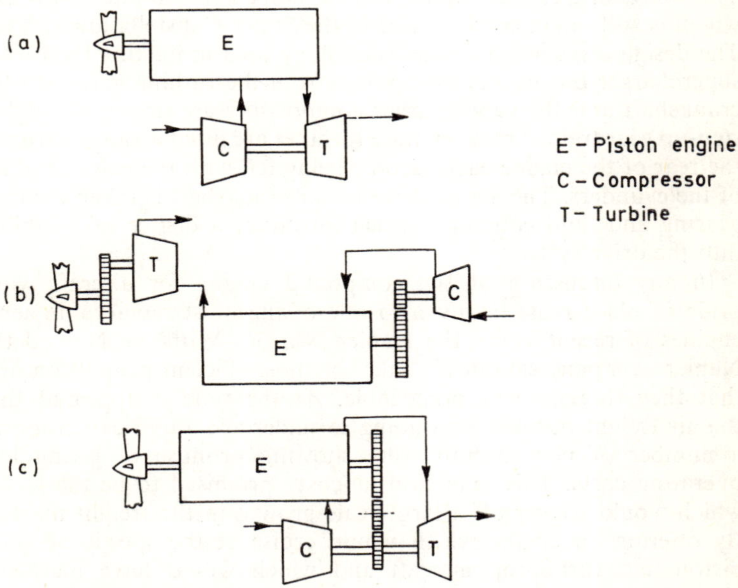

FIGURE 11.6 *Arrangements of compound engines*

(*a*) The straightforward turbo-supercharger where a free-running compressor/turbine boosts the piston engine.

(*b*) An arrangement where the supercharger is driven mechanically by the engine and the entire turbine output goes into the crankshaft.

(*c*) A combination of these two with the turbine giving sufficient power to drive the compressor and to contribute some shaft power at the engine crankshaft.

Brief mention has already been made of the turbo-supercharger arrangement and of the difficulties that had to be overcome in making a successful unit. As both the centrifugal compressor and the turbine are fundamentally high speed machines the matching problem is not severe and the wide range of mass flows possible with a centrifugal compressor meant that the air-breathing characteristics of the piston engine could be met without too much risk of compressor surge. The main problems that had to be overcome were those of finding suitable materials for the turbine blades and of avoiding back pressure effects on the piston engine.

Arrangement (*b*) was that adapted by Curtiss Wright for their Turbo-Cyclone engine, which was developed in the late 1940s and which is still in extensive service in the Super-Constellation airliner. The design owes much to the technology used in making the turbo-supercharger but in this case power from the turbine goes into the crankshaft and the supercharger compressors are driven through a step-up gear train. Three separate turbines are fitted arranged around the rear of the engine each turbine being fed with the exhaust of six of the cylinders. The actual drive to the crankshaft is taken through gearing and fluid couplings which introduce a degree of flexibility into the drive system.

In any discussion of the compound engine for aircraft work pride of place must be given to one of the most remarkable aero-engines of recent years, the Napier Nomad. With the Nomad the Napier Company set out to make the most efficient propulsion unit that they thought was practicable. At the time it appeared that the air freight business was going to undergo a rapid expansion as a number of new aircraft were showing promise of giving low operating costs. This reduction in costs promised to be the factor which would trigger off a large-scale growth in the freight market. By offering an engine which would cruise at the speeds of good piston and turboprop aircraft and which would have markedly lower fuel costs it was hoped to cash in on the boom.

The best cycle which seems feasible for propeller driven aircraft appeared to be a combination of the two-stroke diesel engine and the gas turbine. Before the Second World War a number of compression-ignition engines had been made for aircraft work, notably the Junkers Juno 205E in Germany and the Bristol Phoenix in this country. This development of the diesel had looked particularly promising for long-range aircraft where the reduced fuel consumption would, over a long flight, balance out the heavier installed weight of the power plant. Proponents of the diesel could point to the higher thermal efficiency and could show that diesel fuel was a very much safer fuel than petrol if the aircraft should crash. The disadvantage of a much higher freezing point for the diesel fuel could be overcome by providing some form of heating in the fuel tanks—a difficult problem but not insurmountable.

Napier approached the job of compounding the two-stroke diesel and gas turbine with a considerable background of experience. On the high speed piston engine side the company had a long history of successes culminating in the Sabre, in its day one of the most powerful engines in the world; on the diesel side development was proceeding of the Deltic which is a highly successful engine and which has set new standards of compactness in the medium power range; on the gas turbine side the first Napier turboprops had been successfully run and flown. All these factors pointed to the company being in a very strong position to tackle this highly efficient advanced technology engine, the Nomad.

The actual layout used in the Nomad is shown in Figure 11.7 while a full and detailed discussion of the engine was given by

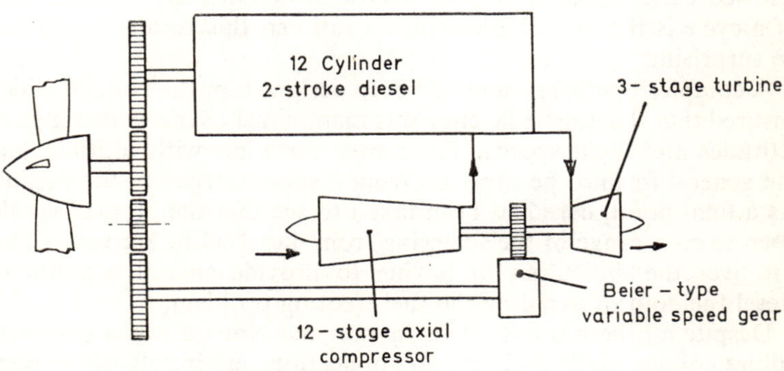

FIGURE 11.7 *The layout of the Napier Nomad*
(From ref. 5, Courtesy the Royal Aeronautical Society)

Chatterton (ref. 5). Air entered a 12-stage axial flow compressor which supercharged the 12-cylinder horizontally opposed two-stroke diesel. At the forward end of the crankshaft a reduction gear was fitted for the propeller drive. The exhaust gases from the diesel were carried to a three-stage axial flow turbine which rotated at the same speed as the compressor and supplied all the compressor power.

Surplus turbine power to that required by the compressor was taken off and passed through a variable speed gear on its way to the propeller shaft. By using this it was possible to vary the axial compressor/turbine speed independently of propeller and diesel crankshaft speed and so get better performance.

Full details of the performance of the Nomad over the whole range of flight conditions were quoted in Chatterton's paper. The take-off power of the basic engine was 3 135 ehp (2 340 kW) and this could be increased by using water injection to 3 580 ehp (2 600 kW) and by further adding a small measure of combustion before the turbine to 4 100 ehp (3 060 kW). (Powers quoted are equivalent horsepower and make allowance for the small amount of jet thrust.) These powers are of the same order as those being produced at the same time by the Bristol Siddeley Proteus, which was in many ways a competitor of the Nomad. Of much greater interest than the power was the specific fuel consumption which at ISA take-off was 0·345 lb/eph hour (0·21 kg/kWh), corresponding to a net thermal efficiency of about 41 %. This incredibly high thermal efficiency—half as much again as the Proteus—was a result of the very high expansion ratio. If a combined expansion ratio for diesel and turbine is calculated it found to be equal to a volume expansion ratio of 27:1. It has been stressed earlier that the crucial factor in determining the efficiency of a cycle is the available expansion ratio so this result should not be surprising.

A complex control system had been designed for the Nomad which ensured that the good efficiency was maintained over a wide range of altitudes and flight speeds. The power variation with altitude had the general form to be expected from a supercharged piston engine. As a final point, care had been taken to see that the Nomad would operate on a range of fuels varying from diesel oil to kerosene. This got over the difficulties of having to provide special supplies of diesel fuel and of avoiding the fuel freezing problem.

Despite all these points in its favour, the Nomad was a complete failure commercially and had no applications in aircraft which were sold to operators—its only flights were in a Lincoln flying test bed as part of its development programme. Looking back it is of interest to try and decide just *why* there should have been this utter lack of success. Four points may be put forward for consideration:

(i) It is extraordinarily hard for a new type of engine to break into an established market. This problem comes up repeatedly in engineering and at present, for example, we are witnessing the attempted break-in to the automotive market by the

Wankel and the gas turbine engine. Each of these has some decided advantages over the well established piston engine but neither offers a breakthrough as complete as did the turbojet compared with the piston engine. The Nomad arrived to compete against turboprops which were themselves improving in efficiency all the time. Any initial margin of advantage it might have had would be continuously eroded away.

(ii) The expected explosive growth of the freight market did not materialize at the time expected. The commercial estimates had been too optimistic.

(iii) Fuel cost are only one item in the direct operating costs of aircraft and, though important, they are by no means the dominant factor. Other terms such as utilization and block-speed can have a far greater influence on costs. The pattern expected therefore of the freight market being dominated by propeller aircraft probably gave a false picture of what was to happen. It now seems that pure jet aircraft used on freight can offer lower costs (*with* higher speed) and the mixed traffic airliner—easily convertible from passengers to freight—is one of the most important types.

(iv) The Nomad was too ingenious and too complex. This may have been the crucial point. As an example of aiming for the ultimate it was excellent. As an example of engineering skill to make it all work, it was a magnificent achievement. But airline engineers and maintenance men want reliability and, for this, simplicity is a very great advantage. The service engineers were metaphorically breathing sighs of relief to see the end of piston engines with all their mechanical problems and they did not want to see a new highly rated engine coming into service.

Those points may suggest why the Nomad did not succeed. The verdict by an outsider must be that it was a magnificent experiment which came at the wrong time to be a commercial success.

11.5 THE PLACE OF THE PISTON AERO-ENGINE TODAY

Many hundreds of airliners using high power piston aero-engines are still in service but they are being relegated to the minor routes of the World's airways and comparatively little new design and development work is being carried out on the engines. Instead, the mainstream of design effort has turned to the low power range of engines as applied to light aircraft and the smaller business machines. For these types the piston engine has advantages of lower first cost and good fuel consumption, while the utilization is comparatively low so that the gas turbines reliability and long life do not really offer real benefits. No doubt small gas turbines *will* gradually eat into the market but the piston engine will almost certainly predominate for many years to come.

The market for light engines is so large in the United States compared with that in other countries that it is not surprising that the American manufacturers dominate the field. Two companies in particular, Avco Lycoming and Continental, produce the great bulk of the engines. Between them they offer over fifty distinct types, ranging in power from 65 to 400 hp (50–300 kW). Since 1960 the Rolls-Royce Light Engine Division has been building a number of Continental designs under licence for use in aircraft built outside the United States

A classic pattern of light piston engine has evolved over the years, namely a four or six cylinder horizontally opposed air cooled engine of the layout shown in Figure 11.2(d) (p. 312) and the great majority of the new designs are of this layout. It allows a compact installation with the cylinder heads readily accessible to the cooling air. Within this general layout though there is a large number of variants. Some engines are supercharged and some are naturally aspirated; some use conventional carburation and some employ fuel injection; the great majority use a direct drive from crankshaft to propeller but a few of the larger engines have a reduction gear to allow the engine to run faster and so give more power. Despite this wide complexity of possible arrangements, the makers try to standardize as far as they can over their ranges and use a number of common components in different engines.

The other main layout of light engine which has been used extensively is the four or six cylinder in line inverted air cooled type (Figure 11.2(c)). Very many engines of this construction were made by the Blackburn Company as the "Cirrus" range and by the de Havilland Company as the "Gipsy Queens" the engines powering such famous aircraft as the Tiger Moth, Dove and Heron. Compared with the horizontally opposed, the inverted inline is much slimmer—although it must be deeper—and it was better suited to light aeroplanes where pilot and passenger sat one behind the other. When the American light plane with its side by side seating came into use the fuselage was wider and so could accommodate the horizontally opposed layout so that the incline lost one of its main advantages. The only inline type still in production in Britain is the Rolls-Royce

Gipsy Queen Six, which is supercharged and which has a take-off rating of about 400 hp (300 kW).

Occasionally designers of light aircraft look to the possibilities of using automobile type engines as power units. At first sight this would seem to be an attractive proposition as the motor industry produces engines in such numbers that unit costs are bound to be low. When a detailed study is made however, it is found that the conventional water cooled automobile engine is ruled out on the grounds of excessive weight. Of the air cooled car engines in current production, the one that has attracted most attention from the aircraft people has been the standard Volkswagen flat four of about 1·2 to 1·5 litre capacity. Modifications made to improve its suitability for aircraft work include replacement of battery and coil ignition by a magneto (which is lighter and somewhat more reliable) and the fitting of two sparking plugs per cylinder. This latter change is in the interests of reliability and a rather higher power output.

It has been pointed out earlier that the big improvements in petrols over the years have been made largely as a result of adding dopes such as tetra-ethyl-lead but that these dopes are liable to cause damage to the sparking plugs from deposition of lead. For this reason very many engines for light aircraft are designed with conservative values of compression ratio—lower than on most modern cars—and are operated on low octane fuels which are free of lead. Here again the aim is to improve reliability at some cost in performance.

12 Mechanical and Material Considerations in the Design of Turbine Engines

12.1 INTRODUCTION

The purpose of this chapter is to highlight some of the main stressing and materials problems which arise in the design of turbine engines. Only a few of the main components are dealt with as many parts of an engine, such as accessory drive shafts, will be designed in a conventional manner. As a rule, though, even conventionally designed components will be worked "closer to the bone" than would be customary in non-aircraft practice.

Of the work that has been done on aero-engine stressing over the years, the greater part has been concerned with the stressing of rotating parts. This is inevitable because the results of failure of moving parts will usually be more serious than those of stationary parts—the dissipation of kinetic energy is liable to cause extensive secondary damage. Although this emphasis is understandable, it must be appreciated that static parts add up to more than half the total weight of an engine. It might well prove worthwhile from the point of view of savings in overall engine weight to concentrate more on the stationary components.

In the work that follows the problems of compressor blade breakages are first considered. It will be seen that the most serious danger arises from blade vibration or impact damage. The compressor casing and combustion chambers are then mentioned briefly—the problems here are those arising from the very high peak pressures and temperatures. Then the turbine is considered both from the points of view of high temperature blading and disc integrity. Finally mention is made of the design of interconnecting shafts between compressor and turbine.

12.2 THE COMPRESSOR

From the earliest days of axial flow compressors it has been appreciated that, with the multitude of slender blades, it is a very vulnerable machine. Foreign body damage is liable to cause secondary failures and the expression "to give a compressor a hair-cut" describes the most serious form of failure.

Compressor blade failures seldom occur due to straightforward overstressing under normal steady running conditions as the design problem in this case is well understood. Instead, the most likely causes of damage are the fluctuating stresses due to vibrations causing fatigue failure of the blade, or straightforward impact damage resulting from ingestion of a foreign body. These will now be considered in turn.

Compressor Blade Vibration Problems

For a review of the problems of compressor blade vibration reference should be made to the papers by Armstrong and Stevenson (ref. 1) and by Armstrong alone (ref. 2). These papers contain much information on the vibration characteristics of blades and on the techniques used in studying the problem.

The fatigue properties of a material are generally plotted in the form shown in Figure 12.1 where stress to cause failure is plotted

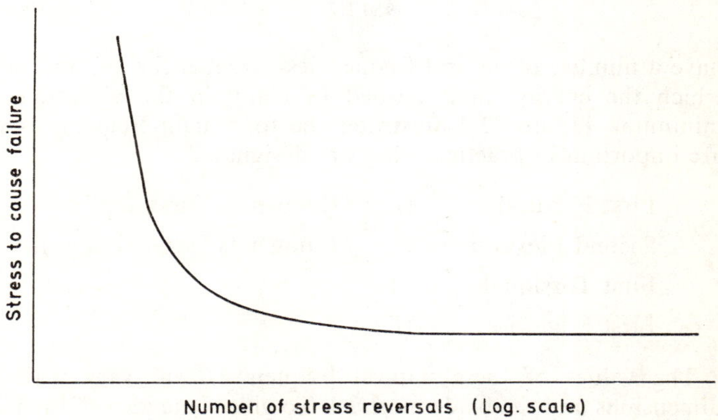

FIGURE 12.1 *Typical fatigue characteristics of a metal*

against the number of reversals of a cyclically applied stress. At very high stresses only a small number of reversals are needed to break the material but below a certain stress level the curve levels off and the material will stand any number of reversals without failure. A design aim will be to work with the highest stress which will give some required life, so knowledge of the frequency of vibration is needed. Great care must be taken in the use of fatigue data as there is generally an appreciable scatter between specimens tested and a small percentage of components will fail in a much shorter time than the average.

A compressor blade will have vibration properties similar to those of a cantilever rigidly mounted at one end. In particular it will

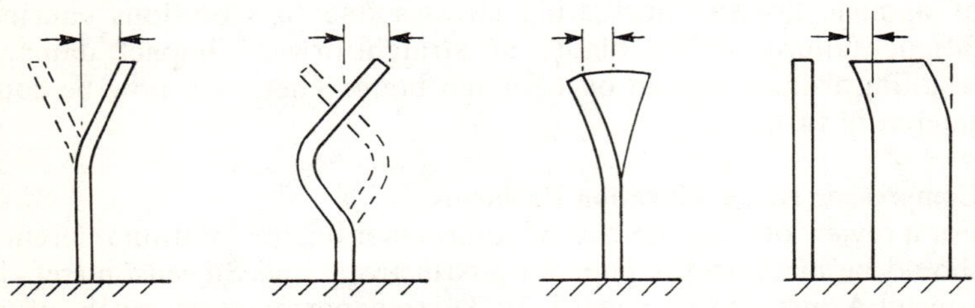

FIGURE 12.2 *Blade vibration characteristics*

(From ref. 1, Courtesy Royal Aeronautical Society)

Values for a typical blade:

1st flexural	2nd flexural	1st torsional	1st edgewise
250 Hz	950 Hz	1 570 Hz	1 940 Hz

have a number of natural frequencies—frequencies of vibration for which the energy input needed to maintain the vibration are a minimum. Figure 12.2 illustrates the four main frequencies which are important in practice. They are designated

First Flexural	1F	(known as "first flap")
Second Flexural	2F	(known as "second flap")
First Torsional	1T	
First Edgewise	1E	

The values of these natural frequencies will vary with blade dimensions and the blade material. To give some idea of the relative frequency values for a typical blade, figures from ref. 1 are quoted on Figure 12.2 (the International unit for frequency is the hertz, 1 Hz being 1 cycle per second).

For the particular case of a vibrating cantilever it can be shown that the maximum stress σ occurs at the root and is given by the expression

$$\sigma = 2\pi \frac{y}{k} \sqrt{(E\rho)} \, . \, af \tag{12.1}$$

banning steady running at 8 400 rev/min—a drastic step—or by redesigning the nose-cone support with three struts.

Compressor Blade Impact Failure Problems

Compressor blades are designed to move at high speed—tip speeds of the order of 300 m/s are possible—and to handle air with a maximum density of say 2·4 kg/m^3 at inlet conditions. Clearly any foreign body entering the intake will cause a sudden increase in local density and an impact load on the blades. If failure of a blade occurs then the broken parts are liable to cause further secondary damage.

Apart from the fairly remote possibility of an oversight in servicing when some tool is left in the intake of an engine, the most common causes of impact failure are bird strikes and the ingestion of debris from the ground. The bird strike problem is a really intractable one with no clear solution yet in sight except that of making the first stage blades of a compressor sufficiently strong to withstand an impact. Airworthiness requirements call for tests on all new engines involving the firing of dead birds into the intake at maximum compressor speed and the engine must be able to digest multiple impacts without serious damage or excessive power loss. The debris ingestion problem is becoming more serious with the advent of larger diameter engines which when mounted under the wing have low ground clearance. A miniature cyclone can be set up between the cowl lip and the ground which is able to suck up quite substantial objects.

Clearly with all these requirements to be met by the blade material, the choice of what is best will not be a straightforward one but will depend on the particular application. For example with a lift engine, light weight will be more important than long fatigue life, while for an engine for civil use impact resistance may have the highest priority. As a general rule though both high strength and low density will be wanted and the specific strength will give some measure of the relative merits of different materials. Table 12.1 shows the values of specific ultimate strength for a number of possible compressor blade materials.

Consider each of these materials.

(1) *Aluminium Alloys.* A very wide range of aluminium alloys has been developed over the years and RR58 is a typical successful type. This alloy has satisfactory strength and good machining properties so has been used extensively for making blades. However the specific strength falls fairly quickly with temperature, making it

unsuitable for the high pressure end of compressors. It is also less rugged than steel so is inferior in applications where resistance to impact or erosion is particularly required.

(2) *Steel.* Steel, a well tried and inexpensive engineering material, has comparable specific strength to aluminium alloy at low temperature and it keeps its strength as temperature increases. It is a good

TABLE 12.1 *Properties of Some Compressor Blade Materials*

Material	Density ρ Mg/m^3 (lb/in^3)	At 15°C: Ultimate tensile strength σ kN/mm^2 (lbf/in^2)	At 15°C: Specific strength σ/ρ kN-m/kg	At 250°C: Ultimate tensile strength kN/mm^2 (lbf/in^2)	At 250°C: Specific strength σ/ρ kN-m/kg
Aluminium alloy RR58	2·71 (0·098)	0·414 (60 000)	153	0·244 (35 400)	90
Steel FV 535	7·76 (0·28)	1·080 (156 000)	139	1·010 (146 000)	130
Titanium alloy	4·60 (0·16)	1·010 (146 000)	220	0·770 (112 000)	168
Beryllium alloy	1·86 (0·067)	0·610 (88 500)	328	0·510 (74 000)	274
Glass fibre reinforced plastic	2·05 (0·014)	1·130 (164 000)	552	0·670 (98 000)	330

(Based on Glenny (Ref. 3))

general purpose blade material which can be used for the high pressure stages and is strong enough to withstand impact at the low pressure end.

(3) *Titanium Alloys.* Titanium alloys have been developed over the years particularly to meet the needs of the aircraft industry where its light weight together with a strength comparable to that of steel makes it particularly attractive. By now its manufacturing properties are well understood.

Surprisingly, titanium introduces a fire risk in compressors. This is because the metal has a very low thermal conductivity—about 20% of that of steel—so that if a rub should occur the heat generated will not be carried away but will raise the blading to its ignition point.

(4) *Beryllium.* Beryllium is another of the less common metals which have been investigated in the search for strength and lightness. Although outstanding in this way beryllium alloys have the major disadvantage of a very low impact strength. Until some means of overcoming this is found there is little likelihood of these alloys being applied to practical compressors. At present the alloys are very expensive and the metal itself is poisonous.

(5) *Glass Fibre Reinforced Plastic.* A new branch of Materials Technology, that concerned with composite materials, has grown up in recent years and the figures quoted in Table 12.1 are for a fairly early composite. A striking feature is its high specific strength at normal temperatures although it will be noticed that this strength falls off rapidly with increase in temperature. The strength/temperature characteristics depend on the mix between the bonding material and the reinforcement so designers have a fair measure of freedom in choosing required characteristics.

Although the composite material referred to in Table 12.1 has high specific strength, its modulus of elasticity (Young's Modulus E) is low, being only 70% of that of aluminium alloy. Equation 12.1 shows that this will lead to a low stiffness which may be a disadvantage with larger blades. More recent materials are greatly improved in this respect. A Rolls-Royce material known as Hyfil has a density of 1·60 kg/m^3, an ultimate tensile strength of 1·16 MN/m^2 and a modulus of rigidity 60% of that of steel. It is a promising material for the manufacture of fan blades for large turbofans. Even more useful materials are possible if the great strength of fibres and whiskers can be used.

Compressor blades are so vulnerable compared with the remainder of the compressor that it is not surprising that the major part of the overall research and development effort has been concentrated on the blading. The remaining parts of the compressor—the discs and the casing—are comparatively straightforward from the design point of view. The weight of a compressor disc is determined to a large extent by the centrifugal loads applied at the rim by the blades and so is dependent on blade weight. If use can be made of lighter and stronger blade materials, very large savings in weight of the disc are possible which can be added to the direct savings of blade weights. As the blade attachments to the disc are highly stressed and require to be made to a high degree of accuracy it is important that the material selected for the discs should have good machinability and low creep growth with time.

Regarding the compressor casing, two main failures must be guarded against. First there is the need to withstand the direct bursting loads due to the internal gas pressure. The critical operating case here is at high forward speed, low altitude on cold days when the highest compressor pressure ratios are encountered along with the highest intake pressures. Secondly there is the necessity of containing the blades in the event of blade breakages. This is absolutely essential from the requirements of the airworthiness authorities as vulnerable pipes and ducts may pass close to the compressor. The energy to be dissipated in containing a blade will be directly proportioned to the kinetic energy of the blade so will be highest at the low pressure end of the compressor where, fortunately, the gas bursting loads and the metal temperatures will be least. Clearly the loads to be withstood by the casing and the corresponding material strength (which will depend on the temperature) will be difficult to predict exactly.

Besides the direct loadings sustained by the compressor casing, there is a need to provide rigidity in order to ensure accurate location of bearings, etc. Loads arising from this will include those encountered during manoeuvres arising from both inertia effects and precession effects due to gyroscopic couples. In addition the casing may be required to transmit some of the engine thrust to the mounting points.

12.3 THE COMBUSTION SYSTEM

The work in Chapter 6 showed that the classic pattern of chamber which has evolved for aero-engine use has two main components, the inner flame tube structure and the outer casing. Of these two the flame tube is exposed to the very high temperatures (which can be mitigated by use of film cooling techniques) but is very lightly stressed under normal steady running conditions. The outer casing on the other hand is a pressure vessel containing the gases at the maximum differential pressure of the whole cycle but fortunately it sees little of the maximum gas temperatures. Some heat from the gases will come to it by radiation but it will be comparatively well shielded from the main high temperature regions. As engine cycle pressure ratios increase over the years the combustion casing design problem becomes more acute owing to the increased gas temperature which accompanies the higher temperature. The flight conditions of high forward speed at low altitudes are particularly severe. In addition the general tendency to make engines more compact with higher average gas velocities leads to higher heat transfer coefficients between gas and casing.

Trouble can arise with the flame tubes during transient operations when there are large and rapid variations in temperature. Under these circumstances thermal shock can be a problem as stresses are set up due to differential expansions in the metal resulting from variations in heat capacity. Similarly any coking—depositions of carbon on the metal—may cause trouble, again due to variations in metal temperatures. These problems can only be overcome by patient development work and reshaping of the air holes. It is clearly important therefore that the material which is used for manufacture of flame tubes should have good machining and forming properties. In particular flame tubes are generally of welded construction so good weldability is essential. The "Nimonic" alloys produced by Henry Wiggin & Co. have been used extensively for flame tube manufacture in Britain.

12.4 THE TURBINE

When dealing with the mechanical design of compressors (above) the two distinct aspects of blade design and disc design had to be considered, although clearly each has influence on the other. On coming to turbines the same two aspects arise. Whereas on compressors, greater emphasis had to be placed on blade design, with turbines the two are of almost equal importance. There are a number of reasons for this change in attitude. Firstly failure of a turbine blade is less likely to be catastrophic in leading to extensive secondary damage because, with the fewer number of stages, a broken blade will escape more easily. Secondly a turbine disc is generally very much heavier than a compressor disc so it stores a greater amount of kinetic energy which may wreck an engine if it disintegrates.

Turbine blades are exposed to the highest temperatures of the engine cycle and are subjected to centrifugal stresses due to their mass and bending stresses due to the gas loads. The primary cause of failure under steady running conditions is the phenomenon of creep, the continuous growth of a member under load. Creep characteristics of a material are generally plotted in the form of curves of strain against time for constant values of stress and temperature. Figure 12.4 shows the typical form of these curves and it will be noticed that there is a fairly rapid primary creep which is followed by a comparatively long time when the creep rate is constant. This secondary creep is followed by a rapid stretching—the tertiary phase—leading to failure. With higher stresses or higher temperatures failure will occur in shorter time intervals.

Although that brief statement describes the *form* of creep data it does nothing to explain the *mechanism* and it must be admitted that the phenomenon of creep is not yet clearly understood. Many attempts have been made to correlate the extensive data now available on creep and some of the more promising results are described by Howe (ref. 4). In this paper Howe also makes a new attempt to explain creep in terms of an equivalent virtual work being carried

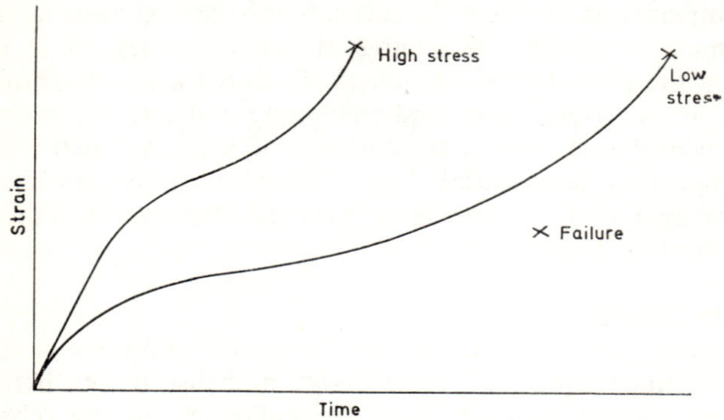

FIGURE 12.4 *Typical creep characteristics of a metal*

out by the thermal loading. There is a possibility that a fundamental understanding may come when the microscopic behaviour of materials can be tied up with the macroscopic behaviour.

During steady running there will be a variation in material temperature along a blade as heat is conducted out through the blade roots into the cooler disc. The highest temperatures will therefore occur at the tips of the blades—that is assuming of course that there are no gross variations in gas temperature leaving the combustion chamber. Blade stresses on the other hand, both due to centrifugal and gas bending loads, will be highest at the roots and will fall steadily with radius. As the factor causing failure will vary with both stress *and* temperature, in some fairly complex manner, it would be expected that the most vulnerable part of a blade would not be either at root or tip. In fact, practical experience of blade breakages shows that very often failure occurs at one third to one half of the distance along the blade from the root.

While the main steady running problems with turbine blading arise from creep, there is a further important loading parameter known as "thermal shock" which is essentially a transient phenomenon. As mentioned previously in Chapter 5, if an engine is shut down suddenly there is more or less a step fall in gas temperature. When this happens the light section and trailing edges of the blade will cool-off very much more rapidly, comparatively speaking, than the main bulk of the blade. The thermal contractions are resisted by the bulk of the blade so that high tension stresses are set up in the leading and trailing edges. With repeated thermal shock, hair-line cracks will open up in the blade leading and trailing edges.

One other necessary property for a suitable turbine blade material besides the resistance to creep and fatigue is that it must withstand corrosion. Corrosion is liable to lead to a special form of fatigue failure which cannot be predicted as easily as straightforward fatigue due to cyclic stressing. Corrosion fatigue is more likely to occur at higher operating temperatures although the exact mechanism is not as yet fully understood.

The materials which have best been able to meet these exacting requirements are those based on high melting point metals particularly nickel, chromium and cobalt. By far the most commonly used alloys in Britain have been the "Nimonic" series made by Henry Wiggin & Co. A whole series of these has been developed each containing about 60% nickel but with varying amounts of chromium, cobalt and other elements. In the search for newer materials much work is being done on chromium based alloys (which have high resistance to corrosion) and on those with the rarer metals such as niobium, molybdenum, tantalum and tungsten. Some properties of representative alloys are given in Table 12.2; it should be noted that the strength of alloys depends to some extent on the method of forming (e.g. whether material is wrought or cast), so the data of Table 12.2 is for comparative purposes only.

Besides the development of new alloys there has always been an interest in the use of ceramics for turbine blades. They have outstanding temperature properties, low density and low thermal expansion but, all in all, are too brittle to be really practicable as yet. Where ceramics *may* be of use is in forming protective coating for metal blades. Because of their low thermal conductivity even thin coatings may reduce the bulk blade temperature appreciably with the added advantage that some corrosion protection will be provided.

The very properties which make high temperature alloys suitable for turbine blades can lead to manufacturing difficulties as they are so tough to machine. Exact profiles are needed and for cooled blades the cooling passages must be formed, so use is made of such advanced production methods as spark erosion and electrochemical machining.

Turning now to the design of the turbine disc, absolute integrity is needed here because it is impractical to design a casing to contain a disintegrating disc. Failure of a disc is likely to occur if there is a

TABLE 12.2 *Properties of Turbine Blade Materials*

Alloy	Density kg/m^3 (lb/in^3)	Temperature for rupture in 100 hrs (°C)	
		At stress of 140 MN/m^3	At stress of 210 MN/m^3
Nimonic 90	7·76 (0·280)	840	800
Nimonic 105	7·88 (0·284)	900	860
Nimonic 115	7·88 (0·284)	970	910
Chromium alloy E*	7·21 (0·260)	1 030	960
Niobium alloy AS 30†	8·60 (0·310)	1 310	1 180

* Chromium alloy E has percentage composition by weight of
Carbon 0·01, Titanium 0·1
Silicon 0·45, Tantalum 2·0
Balance Chromium.

† AS 30 has percentage composition by weight of
Carbon 0·1, Wolfram 20
Z
Balance Niobium

(Based on Glenny, ref. 3)

gross overspeed of the turbine (such as would result from a shaft failure) or a loss of cooling air. All the heat which is conducted through the blade roots must be removed from the turbine disc if it is to be kept at an acceptable temperature so the cooling air is very important. There may also be vibration problems if the gas distribution from the combustion chambers has a peaky profile but generally the natural frequency of a disc will be very high.

The bulk of experience with turbine discs so far has been with steel alloys as these have had acceptable strength properties at representative disc operating temperatures. With more advanced engines there is a need to cater for higher disc temperatures and much work is now being done on new alloys based on nickel. Glenny (ref. 3) describes these new materials and lists their properties.

12.5 THE DESIGN OF SHAFTING

As for the turbine discs, a very high standard of integrity is required for the shafts connecting compressor and turbine of an engine. The consequences of a shaft failure would be a major overspeed of the turbine and the risk of a broken shaft flailing around inside the body of the engine. Almost certainly complete failure of the engine with a strong chance of the loss of the aircraft would result.

In order to achieve lightness, connecting shafts are generally of hollow construction and in many cases they are really quite thin-walled tubes. (Outer diameter is limited by the requirements of the combustion chamber area.) A tube like this, transmitting torque may fail due to excessive shear stress in the material either under steady running or with fluctuating stresses. There is therefore the primary material requirement that the material to be used should have a high allowable specific shear strength—shear stress divided by density—at operating conditions. Most shafts will be designed on the basis of an acceptable life at normal steady running conditions but if reheat is used there is a chance of impact or fluctuating loads particularly at light-up of the reheat system. In addition, if a compressor should surge there will be violent and rapid fluctuations of compressor torque, although surge at full power will be most unlikely to occur.

A hollow shaft will be liable to fail in buckling, the relevant material property in this case being $E/(1 - \nu)^2$ where E is Young's Modulus and ν is Poisson's Ratio for the material. The form of this parameter shows that buckling is related to both tension and compression stresses.

Any shaft in a rotating machine has a critical speed at which whirling occurs. When this happens the shaft centre-line is curved (rather like a rope in skipping) giving appreciable deflections at the mid-point of the shaft. The critical speed must be kept well clear from the normal operating speed of an engine. A designer may design a shaft with a very high critical speed or use a low speed and allow it to run through the critical speed at something near idling conditions. If this second course is chosen the centrifugal stresses are so low at the critical speed that the shaft deflections are acceptably small. From the materials point of view the critical speed for a shaft of given dimensions mounted in bearings of fixed stiffness will vary with E/ρ^2 where ρ is the material density and E, as before, is Young's Modulus. Beryllium is good from this aspect and if it could be used as a shaft material considerable weight savings would be possible compared with shafts made of steel or titanium.

Little need be said about the materials and design of the other shafts in an engine as, compared with the main turbine shafts, they are only required to transmit very small powers. Naturally, of course, drives to fuel pumps, oil pumps and other important accessories must be of high integrity but this can be achieved by working at low stresses and accepting the resultant small weight penalties. The one shaft which is often deliberately made weak is that from the starter. A shear section is usually incorporated which will fail if there is any excessive load on the starter motor. In this way secondary damage can be avoided.

13 Basic Flight Vehicle Performance

13.1 INTRODUCTION

The disciplines and techniques used for making aero-engines are so different from those used in making airframes that two distinct industries have grown up dealing with these two main components of the complete aircraft. This is in many ways unfortunate because the overall vehicle economics depend so much on efficient matching of airframe and power unit. Some basic knowledge of elementary aircraft performance can help the engine man to appreciate what he should do.

With aircraft coming in such a range of sizes and configurations and with such diverse functions to perform it is not feasible here to cover all aspects of engine and aircraft matching. The aim in this chapter is to introduce the subject by looking at the characteristics of conventional fixed wing aircraft and to see how these characteristics can most conveniently be related to those of the engine. This matching exercise will explain how aircraft operating conditions are determined and will answer such questions as why jet aircraft fly high and why extra thrust is needed for manoeuvring. The specific problems of helicopter operation are considered before dealing briefly with the particular needs of hovercraft.

There are a number of excellent textbooks such as refs. 1 and 2 dealing with aircraft performance but for the non-specialist reader a book such as A. C. Kermode's *Mechanics of Flight* (ref. 3) gives an excellent background to the whole subject of aircraft performance and operation.

13.2 THE PERFORMANCE OF FIXED WING AIRCRAFT

When an aircraft is flying at constant speed and altitude it is at steady conditions under the action of two equal and opposite pairs of forces as shown in Figure 13.1. The weight W is exactly balanced by the lift L while the drag D is exactly balanced by the net thrust of the engines X. If the four forces do not pass through one point then there will also be turning couples which must be balanced out by equal and opposite couples set up by the control surfaces. These control forces will alter the effective values of lift or weight but this

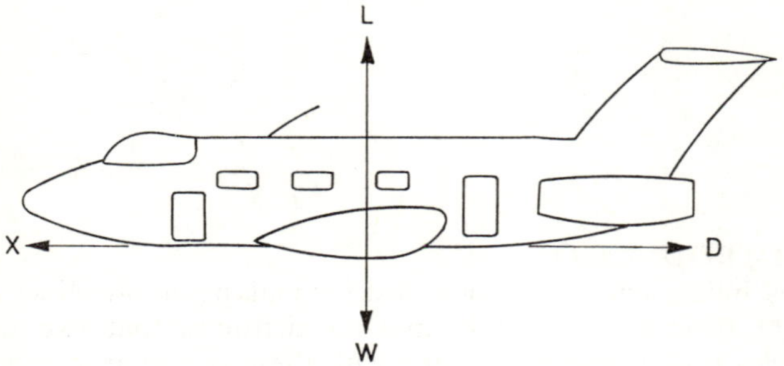

FIGURE 13.1 *Forces on an aeroplane in steady level flight*

will generally be a second order effect. (In the case of Concorde the out-of-balance couples are trimmed out by shifting fuel between tanks as by this means aerodynamic forces can be avoided.) The balance of forces applies only under steady conditions and if, for example, the thrust is greater than the drag the aircraft will accelerate. Similarly an excess of lift over weight will cause it to rise. Finally it should be noted that in some cases there may be appreciable vertical components of thrust which will help to overcome the weight force.

Before looking in more detail at the variation of lift and drag with flight speed it is necessary to introduce the terms *Equivalent Air Speed* and *True Air Speed* written as EAS and TAS respectively. The aerodynamic forces produced on a wing as it passes through the air will be proportional to the dynamic pressure of the airflow, that is on the term $\frac{1}{2}\rho V^2$ for the flow. A given true velocity through the air at altitude where the density is lower will give a lower aerodynamic force than the same velocity would produce at sea level. The EAS at any flight condition is defined as the flight speed which at ISA sea level conditions of air density would produce the same aerodynamic forces. Writing V_T as the true air speed, ρ as the actual air density at same flight condition, V_E as the equivalent air speed and ρ_0 as the air density at ISA sea level conditions then

$$\tfrac{1}{2}\rho {V_T}^2 = \tfrac{1}{2}\rho_0 {V_E}^2$$

Therefore

$$EAS = \sqrt{(\rho/\rho_0)} \times TAS \tag{13.1}$$

The structure of an aeroplane is designed to withstand certain aerodynamic forces and a limitation is placed on the permissible EAS—it is quoted as V_{NE}, the "never exceed" speed. It is surprising how low the EAS may be for high speed aircraft. For example an aircraft cruising at a flight Mach number of 2·0 at 22 km altitude ISA (about 70 000 ft) has an equivalent air speed of 136 m/s or 265 kt. The true air speed at this condition is 575 m/s or 1 150 kt. One of the great advantages of flying at high altitude is that high true speeds can be attained with acceptably low structural loads.

One other air speed frequently encountered is the *Indicated Air Speed* or IAS. This is very nearly the same as Equivalent Air Speed but takes account of what are known as "position errors" which arise from the siting of the air speed detecting instruments. From a pilot's point of view the IAS is the only reading of interest as this is what he uses in driving his aircraft.

Lift and Drag

Having defined these terms we can now return to look at the forces acting on the complete aeroplane. Considering first the lift it is usual to write a lift equation for a wing in the form

$$L = \tfrac{1}{2}C_L\rho V^2 S \tag{13.2}$$

where ρ, V are the density and true air speed as before, S is the wing plan area and C_L is the lift coefficient for the wing. C_L for a wing will vary with the angle between wing chord and the air flow direction—the angle of attack—and will normally vary linearly with angle until stalling of the wing occurs at an angle of about 15° (Figure 13.2). Modern airliners are designed with a multitude of wing flaps, sometimes on leading as well as the trailing edge of the wing which help to give very high lift coefficients.

A similar equation can be written for the drag of an aircraft namely

$$D = \tfrac{1}{2}C_D\rho V^2 S \tag{13.3}$$

but in this case things are rather more complicated. For one thing all parts of the aircraft are contributing to the overall drag force whereas the lift comes almost entirely from the wing. (It is still acceptable though to quote a drag coefficient based on the wing plan area.) For another, it is customary to divide the overall drag coefficient into two components, an induced drag coefficient C_{Di}

and a profile drag coefficient C_{D0}. Reference to any textbook on principles of flight will show that the induced drag is related to the lift of a wing.* At high lift C_{Di} will be high while at high EAS where a given lift is obtained at low C_L the induced drag will be correspondingly low. The profile drag depends on the shape and

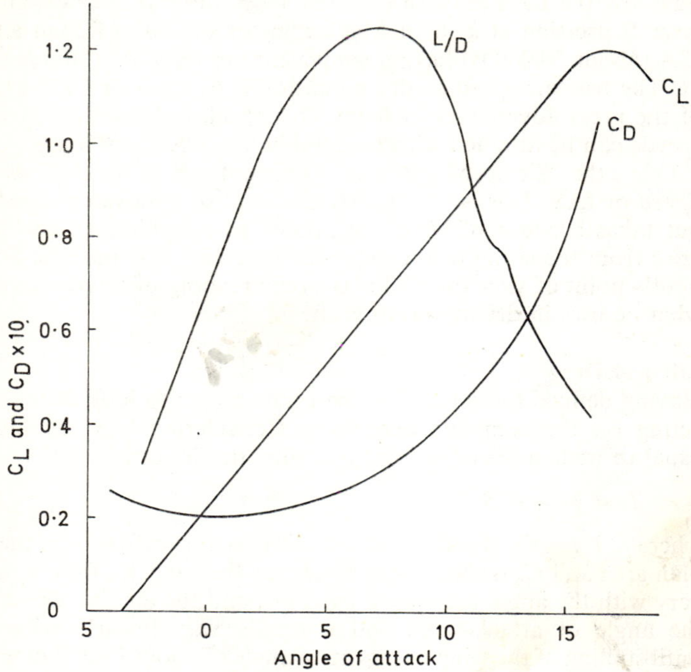

FIGURE 13.2 *Aerofoil characteristics*

surface finish, etc. It is sufficient for our immediate purposes to know that the form of drag coefficient variation will be as shown in Figure 13.2.

The third curve on Figure 13.2, that of the lift/drag ratio, gives a measure of the efficiency of the wing in performing its function of

* In fact the induced drag coefficient C_{Di} can be shown to be related to the lift coefficient C_L by the equation

$$C_{Di} = C_L^2/\pi A$$

where A is the aspect ratio of the wing (this assumes an ideal lift distribution).

producing the highest possible lift for the least possible drag. Normally the best L/D ratio will occur at about 8° angle of attack.

The plot of greatest interest when choosing an engine to power an aircraft will be that of drag against flight speed. Such a graph for an aeroplane at a given operating weight is shown in Figure 13.3 where drag is plotted against EAS. (Strictly speaking, allowance should be made for Mach number effects but these only become important when an aircraft is near its critical Mach number.) Looking at the

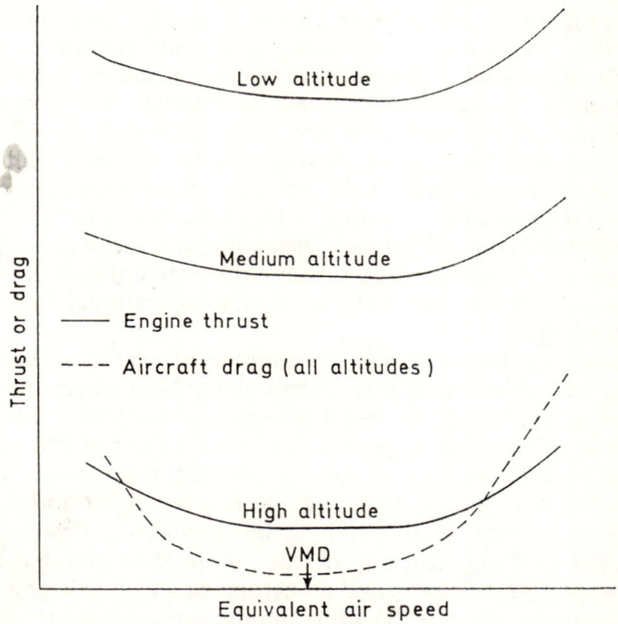

FIGURE 13.3 *Matching of thrust and drag curves*

graph it is seen that as EAS increases from the minimum stalling speed the drag initially falls but that after passing through V_{MD}, the minimum drag speed, it then increases progressively to the limiting aircraft speed. Each point on the graph corresponds to a specific value of lift coefficient which can be calculated from lift, EAS and wing area, that is it corresponds to an angle of attack of the wing.

Clearly the best value of L/D will correspond to the minimum point of the drag curve and this will represent the most efficient point for operating the airframe. However the aeroplane is liable to be rather "mushy" in this region and difficult to control so it is normal to operate it at a higher EAS, of the order of $1{\cdot}3\ V_{MD}$.

Matching for Speed and Altitude

Consider now the matching of airframe and engine. Chapter 3 showed how the net thrust of a jet propulsion engine could be calculated for any set of operating conditions. If a series of calculations is done for an engine at constant rating and constant altitude but with varying flight speed, it is found that curves of the type shown in Figure 13.3 are obtained. Initially as flight speed increases, the net thrust falls due to increasing intake momentum drag but eventually this effect is outweighed by an increasing intake ram pressure ratio. The calculations can be repeated for a range of altitudes and it is found that the same general form of curve is obtained with, of course, lower thrusts at higher altitudes where the air is less dense. (The position of the minimum of the curves will be primarily a function of flight Mach number and this is why when plotted on an EAS basis the minimum varies with altitude.) Eventually an altitude is reached where the engine thrust line intersects the aircraft drag curve at the required operational EAS. This is the matching point.

It is now possible to discuss the question of why jet airliners fly high. An instinctive thought is that by flying high where the air is thin the drag will be reduced, but a moment's consideration will show that this is a fallacious argument. In fact the aircraft will operate at an EAS which will give a certain lift/drag ratio which will be close to the optimum value and so, at a given flight weight, the drag will be constant at all altitudes. But at altitude, this drag will be achieved by flying at a higher true air speed so giving a better engine propulsion efficiency and an aeroplane which will be a more useful transport machine. There is another very important engine aspect which will be apparent from a study of the engine performance curves in Figure 13.3. If the aircraft is to fly at low level then the engine will give far too much thrust if operated at its normal ratings and must be throttled back appreciably. For a turbine engine this operation at part-load conditions will give a serious fall in engine thermal efficiency.

The maximum speed of an aircraft is limited by the thrust available from the engines and the maximum altitude or ceiling may also be determined by thrust limitations. For an aircraft to be able to climb there must be an excess of thrust over drag which can either lift the aircraft directly or allow it to fly under conditions where the lift is greater than the weight. Eventually an altitude will be reached where the net thrust available is just equal to the minimum drag. For many aeroplanes designed for subsonic use an aerodynamic limit may be encountered before this engine limit. The limit arises when the true air speed at stalling has risen to such a value that the corresponding flight Mach number is that at which buffeting of the airframe commenced. Aircraft designed for supersonic operation do not meet this altitude limitation.

Range

Besides speed and altitude the operator will also be interested in the range of his aircraft. The classic equation for still air range is known as the Breguet Range Equation, which states that the range in still air of an aeroplane is given by

$$R = C\frac{V}{SFC}\frac{L}{D}\ln\left(\frac{W_1}{W_2}\right) \tag{13.4}$$

where V is the true air speed, SFC is the engine specific fuel consumption, L/D is the lift/drag ratio, W_1 and W_2 are the all-up weights at start and end of cruise; C is a correction factor to take account of inconsistency of units. (This equation is derived by assuming $(W_1 - W_2)$ results from the consumption of fuel while the fuel flow rate is the drag times the engine specific fuel consumption.) The Breguet equation emphasizes that range varies inversely with engine SFC and directly with the L/D ratio. It is useful in calculating trade-offs between various design variables at the planning stage of a new aircraft.

There is such great interest at this time in the supersonic civil transport (SST) that some mention of its special characteristics is warranted. In many ways the first-generation SST operating at flight Mach numbers in the 2·0 to 3·0 range is comparatively conventional and all the above work on aircraft characteristics can be applied to it. Where it does differ substantially is that the cruising lift/drag ratio is very much lower—of the order of 8 compared with about 15 for a subsonic airliner. Against that can be set a greatly improved engine thermal efficiency which results from a high ram pressure ratio and a good propulsive efficiency.

The other problems arise when the SST is accelerating through the transonic speed range. At about sonic speed the drag coefficient rises very sharply before falling again to a value of two or three times the subsonic figure. A large thrust is needed to push the aircraft rapidly through this speed range and so the transonic acceleration must be carried out at comparatively low altitude—say 11 km—where the air is still dense and the engine thrust is high. Also, in the transonic range, there are large movements in the position of the lift forces on the wing of the aircraft which introduce stability and control problems. As mentioned earlier these out-of-balance couples are trimmed out in Concorde by shifting fuel between tanks and so altering the position of the centre of gravity of the aircraft.

Manoeuvring

The elementary aircraft performance theory considered so far has been for what is essentially steady state operation—the pairs of forces acting on the aircraft have been considered as being more or less in balance so that there are no violent accelerations. As soon as an aeroplane starts manoeuvring additional forces come into operation. To take the simplest case of a turn or the pull out from a dive the apparent weight of an aeroplane will be augmented by the centrifugal forces. In a 2*g* turn for example—and this is a fairly gentle manoeuvre—the apparent weight of the aeroplane is doubled. At an unchanged EAS the lift-drag ratio will remain the same so the drag will be doubled. It is clear that there must be a corresponding change in thrust if the desired flight path is to be followed.

Clearly for a highly manoeuvrable aeroplane big reserves of thrust must be available. These reserves will be whittled away as the aircraft climbs into the regions of reduced air density. When eventually there is insufficient excess thrust over drag the turn can only be carried out at the expense of forward speed and the aircraft will suffer a longitudinal deceleration. This itself may be an embarrassment because a high lift is being called for in a tight turn so the stalling speed will rise. It is all too easy to lose control of the aircraft under these conditions.

One particular type of aircraft which requires high manoeuvrability is the intercepter fighter and it is necessary that this type should also have a high rate of climb. Both these requirements call for high installed thrust which may be obtained by using a comparatively large turbojet with afterburning. An extreme case is of the vertical take-off machine like the Hawker Siddeley Harrier which uses a vectored thrust engine with a take-off thrust greater than the all-up weight. An aeroplane of this type is grossly overpowered by conventional standards and the engine must spend much of its life at part load conditions but it is very useful to have such a large excess of thrust available for acceleration, and manoeuvring at high altitude.

In a sub-chapter such as this it is only feasible to touch briefly on some of the main topics in the vast field of aircraft performance.

For anyone interested in learning more about the practical operation of aircraft a book by D. P. Davies, *Handling the Big Jets* (ref. 4) can make fascinating reading. This work, written by the Chief Test Pilot of the Air Registration Board, is aimed primarily at airline pilots but it can give many varied insights into the operational and handling problems of jet transports.

13.3 THE HELICOPTER

An aerofoil only produces lift when it is moving at some velocity relative to the air. If an aircraft is to obtain lift from "wings" while it is as a whole stationary with respect to the air, then the wings on their own must move. This is the basic principle of the helicopter which uses a powered rotor to provide its lift. In a conventional helicopter the whole rotor disc can be tilted about the vertical axis so that the overall lift can produce horizontal propulsive components of the total aerodynamic force in any required direction.

Although the principle of the helicopter is very simple—Leonardo de Vinci sketched a rudimentary machine—the practical difficulties are very considerable and the first reasonably successful helicopters were not flown until the mid-1930s. To understand one of the main difficulties, consider a helicopter moving forwards at a substantial flight speed. The air velocity relative to a rotor blade which is advancing will be very much higher than that relative to a blade on the opposite side of the disc which is retreating. If a large imbalance of lift between the two sides of the rotor is to be avoided, then the angle of incidence of the blades must be altered continuously as the blades move round. It must be increased for the retreating blades and reduced for the advancing ones. A *cyclic* pitch change mechanism is required to accomplish this.

Overall changes in lift are obtained by altering the whole level of angle of pitch of the blades by a *collective* pitch change mechanism. The obstacles which stood originally in the way of designing a successful helicopter centred largely on the difficulty of making a mechanism to accommodate both cyclic and collective pitch changes. Suitable mechanisms could be designed but it was not easy to make them strong enough to withstand the fluctuating loads while being of acceptable size and weight. In addition, for helicopter mechanisms, reliability is of primary importance.

Although there are certain jobs which a helicopter and only a helicopter can carry out it does suffer a number of limitations. The principle drawback from the performance point of view is that there are fundamental limits on the maximum speed at which a rotor-borne aircraft can fly. This limitation arises again from the fact that at a fixed rotor speed the velocity relative to the air depends on whether the rotor blade is advancing or retreating. A position can be reached where the tip speed of the advancing blade has reached its critical Mach number (and so is losing efficiency) while the retreating one is almost stalling in trying to produce lift. Matters can be mitigated by using thin section rotor blades which have high critical Mach numbers or by reducing the rotor loading, through fitting small fixed wings. The name "compound rotorcraft" is applied to machines in which the rotor loading is reduced either by fitting stub wings or by using separate propulsion engines. Even allowing for these possibilities the rotor craft can never compete on speed grounds with a fixed wing aircraft. The comparatively low speed makes it very susceptible to wind speeds and makes scheduling of regular commercial flights difficult.

Coming now to the power requirements of the helicopter a high power is needed at hovering conditions. The reason for this is that with no overall velocity relative to the air a high lift coefficient is required and this leads to a high induced drag coefficient. As soon as the helicopter starts moving through the air the required power will fall until some minimum is reached—probably at about a third to two thirds of the maximum speed. Thereafter the power for lift will continue to fall but the overall aircraft drag will be rising sharply leading to a net increase in power demand. At maximum speed the power will be of the same order as that for hovering.

In the event of a power failure to the rotor a potentially hazardous situation exists. Due to the drag of the blades the rotor will at once start to slow down. This must be guarded against at all costs as it is the centrifugal forces in the blades which keep them at the correct cone angle and prevent the rotor folding up under the influence of the lift forces—rather like an umbrella being blown inside out. To avoid this the pilot must reduce the collective pitch of the blades as soon as he senses a power loss. The helicopter will descend rather like a sycamore seed and immediately prior to touch-down the pilot can increase collective pitch and use the rotor kinetic energy to cushion the loading. Helicopter pilots, especially those who fly small machines with low inertia rotors, develop rapid responses!

Early helicopters used conventional piston aero-engines but it was soon clear that the gas turbine could offer much in the way of lightness and compactness with the admitted disadvantage of a poorer thermal efficiency.

Chapter 3 dealt with turboshaft engines and showed that there are two main classes, the single shaft or the free turbine engines. For helicopter work the free turbine offers distinct advantages as it is possible to eliminate a clutch from the transmission system to the rotor. (With a piston engine or a single shaft turbine this clutch is needed both for starting and to disengage the rotor from the engine in the event of power failure.) In addition if the design power turbine operating speed is rather higher than the optimum speed—see Figure 3.12—any momentary increase in rotor drag will reduce the turbine speed and automatically give more power. This is a useful factor in stabilizing rotor speed and it may be used in conjunction with a control system which acts to keep rotor speed constant.

In many helicopters the space available for the power plant is limited and of an awkward shape so an engine designed for this work should be mountable with its axis vertical. Special attention must be paid then to bearing and lubrication problems in the mechanical design.

Apart from piston and turboshaft turbines the only other type of power unit used to any extent with helicopters has been reaction propulsion at the rotor tips. Various schemes have been tried or proposed including subsonic ramjets, air bleed from the main engine compressor (either with or without burning of fuel in the air ducts), and small-size turbojet or turbofan engines. Most of these arrangements are noisy and comparatively inefficient and those using air bleed bring in structural problems of housing the air ducts in the rotor blades. They do, however, eliminate the need for a torque reaction at the tail and, by providing a mass at the rotor tip where the centrifugal forces are highest, they help to reduce coning of the rotor blades. Looking to the future of tip drive, it is possible to envisage turbofans being designed which would have good efficiency, but the problem of the gyroscopic precession couples of the engine rotating parts will always be considerable.

13.4 THE HOVERCRAFT

In the years since 1959 when the first practical hovercraft was built a great family of Air Cushion Vehicles have been designed and tested. They range from little single-seat sporting machines which can barely hover under the weight of the pilot to the big Mountbatten class Cross-Channel ferries which operate at 160 Mg (160 tons) while carrying 30 or more vehicles and their passengers. Not all are true flight vehicles as some such as the side wall hovercraft must remain in contact with the water while others require specially prepared tracks.

Conventional hovercraft are again vehicles which depend on power units for both lift and propulsion. There is however one fundamental difference between these and the other flight vehicles we have looked at. All the others depend for their lift on Newton's second law of motion as the wings or rotor produce lift by accelerating downwards a mass of air. The hovercraft on the other hand rides on a cushion of air which is held in place by an air jet. Figure 13.4 which shows the cross-section of a simple hovercraft illustrates

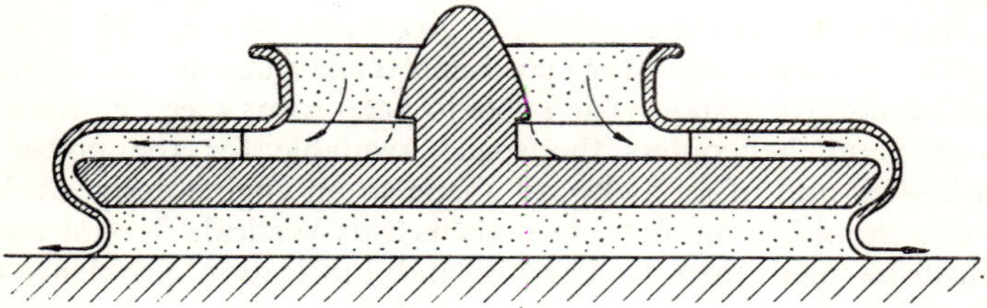

FIGURE 13.4 *Layout of a hovercraft—single annular jet type*

the principle. An air jet from the fan emerges from a slot around the periphery of the hovercraft so forming a curtain of air. The air leaves the jet directed inwards at an angle of about 45° to the vertical and is then turned outwards. Because of the higher pressure under the cushion the cushion air is directed outwards and a force balance equation can be written that

Radial outward component of rate of change of momentum of cushion air

= (Cushion pressure − Ambient pressure) × Hover height × periphery.

Once a cushion pressure and hover height have been determined for a hovercraft it is possible to design the blowing system with its associated ducting and to calculate the power required to hover.

Although the air cushion forms a virtually frictionless support for the craft, as soon as it gets under way air resistance starts and there is an intake momentum drag effect for the cushion air—the large quantities of cushion air represent appreciable drags even at the low speeds of hovercraft. In addition, if the vehicle is operating over water, it experiences an appreciable drag at low speed until it gets "over the hump" and properly into its element. At speeds of the order 40 m/s (80 kt) the lift/drag ratio for a conventional hovercraft will be in the range 5–10 but it will be higher with larger sizes of vehicle.

As weight and space are at a premium in hovercraft most of those built so far have used gas turbine engines which drive centrifugal fans for lift and relatively conventional aircraft type propellers for propulsion. The most suitable engines available have been turbo-props such as the Rolls-Royce Proteus and Gnome. These engines have come into hovercraft use after being fully developed aero-dynamically in the aircraft field. The main modifications needed have been mechanical ones concerned with the bearings and with alterations in materials to make them more suitable to work in the corrosive atmosphere of sea-water spray. One exception to this pattern has been with the Britten-Norman Cushioncraft which used a Rolls-Royce piston engine driving a centrifugal fan which provided both lift and propulsion.

It is always difficult to predict what the future holds in a rapidly changing technological field like that of air cushion vehicles but there would seem to be no insurmountable difficulties in designing much larger craft to operate as freighters at speeds of the order of 40 m/s (80 kt). Problems will arise for propulsion as the number and size of propellers operating in air becomes excessive so the very large vehicles will probably use water-jet propulsion.

Finally on hovercraft, it should be said that of all the flight vehicles we have considered this is the one which is most easily homebuilt. Above all it is a comparatively safe but sporty and exciting type of machine which can be designed and made without expensive facilities. To cater for this interest, an excellent series of articles on Homebuilt Hovercraft by G. H. Williams (ref. 5) goes into all aspects of the design, construction and performance of small vehicle.

Exercises

1.1 A compressor takes in 50 kg/s of air at 25°C and 200 m/s and delivers it at 200°C and 150 m/s. There is a heat loss from the compressor equivalent to 0·25% of the work input. Calculate the power required to drive the compressor and the actual value of the heat loss.

[88·6 MW, 222 kW]

1.2 Air flowing in a horizontal duct decelerates from a velocity of 250 m/s and a static temperature of 600 K to a velocity of 50 m/s. Neglecting heat losses find the final static temperature. If the static pressure ratio between the two stations is 1·15 find also the specific entropy change of the air.

[629 K; 0·0086 kJ/kg-K]

1.3 A supply of hot gas is to be produced by burning methane, CH_4, in air. What is the maximum quantity of methane which may be burnt if the the percentage of carbon dioxide in the products is not to exceed 3% by volume?

[Limiting ratio of CH_4 to air, 0·31 by volume]

1.4 The flow pattern around a model in a gas stream at a Mach number of 1·3 is to be demonstrated in a "wind tunnel". Find the gas velocity past the model if the working fluid in the tunnel is

(*a*) Helium of molecular weight 4 and specific heat ratio γ of 1·63.

(*b*) Freon-12 which has a molecular weight of 121 and a specific heat ratio γ of 1·15.

Assume that in each case the working fluid expands from a reservoir where the temperature is 400 K.

[1 220 m/s; 218 m/s]

1.5 A high pressure gas main carries air at a pressure of 1 MN/m^2, a temperature of 400 K and at negligible velocity. The main springs a leak which is equivalent in area to a circular hole of 25 mm diameter. Use the curves of Figure 1.11 (p. 36) to estimate the maximum side thrust produced on the main by the leaking air.

Hint: Treat the leakage in the limiting case as an isentropic expansion to atmosphere.

[620 N]

1.6 Heat is transferred between the surroundings and air flowing supersonically through a frictionless constant area duct. Initially the air is at a Mach number of 1·60 and a stagnation temperature of 600 K and heat transfer continues until the stagnation pressure has increased by 10%. Calculate the final Mach number of the air and the magnitude and direction of the heat transfer per unit mass of air.

[1·77; – 30·3 kJ/kg]

1.7 The friction factor of an insulated pipe of 60 mm bore is to be estimated by passing air through it and taking readings at two stations 20 m apart It is found that the static pressure falls from 200 kN/m² to 134 kN/m² while the stagnation temperature remains constant at 400 K. If the air flow rate is 0·80 kg/s estimate the mean value of the friction factor *f* based on mean air velocity and density.

[$4{\cdot}2 \times 10^{-4}$]

1.8 Air is to be decelerated from a Mach number of 2·0. Compare the loss in stagnation pressure and the overall static pressure ratio if this deceleration is accomplished

(*a*) through a single plane normal shock wave;

(*b*) through a plane oblique shock and a plane normal shock each each shock having the same pressure ratio.

A trial and error process is advised.

[Loss in stagnation pressure 10% compared with 28%; static pressure ratio increased by 10%]

1.9 An emergency air turbine is required to produce 50 kW of power when supplied with air from a reservoir at 300 kN/m² and 420 K. The air leaves the turbine with stagnation pressure and temperature of 120 kN/m² and 345 K and at a velocity of 120 m/s. Calculate the volume flow rate of air leaving the turbine assuming adiabatic flow.

[0·571 m³/s]

1.10 A convergent-divergent nozzle is designed to expand air isentropically to produce an outlet Mach number of 2·2. What outlet Mach number will result if the nozzle is operating subsonically with isentropic expansion and a Mach number at the throat of 0·90? (Assume the air has a specific ratio $\gamma = 1.4$).

[0·303]

1.11 Calculation of the approximate leakage losses from high pressure air mains can often be simplified by assuming choked flow with a value of $\dot{m}\sqrt{T_0}/Ap_0$ taken as 40. The units are as quoted for Figure 1.11 (p. 36).

Use this expression to estimate the leakage rate through a 2 mm diameter hole in an air duct carrying air at 300 kN/m^2 and 400 K.
[1·88 gm/s]

2.1 Two engines A and B are each designed to produce 25 MN thrust. Engine A has a thrust to weight ratio of 8:1 while engine B has a value of 18:1. The corresponding specific fuel consumptions are 0·09 kg/N-h and 0·2 kg/N-h. Find the break-even time when the total weight of engine plus fuel for engine A is equal to that for engine B.
[3·48 min]

2.2 Find the capital costs per flying hour for an airliner costing £5m operating for 3 000 h/year assuming:
(i) Depreciation to 15% of its initial value over 7 years.
(ii) Insurance at 5% per annum based on original cost.
(iii) An interest rate of 6% per annum based on original cost.
[£385]

3.1 Calculate the overall thermal efficiency of the turbofan of Example 3.3 (p. 109) and the ramjet of Example 3.5 (p. 112).
[0·195, 0·256]

3.2 Compare the increase in specific output possible for a turbine operating with a turbine entry temperature of 1 200 K and an expansion ratio of 4·0 if
(*a*) the turbine entry temperature is increased to 1 400 K at the same expansion ratio;
(*b*) the expansion ratio is increased to 8·0 at the same inlet temperature;
(*c*) turbine entry temperature and expansion ratio are *both* increased by these amounts. Assume constant efficiencies.
[+ 20%; + 41%; + 69%]

3.3 Gases expand in a turbine at a rate of 75 kg/s from 870 kN/m^2 and 1 200 K to produce an output of 24 MW and then pass to a nozzle where expand isentropically to a pressure of 70 kN/m^2. Calculate the gross thrusts produced for turbine isentropic efficiencies of 0·90 and 0·85 assuming that the gases approximate to air in their properties.
[57·4 kN, 55·7 kN]

3.4 Compare the total thrusts produced by expanding gases isentropically to an atmospheric pressure of 101·3 kN/m^2,
(i) where individual expansions take place of 1 kg/s from 205 kN/m^2, 400 K and 1 kg/s, 205 kN/m^2, 800 K. (Stagnation conditions quoted.)
(ii) when the gases are first mixed at constant stagnation enthalpy but with a 5% loss in stagnation pressure before being expanded in a single nozzle.

Treat the gases as air with values of specific heat and specific heat ratio given in Table 1.1 (p. 11).

[Thrust falls in second case by 2%]

3.5 Under what circumstances would it be worthwhile to fit a heat exchanger to a turboshaft engine which operated with a compressor pressure ratio of 14·0 and a turbine entry temperature of 1 100 K?

[Never—compressor delivery temperature will be above turbine outlet temperature]

3.6 As an exercise in crystal-gazing calculate the specific power and thermal efficiency for a turboshaft engine operating at ISA Sea Level with the following conditions:

Compressor pressure ratio 20·0 and efficiency 0·95.
Turbine entry temperature 1 900 K and efficiency 0·97.
Combustion pressure loss 2% and efficiency 0·99.

Neglect all other pressure losses in the cycle.

[720 kW-s/kg; 0·49]

3.7 During starting the compressor and turbine of a turbojet are operating at low pressure ratio and low efficiencies so high turbine entry temperatures are needed to reach self-sustaining speed. Calculate what temperature is needed at ISA sea level for a compressor pressure ratio of 2·0 and isentropic efficiency of 0·60 if the turbine efficiency is 0·65. Neglect pressure losses in combustion chamber and jet pipe/final nozzle and assume for compression $c_p = 1{\cdot}00$ kJ/kg-K, $\gamma = 1{\cdot}4$ and for expansion $c_p = 1{\cdot}17$ kJ/kg-K, $\gamma = 1{\cdot}333$.

[910 K]

3.8 In a nuclear rocket operating at part load conditions, hydrogen gas is heated at a stagnation pressure of 1·5 MN/m² to a stagnation temperature of 2 000 K and is then expanded through a nozzle of isentropic efficiency 0·90 to a static pressure of 100 kN/m². Calculate the nozzle outlet velocity assuming for the hydrogen a mean specific heat c_p of 16 kJ/kg-K and specific heat ratio γ of 1·333. Find also the outlet area for a hydrogen mass flow rate of 600 kg/s. What approximately will be the thrust at these conditions? (The gas constant for hydrogen may be calculated from the Universal Gas Constant and the molecular weight of hydrogen is 2—see p. 19.)

[5 310 m/s; 5·22 m²; 3·19 MN]

4.1 Air enters the first stage of an axial flow compressor axially at a stagnation pressure and temperature of 101·3 kN/m² and 288 K. At mid-radius the rotor blade speed is 200 m/s and the blades deflect the air through an angle of 20° towards the axial direction. The stator blades deflect the air so that it leaves the stage at an angle of + 15° to the axial direction. If the axial component of the air velocity through the stage is constant at 170 m/s calculate

(i) The air angles through the stage;
(ii) The specific power required for a work done factor of 0·90;
(iii) The percentage reduction in air velocity relative to the blades in passing through the stators.

[$\alpha_1 = 49{\cdot}6°$, $\alpha_2 = 29{\cdot}6°$, $\alpha_3 = 31{\cdot}4°$; 18·6 kW; 12%]

4.2 Calculate for the rotor stage of Example 4.1 (p. 127) what pitch chord ratio is needed if the diffusion factor is to have a value of 0·50.
[1·01]

4.3 Which will be the more potent method of increasing the flow per unit frontal area of an axial flow compressor

(i) by reducing the hub/tip ratio from 0·50 to 0·45 at the same axial velocity and outside diameter, or
(ii) by increasing the axial component of the air velocity from 150 m/s to 170 m/s?

[By increasing velocity]

4.4 The axial compressor stage of Example 4.1 (p. 127) is to be designed with constant reaction blading. Calculate the required rotor and stator blade angles at root and tip if the mean blade radius is 250 mm and the blade height is 100 mm and the axial velocity is constant across the blade height.
[Root: $\alpha_3 = 42{\cdot}8°$, $\alpha_4 = 3{\cdot}0°$;
Tip: $\alpha_3 = 45{\cdot}8°$, $\alpha_4 = 26{\cdot}2°$]

4.5 The first stage rotor of an axial compressor has a hub/tip radius ratio of 0·6 and the blade speed at mid radius is 220 m/s. The air enters the stage axially with a velocity which is constant at all radii. Calculate the value of the axial velocity component if the Mach number of the air relative to the blades at the blade tips is to be 1·2 when the air inlet stagnation temperature is 288 K. (Remember that the velocity of sound depends on the *static* temperature; for this reason a trial and error solution is advisible.)
[266 m/s]

4.6 At entry to a stator blade row air has stagnation pressure and temperature values of 300 kN/m^2 and 410 K and an absolute velocity of 205 m/s. In passing through the blade row the velocity falls to 175 m/s and there is a 2% loss in stagnation pressure. Find the entropy change across the blade row based on static properties of the air.
[4·46 J/kg-K]

4.7 When a turbojet is tested on a day when the intake stagnation temperature and pressure are 25°C and 99·7 kN/m^2 the pressure ratio at 8 400 rev/min is found to be 5·0 and the mass flow is 45 kg/s. Find what rotational speed is needed to give this pressure ratio on an ISA standard day (temperature 15°C, pressure 101·35 kN/m^2) and what the corresponding mass flow would be.

[8 250 rev/min; 46·6 kg/s]

4.8 A centrifugal compressor has an impeller of 0·8 m diameter which rotates at 12 000 rev/min. When the intake stagnation temperature is 288 K the mass flow rate is 20 kg/s and the isentropic efficiency is 0·76. Find the power required to drive the compressor and its pressure ratio assuming the air flow at entry to the compressor is axial and the slip factor is 0·92. Neglect windage losses.

[4·65 MW; 5·25]

5.1 In a turbine of 50% reaction, gases which approximate to air in their properties expand from stagnation conditions of 650 kN/m^2 and 1150 K to a static pressure of 250 kN/m^2 with a stagnation to static isentropic efficiency of 88%. The gas leaves the nozzles at an angle of 20° to the plane of the turbine wheel. Find the blade speed at mid-radius conditions and the magnitude and direction of the gas velocity leaving the stage. (A trial and error solution is advised.)

[295 m/s, 243 m/s at 44·6°]

5.2 A convergent-divergent nozzle is used to expand air at a rate of 100 kg/s from 600 kN/m^2, 600 K to 100 kN/m^2 static pressure. Use the data of Chapter 1 on the isentropic flow of air with $\gamma = 1{\cdot}4$ to estimate the length of the divergent portion of the nozzle if this portion is conical in shape with an included angle of 15°.

[288 mm]

5.3 At the preliminary design stage of a turbine, use is to be made of the loading/efficiency correlation given in Figure 5.6. The mean blade speed is to be limited to 200 m/s and the target isentropic efficiency, neglecting leakage losses, is to be 0·90. In the interests of reducing size (that is, of increasing the ratio V_a/u) a specific output of 5% less than the maximum possible is to be accepted.

Find what value of axial velocity should be used and what mass flow rate will be needed for a power output of 20 MW.

[184 m/s; 228 kg/s]

5.4 An example of the rare case of the purely reaction turbine is when tip drive units are mounted on a helicopter rotor. To achieve reasonable overall efficiency it is necessary to use a comparatively low exhaust velocity as in the following exercise.

The rotor of a helicopter is 16 m in diameter, has three blades and requires a total of 900 kW to drive it at 180 rev/min. Compressed air is taken through the blades to tip drive units where fuel is added before the gases are expanded through the nozzles to atmosphere. At entry to the nozzles the gas stagnation pressure and temperature are 180 kN/m^2 and 900 K and the atmospheric static pressure is 101 kN/m^2. Assuming isentropic expansion in the nozzles calculate the nozzle flow rate to give the required power. Find also the overall thermal efficiency assuming that the air enters the tip drive units at a stagnation temperature of 300 K.

[3·84 kg/s per nozzle; 0·124]

6.1 At the preliminary design stage of making a new turbojet it is predicted that the engine will produce 40 kN thrust at ISA sea level static conditions the corresponding specific fuel consumption being 0·09 kg/N-h with a fuel of calorific value 42 MJ/kg. The compressor pressure ratio is to be 11·0. Estimate the volume required for combustion if the combustion intensity is to be 50 MW/m^3-atmosphere.

[0·0764 m^3]

6.2 In a test of a tubular combustion chamber with inlet diameter 200 mm and maximum diameter 250 mm the following readings were obtained:

Air mass flow rate = 15 kg/s
Weighted mean stagnation temperatures: 600 K at inlet
1 250 K at outlet.
Static pressure at inlet = 1 MN/m^2
Stagnation pressure drop = 50 kN/m^2

Find the fuel flow rate if the fuel has a calorific value of 42 MJ/kg and the combustion efficiency is 0·975; find also the pressure loss factor.

[960 kg/h; 2·62]

6.3 The gases at entry to the reheat section of a turbojet are at stagnation pressure and temperature of 200 kN/m^2 and 900 K while combustion in the reheat section raises the stagnation temperature to 1 600 K. Expansion in the final nozzle takes place isentropically to a pressure to 70 kN/m^2. Neglecting the baffle loss in stagnation pressure find the increase in gross thrust resulting if the Mach number at entry to reheat is reduced from 0·40 to 0·25. (It is acceptable in calculating a relative change to use the data for air with $\gamma = 1{\cdot}4$ given in Chapter 1.)

[+ 4%]

7.1 Calculate the kinetic efficiency of a pitot type intake operating at a flight Mach number of 1·8 assuming that the only loss in stagnation pressure is that occurring through the plane normal shock.

[0·845]

7.2 The intake for a turbojet designed for operation at a flight Mach number of 2·5 at 15 km, ISA conditions, is of the purely internal compression type. Air decelerates from the free stream conditions to the intake throat where the Mach number is 1·2 and then accelerates to a Mach number of 1·4. It then passes through a plane normal shock before decelerating subsonically to the compressor inlet where the Mach number is 0·50.

Assuming that the flow is isentropic except through the shock calculate the ratio of intake throat area to compressor inlet flow area. Find also what percentage reduction in compressor inlet non-dimensional mass flow can be accommodated before the shock is expelled.

Hint: In calculating the area ratio treat as two isentropic flows, from throat to shock and from shock to inlet which meet at the constant area shock plane.

[1·360; 3·5%]

7.3 For the intake of Example 7.1 (p. 222) calculate the flow area required at the intake lip, when operating at the required design conditions as stated in the example. Hence, recalling that the intake is of constant 1 m width, find the perpendicular distance from the cowl lip to the lower surface of the intake.

[0·93 m^2; 1·52 m]

7.4 A single wedge external compression supersonic intake is designed so that at a flight Mach number of 2·0 with a ramp angle of 16° the oblique shock impinges on the cowl lip (as shown in Figure 7.10). The ramp angle is made variable so that the "shock-on-lip" condition can be maintained at a range of flight Mach numbers between 1·5 and 2·5. Find what range of ramp angles must be provided to cover this Mach number variation.

[4·2° to 22·8°]

8.1 The rotor of a turbojet has a moment of inertia of 30 kg-m^2. When accelerating past a rotational speed of 7 000 rev/min. the turbine is producing 4 MW more than the compressor is absorbing. Calculate the instantaneous acceleration rate in rev/min per second.

[1 735]

8.2 Draw block diagrams to illustrate the control systems of

(i) A car driver being followed by a police car through an area with a speed restriction.

(ii) An office heating system where the heating is turned on on 1st October and off on 30th April no matter what the true weather may be.

(iii) A turboshaft power unit for a helicopter which keeps rotor speed constant and protects the engine from exceeding safe temperatures.

8.3 Estimate the percentage increase in the time constant of a turbojet flying at a flight Mach number of 0·3 at 11 km altitude ISA conditions compared with the time constant at ISA sea level static. (At 11 km, ISA $p = 22{\cdot}7$ kN/m^2. $T = 217$ K.)

[370%]

9.1 A ramjet is flying steadily at a Mach number of 3·0 where the atmospheric pressure and temperature are 7·2 kN/m^2 and 217 K. The overall isentropic efficiency of the intake and diffuser based on stagnation pressure recovery is 0·80 and the static pressures before and after combustion are 203 kN/m^2 and 106 kN/m^2. Assuming that the flow through the combustion chamber is frictionless and of constant area find the Mach number at entry to the final nozzle and the nozzle inlet stagnation temperature. Neglect the effect of fuel addition to the mass flow rate and use the tables in Chapter 1 for air with $\gamma = 1{\cdot}4$.

[0·88; 2 345 K]

9.2 Power for a helicopter rotor is to be provided by fitting subsonic ramjets at the rotor tips. Calculate the maximum possible specific thrust (net thrust per unit air flow rate) at ISA sea level conditions if the tip Mach number is 0·70 and the combustion temperature is 1 200 K. Neglect the effect of fuel addition to the mass flow rate through the ramjet.

[230 N/kg]

10.1 Develop an expression for the ideal specific impulse of a rocket operating in a vacuum in terms of the temperature of the propellants entering the final nozzle, their molecular weight and the universal gas constant. Hence find to what temperature monatomic hydrogen should be raised to give an ideal specific impulse of 12 MN-s/kg. (For monatomic hydrogen assume a molecular weight of 1 and a specific heat ratio γ of 1·67.)

[3 470 K]

10.2 A liquid fuel rocket uses hydrogen as the fuel and oxygen as the oxidizer. What is the exhaust gas composition, neglecting dissociation when operating:

(*a*) With 15% excess oxygen

(*b*) With 15% excess hydrogen,

compared with the stoichiometric value? Assuming the same combustion temperature and expansion ratio in each case which will give the higher specific impulse?

[(*a*) 7% oxygen; (*b*) 13% hydrogen with remainder H_2O]

11.1 The supercharger of a piston engine is a centrifugal compressor with an impeller of 0·5 m diameter driven at 6·33 times crankshaft speed. Would you expect the full throttle height at 3 000 engine rev/min and zero boost (induction manifold pressure 101·35 kN/m^2) to be above or below 6 000 m when flying under ISA conditions?

[Probably above. Pressure ratio needed is 2·16 which would be obtained with compressor efficiency of less than 40%]

References

Chapter 1

1 Keenan, J., *Thermodynamics* (Wiley, 1941).

2 Rogers, G. F. C. and Mayhew, Y. R., *Engineering Thermodynamics, Work and Heat Transfer* (Longmans, 1967)

3 Wallace, F. J. and Linning, W. A., *Basic Engineering Thermodynamics* (Pitman, 1970)

4 Tribus, M., *Thermostatics and Thermodynamics* (Van Nostrand, 1965).

5 Shapiro, A. H., *The Dynamics and Thermodynamics of Compressible Fluid Flow* (The Ronald Press, 1953)

6 Cheers, F., *Elements of Compressible Flow* (Wiley-Interscience, 1963).

7 Jamison, R. R. and Mordell, D. L., *Compressible Flow of Fluids in Ducts* (Aeronautical Research Council R. & M. 2031, 1945).

8 Shapiro, A. H. and Hawthorne, W. R., Mechanics and thermodynamics of one-dimensional gas flow, *Trans. A.S.M.E. J. App. Mech.*, vol. 69 (1947).

9 Schlichting, H., *Boundary Layer Theory* (McGraw-Hill, 1955).

Chapter 2

1 Braddick, H. J. J., *Physics of Experimental Method* (Chapman and Hall, 1963).

2 Hafner, R., Safe mechanisms, *J. Roy. Aero. Soc.*, vol. 63 (May 1959).

3 Eden, J. J., Overhaul life development of gas turbines, *Aircraft Engineering*, vol. 34 (1962).

4 Greatrex, F. B. and Bridge, R., The evolution of the engine noise problem, *Aircraft Engineering*, vol. 39 (Feb. 1967).

5 Richards, E. J., Aircraft noise—mitigating the nuisance, *Aircraft Engineering*, vol. 39 (Feb. 1967).

Chapter 3

1 Mallinson, D. H. and Lewis, W. G. E., Part load performance of various gas-turbine engine systems, *Proc. Inst. Mech. Eng.*, vol. 145, W.E.P. 41 (1948).

2 Tharratt, C. E., The propulsive duct, *Aircraft Engineering*, vol. 37 and vol. 38 (Nov. 1965 to Feb. 1966).

3 Gozlan, A., Achievements and prospects with composite air breathing engines, *Aircraft Engineering*, vol. 38 (Feb. 1966).

4 Lombard, A. A. and Keenan, J., The turborocket, *Flight International*, vol. 11 (29th Oct. 1964).

5 Mordell, D. L., Eyre, F. W. and Sreenath, A. W., The inverted turbojet, *Proc. Eighth Anglo-American Aeronautical Conference* (Roy. Aero. Soc., 1961).

Chapter 4

1 Howell A. R. and Bonham R. P., Overall and stage characteristics of axial flow compressors, *Proc. Inst. Mech. Eng.*, vol. 163, W.E.P. 60 (1950).

2 Lieblein, S., Schwenk, F. C. and Broderick, R. L., *Diffusion Factor for Estimating Losses* RME, 53001 (N.A.S.A., 1953)

3 Lieblein, S and Johnsen, I. A., Résumé of transonic compressor research at N.A.C.A. Lewis Laboratory, *Trans. A.S.M.E., J. of Engineering for Power*, vol. 83 (July 1961).

4 Lakshminarayana, B. and Horlock, J. H., Review: secondary losses in turbomachines. *Int. J. of Mech. Eng. Sc.*, vol. 5 (1963).

5 Horlock, J., *Axial Flow Compressors* (Butterworth, 1958).

6 Carter, A. D. S., Blade profiles for axial flow fans, pumps, and compressors, *Proc. Inst. Mech. Eng.*, vol. 175, no. 15 (1961).

7 Capon, R. S. and Brooke, G. V., *Application of Dimensional Relationships to Air Compressors* (Aeronautical Research Council R. & M. 1336, 1930).

8 Stenning, A. H. and Kriebel, A. R. *Stall Propagation in a Cascade of Aerofoils*, A.S.M.E. paper No. 57-SA-29 (1957).

Chapter 5

1 Smith, S. F., A simple correlation of turbine efficiency, *J. Roy. Aero. Soc.*, vol. 69 (July 1965).

2 Stannitz, J. D., *Design of Two-Dimensional Channels with prescribed Velocity Distribution along the Channel Walls*, N.A.S.A. Report 1115.

3 Ainley, D. G., The high temperature turbojet, *J. Roy. Aero. Soc.*, vol. 60 (Sept. 1956).

4 Halls, G. A., Air cooling of turbine blades and vanes, *Aircraft Engineering*, vol. 39 (Aug. 1967).

5 Migdal, D. and Horgan, J. J., Thrust nozzles for supersonic transport aircraft, *Trans. A.S.M.E., J. of Engineering for Power*, vol. 86 (April 1964).

6 Herbert, M. V., Centre-body nozzles for supersonic transport aircraft, *J. Roy. Aero. Soc.*, vol. 71 (Jan. 1967).

7 Greatrex, F. B. and Bridge, R., The evolution of the engine noise problem, *Aircraft Engineering*, vol. 39 (Feb. 1967).

Chapter 6

1 Spalding, D. B., *Some Fundamentals of Combustion* (Butterworth, 1955).

2 Spalding, D. B., Performance criteria of gas turbine combustion chambers, *Aircraft Engineering*, vol. 28 (1956).

3 Clarke, J. S., A review of combustion problems for the aero gas turbine, *J. Roy. Aero. Soc.*, vol. 60 (April 1956).

4 Clarke, A. E. and Odgers, J., The spherical reaction chamber and some practical applications, *Lucas Engineering Review*.

5 Edwards, J. L., Reheat for gas turbines, *J. Roy Aero. Soc.*, vol. 59 (Feb. 1955).

Chapter 7

1 Hewson, C. T., Development of a VTOL power plant, *Proc. Inst. Mech. E.*, vol. 178, pt.1, no. 10 (1963–64).

2 Wyatt, D. D., A review of supersonic air intake problems, *Agardograph* 27, Section 2, Ed. J. Fabri (Pergamon Press, 1958).

3 Shapiro, A. H., *The Dynamics and Thermodynamics of Compressible Fluid Flow* (The Ronald Press, 1953).

4 Williams, J. and Butler, S. F. J., Aerodynamic aspects of boundary layer control, *J. Roy. Aero. Soc.*, vol. 67 (April 1963)
5 Davidson, I. M., The jet flap, *J. Roy. Aero. Soc.*, vol. 60 (Jan. 1956).

Chapter 8
1 Otto, E. W. and Taylor, B. L., *Dynamics of a Turbojet considered as a Quasi-static System*, N.A.C.A. Report 1011 (1951).
2 Lawrence, O. N., Fuel systems for aero gas turbines, *J. Roy. Aero. Soc.*, vol. 59 (Nov. 1955).
3 Carey, F. H., Development of a spill burner fuel system, *J. Roy. Aero. Soc.*, vol. 58 (Nov. 1954).

Chapter 9
1 Bragg, S. L., Expensive noise, *Flight International*, vol. 13 (20th Feb. 1964).
2 Jamison, R. R., Ramjets, *J. Roy. Aero. Soc.*, vol. 61 (June 1957).
3 Shapiro, A. H., *The Dynamics and Thermodynamics of Compressible Flow* (The Ronald Press, 1953).
4 Woods-Ballard, W. R., Hypersonic propulsion, *Science Journal*, vol. 1 (April 1965).
5 Ferri, A., Review of problems in application of supersonic combustion, *J. Roy. Aero. Soc.*, vol. 68 (Sept. 1964).

Chapter 10
1 Sutton, G. P., *Rocket Propulsion Elements* (Wiley, 1963).
2 Bragg, S. L., *Rocket Engines* (Newnes, 1962).
3 Rao, G. V. R., Recent developments in rocket nozzle configurations, *A.R.S. Journal*, vol. 31, no. 11 (1961).
4 Evered, D. S. and Kit, B., *Rocket Propellant Handbook* (Collier-Macmillan, 1960).
5 Pearce, K. W., Thrust control methods for solid propellant rockets, *J. Roy. Aero. Soc.*, vol. 65 (Oct. 1961).
6 Gudmondson, G., Engine design and manufacturing in Sweden, *J. Roy. Aero. Soc.*, vol. 71 (June 1967).
7 King-Hele, D., *Observing Earth Satellites* (Macmillan).
8 Finger, H. B., Nuclear rockets for space flight, *Flight International*, vol. 91 (30th March and 6th April 1967).

Chapter 11
1 Banks, F. R., The aviation engine, *Proc. Inst. Mech. E.*, vol. 162, No. 4 (1950).
2 Pye, D. R., *Internal Combustion Engines* (Oxford U.P., 1937).
3 Bass, E. L., Piston engine fuels, *J. Roy. Aero. Soc.*, vol. 70 (Jan. 1966).
4 Ricardo, H. R., The supercharging of internal combustion engines, *Proc. Inst. Mech. E.*, vol. 162, No. 4 (1950).
5 Chatteron, E. E., Compound diesel engines for aircraft, *J. Roy. Aero. Soc.*, vol. 58 (Sept. 1954).

Chapter 12
1 Armstrong, E. K. and Stevenson, R. E., Compressor blade vibrations, *J. Roy. Aero. Soc.*, vol. 64 (March 1960).
2 Armstrong, E. K., Recent blade vibration techniques, *Trans. A.S.M. J. of Engineering for Power*, vol. 89 (July 1967).

3. Glenny, R. J. E., *Gas Turbine Materials*, Bulleid Memorial Lectures, vol. 4 (University of Nottingham, 1967).
4 Howe, P. W. H., *Thermal, Mechanical and Microscopic Loading*, Bulleid Memorial Lectures, vol. 4 (University of Nottingham, 1967).

Chapter 13

1 Carroll, R. L., *The Aerodynamics of Powered Flight* (Wiley, 1960).
2 Dommasch, D. O. and others, *Airplane Aerodynamics* (Pitman, 1961).
3 Kermode, A. C., *Mechanics of Flight* (Pitman, 1962).
4 Davies, D. P., *Handling the Big Jets* (A.R.B., 1968).
5 Williams, G. H., Homebuilt Hovercraft, series in *Flight International* (1965–66).

Appendix: Getting the Feel of SI Units

Over the years anyone working with engines gets the feeling of a certain set of units and grows familiar with sizes and shapes. For example he will know that a turbojet with a mass flow rate of 100 lb/s will produce about 7000 lbf thrust and the final nozzle will be about 18 in. diameter. It is difficult to ask someone wholly familiar with British Units to make the effort and relearn all his units but the advantages of working in SI are worth the effort—that, at least, is my opinion.

Some exact conversions are quoted below but the object of this note is to suggest the feel of the new units in approximate terms. The basic mass unit, the kilogramme, is about $2\frac{1}{4}$ lb; the unit of length, the metre, is about $3\frac{1}{4}$ ft; the unit of time, the second, is, of course, the same!

The basic force unit is the *newton* which is the force needed to accelerate 1 kg to 1 m/s^2. In magnitude the newton is about $\frac{1}{5}$ to $\frac{1}{4}$ lbf. As someone pointed out, if you pick up a reasonable size apple, it will exert a downward force of about 1 newton on your hand. The work or energy unit, the newton-metre, is called the *joule* and a work rate of 1 joule per second is our old friend the *watt*. 1 kilojoule per second is therefore 1 kilowatt or about $1\frac{1}{3}$ horsepower.

A pressure force of 1 newton per square metre is very small and a suitable practical unit is the kilonewton per square metre. In value this is about one-seventh of a pound per square inch. 100 kN/m^2 is therefore about 1 atmosphere.

The preferred system of multiples in SI is in steps of 10^3, i.e. micro (10^{-6}), milli (10^{-3}), unit (10^0), kilo (10^3) and mega (10^6). For example it is better to write 12 newtons rather than 1·2 decanewtons.

Practice is everything and with a little time one can get used to the SI units.

Conversion factors

1 lb = 0·4536 kg (0·453 592 37 exactly)
1 kg = 2·2046 lb

1 ft = 0·3048 m (exactly)
1 m = 3·2808 ft

1 lbf = 4·4482 N
1 N = 0·2248 lbf

1 lbf/in^2 = 6·8948 kN/m^2
1 kN/m^2 = 0·1450 lbf/in^2

Index